VALORIZATION OF WINE MAKING BY-PRODUCTS

VALORIZATION OF WINE MAKING BY-PRODUCTS

Edited by
Matteo Bordiga
Universita del Piemonte Orientale
Novara, Italy

CRC Press is an imprint of the
Taylor & Francis Group, an **informa** business

CRC Press
Taylor & Francis Group
6000 Broken Sound Parkway NW, Suite 300
Boca Raton, FL 33487-2742

© 2016 by Taylor & Francis Group, LLC
CRC Press is an imprint of Taylor & Francis Group, an Informa business

No claim to original U.S. Government works

Printed on acid-free paper
Version Date: 20150909

International Standard Book Number-13: 978-1-4822-5533-1 (Hardback)

This book contains information obtained from authentic and highly regarded sources. Reasonable efforts have been made to publish reliable data and information, but the author and publisher cannot assume responsibility for the validity of all materials or the consequences of their use. The authors and publishers have attempted to trace the copyright holders of all material reproduced in this publication and apologize to copyright holders if permission to publish in this form has not been obtained. If any copyright material has not been acknowledged please write and let us know so we may rectify in any future reprint.

Except as permitted under U.S. Copyright Law, no part of this book may be reprinted, reproduced, transmitted, or utilized in any form by any electronic, mechanical, or other means, now known or hereafter invented, including photocopying, microfilming, and recording, or in any information storage or retrieval system, without written permission from the publishers.

For permission to photocopy or use material electronically from this work, please access www.copyright.com (http://www.copyright.com/) or contact the Copyright Clearance Center, Inc. (CCC), 222 Rosewood Drive, Danvers, MA 01923, 978-750-8400. CCC is a not-for-profit organization that provides licenses and registration for a variety of users. For organizations that have been granted a photocopy license by the CCC, a separate system of payment has been arranged.

Trademark Notice: Product or corporate names may be trademarks or registered trademarks, and are used only for identification and explanation without intent to infringe.

Visit the Taylor & Francis Web site at
http://www.taylorandfrancis.com

and the CRC Press Web site at
http://www.crcpress.com

To my son, too young to drink wine
old enough to understand its poetry

Contents

LIST OF FIGURES   IX

PREFACE   XIII

EDITOR   XV

CONTRIBUTORS   XVII

CHAPTER 1   OVERVIEW   1

MATTEO BORDIGA

CHAPTER 2   WINE MAKING PROCESS   27

HEND LETAIEF

CHAPTER 3   WINE MAKING BY-PRODUCTS   73

ZHIJING YE, ROLAND HARRISON, VERN JOU
CHENG, AND ALAA EL-DIN A. BEKHIT

CHAPTER 4   TECHNOLOGICAL ASPECTS OF BY-PRODUCT
UTILIZATION   117

ALAA EL-DIN A. BEKHIT, VERN JOU CHENG,
ROLAND HARRISON, ZHIJING YE, ADNAN A.
BEKHIT, TZI BUN NG, AND LINGMING KONG

CHAPTER 5   REGULATORY AND LEGISLATIVE ISSUES   199

MATTEO BORDIGA

CHAPTER 6   SUSTAINABILITY ISSUES   227

MATTEO BORDIGA

CONTENTS

Chapter 7 Marketing Potential   271

JORGE A. CARDONA AND THELMA F. CALIX

Chapter 8 Future Perspectives   293

MATTEO BORDIGA

Index   345

List of Figures

Figure 1.1 Continental evolution as a percentage of global production, 2000–2013.   6

Figure 1.2 Variation of grape production by leading countries.   7

Figure 1.3 Leading grape producers (all uses).   7

Figure 1.4 A modern wine bottling line.   8

Figure 1.5 Forecast for wine production in 2014, in the 10 main producing countries.   11

Figure 1.6 Shift in wine consumption geography.   14

Figure 1.7 Trends of top world exporters: France, Italy, Spain, Australia, and Chile (2010–2013).   17

Figure 1.8 Trends of top world importers: the United States, the United Kingdom, Germany, Canada, Japan, China, France, and Russia (2010–2013).   18

Figure 2.1 Early summer vineyard (Northern Hemisphere).   30

Figure 2.2 Components of a *terroir*.   31

Figure 2.3 Harvested Nebbiolo grapes.   36

Figure 2.4 A flowchart for making white and red wines.   40

Figure 2.5 Direct pressing with three fractions: (first) up to 0.5 bar, (second) up to 0.7 bar, and (third) up to 1.5 bar.   45

X LIST OF FIGURES

Figure 2.6 Stainless steel tanks, at a controlled temperature. 47

Figure 2.7 Stainless steel fermentation vessels. 48

Figure 2.8 Crude wine compounds classified according to their size. 61

Figure 3.1 The vinification process, composition of by-products and potential utilization. 74

Figure 3.2 Wine production of major countries in 2012. 76

Figure 3.3 Pruning grapevine rows. 80

Figure 3.4 Grape stems. 83

Figure 3.5 The solid remains of grapes: stems, seeds, and skins. 85

Figure 3.6 Grape skins. 86

Figure 3.7 Grape seeds. 89

Figure 3.8 The concentration of the total tannins in a model waste system containing catechin: grape-seed extract (50:50 w/w) oxidized with (HRP+) and without (HRP−) horseradish peroxidase. 105

Figure 4.1 Grape stems. 119

Figure 4.2 Grape skins. 120

Figure 4.3 Grape seeds. 121

Figure 4.4 Grape seeds separated from pomace. 131

Figure 4.5 Conversion of grape-seed oil to biofuel. 132

Figure 4.6 Structures of the main constitutive units of flavan-3-ols identified in grape seeds and skin and the general structure of anthocyanin compounds. 159

Figure 4.7 Grape-seed oil extracted using hexane. 172

Figure 5.1 The food waste management hierarchy. 201

Figure 5.2 The Common Market Organization for Wine (CMO) framework for regulations toward different issues. 212

Figure 5.3 Established waste hierarchy as a priority order in waste prevention and management legislation. 215

Figure 5.4 Winery waste constituents showing potential environmental impacts. 219

Figure 6.1 Grape pomace (solid parts separated in skins, seeds, and stems). 229

Figure 6.2 A combination of environmental, economic, and social principles is responsible for the general definition of sustainability. 231

LIST OF FIGURES

Figure 6.3 Grape skins (Muscat). 242

Figure 6.4 Lees from Nebbiolo after fermentation. 257

Figure 6.5 Lees from Nebbiolo after fermentation (overview from the top). 257

Figure 6.6 Sustainability as defined by the three overlapping principles of Environmentally Sound, Economically Feasible, and Socially Equitable. 258

Figure 7.1 Physiology and composition of grapes. 273

Figure 7.2 World grape production: 2008–2013. 275

Figure 7.3 Most important grape producers: 2008–2013. 275

Figure 7.4 World wine production 2008–2013. 276

Figure 7.5 Grape seeds and related powdered extract. 279

Figure 7.6 Lees. 285

Figure 8.1 The structure suggested for xyloglucans reported in wine. 295

Figure 8.2 The structures suggested for (a) Galacto-oligosaccharides and (b) Arabinogalactans reported in wine. 295

Figure 8.3 Positive-mode MALDI-FTICR spectra of an oligosaccharide fraction of Chardonnay wine. 312

Figure 8.4 MALDI-FTICR spectra of GOS with a DP of 8 (m/z = 1337). 313

Figure 8.5 Positive-mode MALDI-FTICR spectra of an oligosaccharide fraction of Chardonnay wine. 315

Figure 8.6 PRE distillation process pomace. 322

Figure 8.7 POST distillation process pomace. 323

Figure 8.8 Acid-catalyzed degradation of proanthocyanidins in the presence of phloroglucinol. 325

Figure 8.9 Seed oil extracted from PRE (left) and POST (right) samples. 334

Preface

This book focuses on the by-products of wine making and their valorization in a broad spectrum. Vine cultivation and wine making processes generate a significant amount of waste and several by-products, including prunings, stems, pomace and seeds, yeast lees, tartrate, carbon dioxide, and wastewater. Generally, only a very small portion of these materials are used (e.g., as fertilizer, as animal feed, or for fuel production). This book provides a comprehensive overview of the by-products of wine making and their conventional and nonconventional utilization with a number of value-adding technologies for the valorization of such products (e.g., as a source of novel functional ingredients). Over the past few years, these by-products, as well as a number of other agricultural wastes from plant origins have attracted considerable attention as potential sources of bioactive phytochemicals, which could be used for various purposes in the pharmaceutical, cosmetic, and food industries. Considering the challenges in the food industry arena, efforts need to be made to optimize food processing technology in order to minimize the amounts of by-product waste. The food industry generates an increasing amount of by-products throughout the chain of food production. The most important means of waste minimization is the application of more efficient production technologies; however, by-product generation is inevitable. Environmental regulations and the high cost of waste disposal have

PREFACE

forced food processors to find better ways to treat, utilize, and process waste. The efficient utilization of food processing by-products is important for profitability in the food industry. By-products and waste of plant-based food processing, which represent a major disposal problem for the industry, are very promising sources of value-added substances, with a particular emphasis on the retrieval of bioactive phytochemicals and technologically important secondary metabolites. In this sense, the huge amounts of by-products from wine making generated by one of the most important agricultural activities in the world, with low economic value but still characterized as being usable, appear to be suitable raw materials to be industrially valorized. The current trend is to utilize and convert waste into useful products and to recycle waste products as a means to achieve sustainable development. Over the next few years, the area of waste management in food processing will expand rapidly. Future legislation regarding industrial waste, including those of wineries, will become even more demanding, thus increasing the cost of waste management. The recovery of value-added products can help in this direction. These recovery processes are part of a new philosophy of sustainable agriculture.

Matteo Bordiga, PhD

Editor

Matteo Bordiga, PhD, earned his PhD in food science from the Università del Piemonte Orientale, Novara, Italy, in 2010. He received his MS in chemistry and pharmaceutical technologies from the same university. Currently, Dr. Bordiga is a postdoctoral fellow working on wine aroma analysis. His main research activity is in the area of food chemistry, investigating the different classes of polyphenols under analytical, technological, and nutritional points of view. More recently, his research interests have shifted toward wine chemistry, specifically the entire production process—from vine to glass. Dr. Bordiga's current research is focused on two major areas: first, the development and application of analytical chemistry techniques to study wine flavor chemistry and the physicochemical interactions of flavors with nonvolatile wine components and, second, the elucidation of the chemical mechanisms for observed health effects of wine and wine components. In the area of wine quality, his current interest is on the effect of oxidation on wine chemistry and how this affects important quality parameters of wine, such as taste and color. Dr. Bordiga has also contributed to the development of general analytical methodologies of interest in wine by-product analysis. Considering this, over the past few years, waste from wine making has attracted a significant amount of attention as a potential source of bioactive phytochemicals, which could be used for various purposes in the pharmaceutical, cosmetic, and food industries. All related research activities have been developed through important collaborations with foreign institutions, such as the Department of Foods Science and Technology, Foods for Health Institute, the University of California, Davis (United States); the Fundación Parque Científico y Tecnológico de Albacete (Spain); and the Instituto Regional de Investigación Científica Aplicada, Universidad de Castilla-La Mancha (Ciudad Real, Spain). Dr. Boriga has published more than 20 research papers in peer-reviewed national and international journals.

Contributors

Adnan A. Bekhit
Department of Pharmaceutical
  Chemistry
University of Alexandria
Alexandria, Egypt

Alaa El-Din A. Bekhit
Department of Food Science
University of Otago
Dunedin, New Zealand

Thelma F. Calix
Department of Food Science
  and Technology
Zamorano University
Francisco Morazán, Honduras

Jorge A. Cardona
Department of Food Science
  and Technology
Zamorano University
Francisco Morazán, Honduras

Vern Jou Cheng
Department of Food Science
University of Otago
Dunedin, New Zealand

Roland Harrison
Department of Wine, Food, and
  Molecular Biosciences
Lincoln University
Dunedin, New Zealand

Lingming Kong
College of Food Science and
  Pharmacology
Xinjiang Agricultural University
Ürümqi, Xinjiang, People's
  Republic of China

Hend Letaief
Department of Viticulture and
  Enology
California State University
Fresno, California

Tzi Bun Ng
School of Biomedical Sciences
The Chinese University of
  Hong Kong
Shatin, New Territories,
  Hong Kong

Zhijing Ye
Department of Wine,
  Food, and Molecular
  Biosciences
Lincoln University
Dunedin, New Zealand

1

OVERVIEW

MATTEO BORDIGA

Contents

1.1 Introduction	1
1.2 Environmental and Economic Challenges	3
1.2.1 Food Waste Issues	4
1.2.2 Wine Production	5
1.3 The Current State of the Vitiviniculture World Market	6
1.4 Vitivinicultural Production Potential	9
1.4.1 The Areas under Vines in European Vineyards	9
1.4.2 Outside the European Union	9
1.5 Wine Production (Resulting from Grapes Harvested in the Autumn of 2014 in the Northern Hemisphere and in the Spring of the Same Year in the Southern Hemisphere)	11
1.5.1 Within the European Union	11
1.5.2 Outside the European Union	12
1.6 Evaluation of World Consumption	14
1.7 Trends in the World Wine Trade	15
1.8 Top Wine Exporters	16
1.9 Top Wine Importers	17
1.10 The Five Largest Wine Markets	19
1.11 Wine Production–Consumption in China	20
References	22

1.1 Introduction

The food industry generates high amounts of solid waste/by-products and high volumes of effluents with an organic component. Considering the challenges in the food industry area, efforts need to be made to optimize food processing technology, minimizing the amounts of by-products and waste. During food processing, the generation of by-products is inevitable, and the amount and kind of waste produced, which consists primarily of the organic residues of processed raw materials, cannot be altered if the quality of the

2 VALORIZATION OF WINE MAKING BY-PRODUCTS

finished product is to remain consistent. Instead of treating these by-products as waste, increasingly food companies are turning them into useful products, promoting the principle of sustainable development. Environmental regulations and high waste disposal costs have forced food processors to find better ways to treat and utilize processing waste. Environmental legislation agencies have significantly contributed to the introduction of sustainable waste management practices throughout the world. In this context, by-product valorization, a relatively new concept in the field of industrial residue management, holds immense potential for the prospective production of biologicals of commercial significance, enzymes, pigments, flavors, functional ingredients, micronutrients, nutraceuticals, active pharmaceutical ingredients, phytochemicals, biofuel, and biomaterials (Patras 2009; Rockenbach 2011; Rondeau 2013). In this sense, it is reasonable to imagine that, over the next few years, the area of food processing waste management will rapidly expand. Plant-based by-products and waste, which represent the main disposal problem for the industry concerned, are very promising sources of value-added substances, with particular emphasis on the retrieval of bioactive phytochemicals and technologically important secondary metabolites (Basalan 2009; Deng 2011; Fontana 2013). The nutritional composition of such food waste appears to be rich in sugars, vitamins, minerals, and various beneficial health phytochemicals such as fine chemicals (antioxidants, polyphenols, etc.) and natural macromolecules (cellulose, starch, lignin, lipids, enzymes, pigments, etc.), which are of great interest to the chemical, pharmaceutical, and food industries (Bordiga 2011, 2013; Pastor del Rio 2006). Moreover, the search for cheap, renewable, and abundant sources of bioactive compounds (e.g., antioxidants) is attracting worldwide interest. Much research is needed in order to select raw materials; those of residual origin are especially promising due to their lower costs. Industrial biotechnology can offer effective strategies and tools for the valorization of by-products and waste of the food industry, thus achieving a significant increase of environmental, social, and economic sustainability (International Organization for Standardization 2004; Kim 2006; Laufenberg 2003). However, extensive research on potential sources, optimization of extraction processes, knowledge of the mechanisms of the *in vivo* action and assimilation are still required.

1.2 Environmental and Economic Challenges

Agriculture generates waste stream (also co-products and by-products) that is not taken care of properly. In plant production, losses take place at the farm and postharvest levels and also at the lower level of the retail sector. Co-products or by-products are generated, which require sustainable use (Torres 2002; Yu 2013). For example, straw has been given a significant amount of attention as a biomass feedstock with potential trade-offs because of its relevance to soil improvement, which needs to be considered. In livestock production, manure and other effluent management appears to be a challenge, particularly for industrial production systems. While these effluents can be used as fertilizer, they can also be important sources of bio-energy or valuable bio-products. However, the impact on the environment needs to be evaluated (e.g., air, soil, and water emissions). It is crucial to consider the entire effluent chain to avoid pollution swapping and health issues (possible transmission of pathogens). Other than the reduction and recycling of agricultural waste, there may be several opportunities for new processes which may enable innovative uses of these materials outside of the agricultural sector. Both awareness and dialogue can be enhanced between sectors towards options for smart use of agricultural waste and by-products through the creation of joint stakeholder platforms and other joint structures. Also, resource efficiency can be improved through the reduction of waste and enhanced waste management in primary production. And different opportunities for the valorization of waste and by-products resulting in environmental and economic benefits for the farming sector (development of new products and processes) have to be increased. In this direction, competitiveness will be enhanced through more varied and new types of sources for bio-products and bio-energy in the agro-food (conventional and organic) and bio-economy sectors. Soil quality and crop productivity, through the optimal use of crop waste (taking into account the need to maintain soil organic matter levels) and nutrient recovery, can be improved. Similarly, water quality can be improved, by reducing pollution and the eutrophication of groundwater, and thus indirectly improving marine waters. Even air quality can be improved by reducing livestock emissions.

4   VALORIZATION OF WINE MAKING BY-PRODUCTS

1.2.1 Food Waste Issues

Food waste is a significant problem for economic, environmental, and food security reasons (Eurostat 2013; FAO 2012, 2014). Efforts made by governments have focused on diverting waste away from landfills through regulation, taxation, and public consciousness. However, efforts to understand why waste occurs have been limited; therefore, detailed investigations are necessary in this field. Environmental pollution problems associated with disposal methods have been the factors behind the search for alternative methods (environment-friendly) to handle food waste. This biodegradable waste can be used to produce industrially relevant metabolites with a great economical advantage (e.g., enzymes, organic acids, flavor and aroma compounds, and polysaccharides). Thus, the cultivation of microorganisms on this waste may be a value-added process capable of converting these materials into valuable products (Beveridge 2005; Gonzalez-Paramas 2004; Molina-Alcaide 2008). However, much remains to be done in this area to develop commercial processes. For example, filamentous fungi are metabolically versatile organisms, which are exploited commercially as cell factories for the production of enzymes and a wide variety of metabolites. It has been possible to control simultaneous production of enzymes (such as pectinolytic, cellulolytic, and xylanolytic) by fungal strains of the genera *Aspergillus, Fusarium, Neurospora,* and *Penicillium* and generate multienzyme activity using a simple growth medium consisting of a solid by-product of the citrus-processing industry (orange peels) and a mineral medium. The green production concept shows a good utilization potential for solid vegetable waste. In addition to accomplishing a reduction of investment and raw material costs, it could contribute to waste-minimized food production. The exploitation of by-products of fruit and vegetable processing as a source of functional compounds and their application in food is a promising field that requires more interdisciplinary research in several aspects. Methods to utilize by-products from food processing on a large scale and at affordable levels should be developed and optimized to minimize the amounts of waste. This approach calls for the active participation of the food and allied industries with respect to sustainable production and waste management. Designing new functional foods,

OVERVIEW

to meet the demands of an innovative and rapidly growing food market, requires careful assessment of the potential risks, which might arise from isolated compounds recovered from by-products. The bioactivity, bioavailability, and toxicology of phytochemicals need to be carefully assessed. Furthermore, investigations on stability and interaction of phytochemicals with other food ingredients during processing and storage need to be initiated and improved. It is still difficult to regulate considering there are some that appear on the borderline between foods and drugs. In any case, consumer protection must be a priority over economic interests, and health claims need to be confirmed by sound and reliable studies.

1.2.2 Wine Production

Humans have cultivated vines for producing wine for thousands of years. The origins of viticulture lie in the region between the Black Sea and the Caspian Sea and date back to 6000 BCE. Only a few products have such a long history with a production process that has remained more or less unchanged. Early examples of viticultural and enological literature date back to more than 2000 years. In *De agri cultura*, the Roman statesman Marcus Cato—also known as Cato the Elder (234–149 BCE)—provides detailed practical advice on how to profitably run a wine farm. For example, he stresses that grapes should be fully ripe when harvested and that all vats need to be perfectly clean to prevent wine from turning into vinegar. In the second book of his scientific treatise, *De re rustica*, Lucius Columella (4–70 CE), one of the most important writers on agriculture of the Roman Empire, discusses many technical aspects of Roman viticulture. He elaborates on topics like which grape variety grows best on which soil type. He lays out many criteria of modern vine training and trellising. Wine production has been traditionally considered as an environmentally friendly process. However, it requires a considerable amount of resources such as water, fertilizers, and organic amendments, and on the other hand, produces a large amount of wastewater and organic waste. Environmental analysis of the wine industry shows that the main effluents of the sector are wastewater and organic solid waste. Innovative solutions must be proposed and tested to develop a real sustainable industry.

1.3 The Current State of the Vitiviniculture World Market

During the *37th World Congress of Vine and Wine* held in Mendoza (Argentina) in November 2014, the International Organization of Vine and Wine (OIV) director general, Jean-Marie Aurand, presented a global overview of the vitivinicultural sector. In 2013, the worldwide area under vines was 7519 thousands of hectares (mha). The global production of grapes has increased to 751 million of quintals (Mql) in 2013, despite the decrease in the vine area (Figure 1.1). 271 millions of hectoliters (Mhl) was the first estimation for the wine produced in 2014. Sparkling wine market has expanded in recent years. The production has increased by more than 40% and consumption by 30% in 10 years. Regarding the area under vines, the world's total vineyard surface area (7519 mha) is decreasing mainly because there are fewer vineyards in Europe now (OIV 2004, 2008, 2012, 2012–2014, 2013). The increase in the planted surface areas in the rest of the world has partially offset this reduction. In China and South America, the total area under vines has continued to increase; these areas are the world's main vineyard growth centers. Meanwhile, the global production of grapes (for all use) is 751 Mql. There is an increased trend in grape production (+17% compared with 2000), despite decrease in the area under vines. This can be explained in part by an increase in yield because of the particularly favorable climate conditions in some

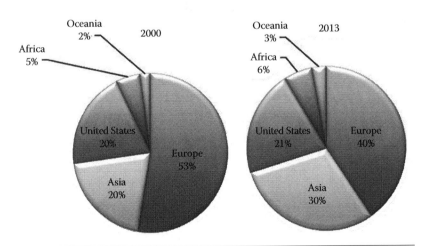

Figure 1.1 Continental evolution as a percentage of global production, 2000–2013. (From OIV, State of the Vitiviniculture World Market, 2014, http://www.oiv.int/oiv/info/en_vins_effervescents_OIV_2014, accessed on December 2014.)

countries and the continued improvements in viticultural techniques (Figure 1.2). China with 115 Mql produces about 15% of the world grape production followed by Italy (79 Mql) and the United States (75 Mql) (Figure 1.3). In 2014, there was an estimated global wine production (excluding juice and musts) of 271 Mhl, which was 6% less than the previous year. The year 2014 was marked by major climatic hazards, which is the reason why there was a fall in production. Europe remains the major producer of wine, despite the decreasing

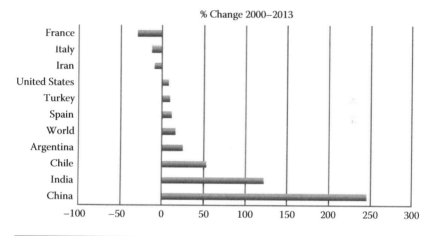

Figure 1.2 Variation of grape production by leading countries. (Created using information from OIV, State of the Vitiviniculture World Market, 2014, http://www.oiv.int/oiv/info/en_vins_effervescents_OIV_2014, accessed on December 2014.)

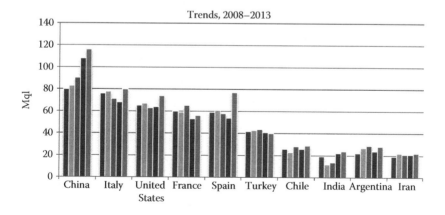

Figure 1.3 Leading grape producers (all uses). (Created using information from OIV, State of the Vitiviniculture World Market, 2014, http://www.oiv.int/oiv/info/en_vins_effervescents_OIV_2014, accessed on December 2014.)

trend. France is the biggest producer (46.2 Mhl) ahead of Italy, which saw a poor harvest (44.4 Mhl); Spain returned to an average level of production (37 Mhl) after a record year in 2013. On the other hand, the production in the Southern Hemisphere and in the United States continued to increase: Argentina 15.2 Mhl (+1% compared with 2013); New Zealand, a new record with 3.2 Mhl (+29%); South Africa 11.4 Mhl (+4%); and the United States registered a high level of production (22.5 Mhl). The data available (about wine consumption) show a consolidation of the global consumption in 2014 (estimated at around 243 Mhl). The data confirms that the wine consumption growth is no longer driven by traditional wine-producing and -consuming countries. A change in wine consumption pattern has characterized the period between 2000 and 2013. Today, about 39% of wine is consumed outside European countries, compared with 31% in 2000. The wine market is now an internationalized sector; the share of wine production traded internationally has nearly doubled. In 2000, 25% of the wine consumed in the world was imported; in 2013, this share reached more than 40% (Figure 1.4). In 2013, the trade in wine

Figure 1.4 A modern wine bottling line.

decreased by 2.2% in volume, representing 98 Mhl, even though the growing prices resulted in an increase in the total revenue of 1.5% up to U.S. $28.5 billion. OIV is focusing more on the sparkling wine market, considering that in 2013, 17.6 Mhl of this wine typology was produced, representing 7% of the global wine production. Its global consumption was 15.4 Mhl in the past 10 years, a 30% rise reaching 8.7 Mhl in 2013, resulting in a revenue of U.S. $4.8 billion. The share of sparkling wine exports accounts for nearly 9% of the volume of wine exports and 18% of their value.

1.4 Vitivinicultural Production Potential

1.4.1 *The Areas under Vines in European Vineyards*

Since the European Union (EU) ended its program to regulate wine production potentially under which the EU introduced permanent abandonment premiums for vineyards (Council Regulation [EC] No. 479/2008; No. 822/1987; No. 823/1987; No. 1493/1999; No. 491/2009; No. 607/2009), the rate of decline of EU vineyards (the EU is now composed of 28 member states) has significantly slowed down (European Landfill Directive [1999/31/EC]; European Court of Auditors 2012; European Commission 2006a,b, 2007a,b, 2008, 2009–2012). Although between 2011 and 2012, the EU area under vines decreased by 54 mha and between 2012 and 2013 EU vineyards declined by only 19 mha overall (Table 1.1). The total vineyards (vines for winegrapes, table grapes, or grapes for drying) even grew by 5 mha in Spain, while Italian, Portuguese, and French vineyards each declined by 6–7 mha. According to the latest data, European vineyards started at 3481 mha in 2013, a decrease of 0.5% between 2012 and 2013.

1.4.2 *Outside the European Union*

The information provided in Table 1.2 shows that vineyards outside of Europe appeared to increase between 2012 and 2013 (+19 mha) (New Zealand Wine 2009; New Zealand Wines 2014; SAWIS 2014; Wines of Argentina 2011). This moderate overall growth is the result of contrasting developments. In China and South America, the total area

10 VALORIZATION OF WINE MAKING BY-PRODUCTS

Table 1.1 Areas under Vines in European Vineyards

1000 ha	2011	2012	2013 FORECAST
Austria	46	44	44
Germany	102	102	102
Bulgaria	83	78	78
Spain	1032	1018	1023
France	806	800	794
Greece	110	110	110
Hungary	65	64	63
Italy	776	759	752
Portugal	240	236	229
Romania	204	205	205
EU—28 total	3554	3500	3481

Source: Created using information from OIV, State of the Vitiviniculture World Market, 2014, http://www.oiv.int/oiv/info/en_vins_effervescents_OIV_2014, accessed on December 2014.

Table 1.2 Total Areas under Vines in Vineyards outside the European Union

1000 ha	2011	2012	2013 FORECAST
South Africa	131	131	130
Argentina	218	221	224
Australia	170	162	158
Brazil	90	91	87
Chile	200	205	207
China	560	580	600
United States	407	407	408
New Zealand	37	38	38
Russia	63	62	63
Turkey	508	497	504
Other African countries	242	239	239
Other American countries	86	88	87
Other Asian countries	615	615	615
Total outside the EU	3955	3936	3955

Source: Created using information from OIV, State of the Vitiviniculture World Market, 2014, http://www.oiv.int/oiv/info/en_vins_effervescents_OIV_2014, accessed on December 2014.

under vines has continued to increase; these areas represent the main vineyard growth centers worldwide. Turkey has seen a break in the downward trend observed in the previous years with a vineyard increase of about 7 mha. On the other hand, Australia recorded a reduction for a second consecutive year, even if only declining by half as much

(–4 mha) between 2011 and 2012 (ABARES 2013; Wine Australia 2013). In 2013, vineyards had to reach 3955 mha outside of the EU, a relatively moderate increase of 0.5% in relation to 2012. Finally, in 2013, the total world area under vines (including the area not yet in production) had to therefore remain stable, similar to that in 2012 at 7436 mha.

1.5 Wine Production (Resulting from Grapes Harvested in the Autumn of 2014 in the Northern Hemisphere and in the Spring of the Same Year in the Southern Hemisphere)

1.5.1 Within the European Union

In the forecast for 2014, the EU vinified production may be described as average. Wine production indeed began at quite a significantly higher level than the very low 2012 production (146 Mhl), nearing production levels for the 2007–2009 period, but significantly lower than 2013 production (170 Mhl). In 2014, the production (excluding juice and musts) should reach about 159 Mhl, a decrease of about 7% compared with 2013 (Figure 1.5). Indeed, compared

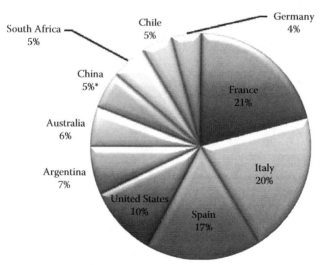

80% of the world's wine is produced by 10 countries. *Report of 2013, 2014 data not yet available.

Figure 1.5 Forecast for wine production in 2014, in the 10 main producing countries. (Created using information from OIV, State of the Vitiviniculture World Market, 2014, http://www.oiv.int/oiv/info/en_vins_effervescents_OIV_2014, accessed on December 2014.)

12 VALORIZATION OF WINE MAKING BY-PRODUCTS

with the extremely low production in 2012 excluding in Germany (8.4 Mhl in 2013, which is −8% more compared with the production in 2012), the expected outcome in 2014 was stable or positive. The significant increase in production in Spain, where a record harvest of 52.5 Mhl (but including juice and musts) was reported, should be noted. This situation led to Spain not vinifying 7.9 Mhl (compared with 5–6 Mhl normally), which is an exception. However, Italy with almost 53 Mhl (excluding juice and musts) became the largest wine producer in 2013. Spain, with 45.6 Mhl vinified in 2013, along with the slow growth of 2013 French production compared with that of 2012 (which in itself was very low at 42 Mhl in 2013 compared with 41.5 in 2012), was the second largest global wine producer after Italy. It should also be noted that without returning to their production levels at the end of the 2000s, Romanian and Hungarian production grew compared with their extremely low 2012 production (+57% and +52% compared with 2012, respectively). Finally, observing the 2014 forecast, it seems that France with 46.1 Mhl (+10% compared with 2013) would be the largest wine producer. Italy and Spain would be second and third, respectively, but both with a negative trend of production (−15% and −19%).

1.5.2 Outside the European Union

In the key Southern Hemisphere countries, together with the United States and China, production levels (excluding juice and musts) reached 88.7 Mhl in 2013, which is a noticeable increase (almost +6%) compared with 2012. This overall trend reflects contrasting developments, as outlined next (Table 1.3). The United States recorded a significant rise in wine production in 2013 at 23.5 Mhl excluding juice and musts, particularly in California, which produced +7% compared with 2012. In South America, Chile reported a record production again reaching 12.8 Mhl (+2% compared with the previous 2012 record). Meanwhile in 2013, Argentina returned to a wine production in line with its potential (14.9 Mhl compared with 11.7 Mhl of 2012, which is +28%). Again in 2013, Brazil showed a decline with a vinified production of 2.7 Mhl, a level that is very close to its low 2009

OVERVIEW **13**

Table 1.3 Wine Production (Excluding Juices and Musts)

1000 hL	2011	2012	2013 PROVISIONAL	2014 FORECAST	2014/ 2013 VOL. VARIAT.	2014/ 2013 % VARIAT.	RANKING
France	50,757	41,548	42,004	46,151	4,147	10	1
Italy	42,772	45,616	52,429	44,424	−8,005	−15	2
Spain	33,397	31,123	45,650	37,000	−8650	−19	3
United States[a]	19,140	21,740	23,500	22,500	−1,000	−4	4
Argentina	15,473	11,780	14,984	15,200	216	1	5
Australia	11,180	12,260	12,310	12,560	250	2	6
China[b]	13,200	13,810	11,780	11,780	0	0	7
South Africa	9,725	10,568	10,980	11,420	440	4	8
Chile	10,464	12,554	12,846	10,029	−2,817	−22	9
Germany	9,132	9,012	8,409	9,725	1,316	16	10
Portugal	5,622	6,327	6,238	5,886	−352	−6	11
Romania	4,058	3,311	5,113	4,093	−1,020	−20	12
New Zealand	2,350	1,940	2,480	3,200	720	29	13
Greece	2,750	3,115	3,343	2,900	−443	−13	14
Brazil	3,460	2,967	2,710	2,810	100	4	15
Hungary	2,750	1,776	2,666	2,734	68	3	16
Austria	2,814	2,125	2,392	2,250	−142	−6	17
Bulgaria	1,237	1,442	1,755	1,229	−526	−30	18
Switzerland	1,120	1,000	840	900	60	7	19
Croatia	1,409	1,293	1,249	874	−375	−30	20
OIV world total	267,243	256,222	287,600	270,864	−16,736	−6	

Countries for which information has been provided with a wine production of more than 1 Mhl.

[a] OIV estimate (USDA basis).

[b] Report for the year 2013; 2014 figures were not yet available.

production. In South Africa, the production reached a consistent level at nearly 11 Mhl (compared with 10.5 Mhl in 2012, which is +4%). While Australian production continued its recovery to reach 12.3 Mhl (+1% compared with 2012), New Zealand production hit a new record of 2.4 Mhl in 2013 after the previous record (of 2.3 Mhl) dating back to 2011. Finally China's production, which with 11.7 Mhl recorded a decrease of 2.1 Mhl compared with the previous year (−15%), should be noted. These developments have resulted in a 2013 global wine production (excluding juice and musts) of 287.6 Mhl, which is +31.4 Mhl compared with the production in 2012. This global wine production may therefore be described as average to high.

1.6 Evaluation of World Consumption

The long-awaited recovery that will mark the end of the financial and subsequently economic crisis, which began in 2008, is still to take place. Furthermore, the 2012 vitivinicultural year was marked by a very low global production that restricted global consumption levels at the end of 2012, and particularly in 2013. Indeed, at the beginning of the 2012/2013 harvest, stocks resulted low due to a succession of modest productions, although there were some regional exceptions to this macroeconomic vision. All this resulted in a 2013 global wine consumption of between 234 and 243 Mhl, corresponding to 238.5 Mhl at the mid-range estimate (2.5 Mhl compared with 2012, resulting –1.0%). However, this number should be noted with caution considering the realistic margin of error with respect to monitoring global consumption. Since the 2008 crisis, global consumption appears to have stabilized overall with 240.9, 240.3, 241.2, and 241.2 Mhl in 2009, 2010, 2011, and 2012, respectively. In Europe, in line with the trends of previous years, traditional consumer countries resumed their decline between 2012 and 2013 (Figure 1.6). A decrease of 2.1 Mhl in France (28.1 Mhl of wine consumed), 0.8 Mhl in Italy (21.7 Mhl), and 0.2 Mhl in Spain (9.1 Mhl) was recorded. In Portugal (4.5 Mhl) and the United Kingdom (12.7 Mhl), consumption remained stable, while in Germany

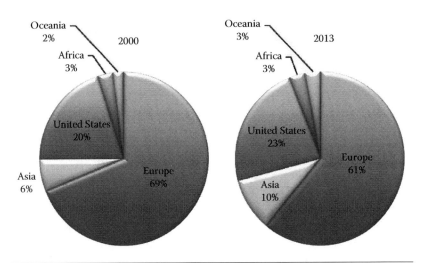

Figure 1.6 Shift in wine consumption geography. (Created using information from OIV, State of the Vitiviniculture World Market, 2014, http://www.oiv.int/oiv/info/en_vins_effervescents_OIV_2014, accessed on December 2014.)

(20.3 Mhl) a slight increase was reported. The United States, with 29.1 Mhl of wine consumed (excluding vermouth and special wines), became the primary market in the world in terms of volume in 2013. However, this rate of growth did not continue between 2012 and 2013 (only +0.1 Mhl; +0.5% compared with 2012 but lower, for example, related to +2% between 2011 and 2012). Meanwhile in China, the rapid increase in consumption since the beginning of the 2000s appears to have come to an abrupt end. The development in consumption between year n – 1 and year n is considered here to be determined in two halves: one by the apparent consumption for the current year (calculated by "Production + Imports – Exports") and the other by the consumption for the previous year. Despite the information on a reduction in production (from 13.8 Mhl in 2012 to 11.7 Mhl in 2013) and imports (from 3.9 to 3.7 Mhl), this approach results in the indirect inclusion of stocks (being at a high level), thereby bringing down the effect of the reduced "supply" (in terms of production + imports) on the estimated level of consumption. Chinese wine consumption is measured at 16.8 Mhl in 2013, showing a decrease of 3.8% compared with 2012. The main South American countries, namely Argentina, Chile, and Brazil, as well as South Africa, recorded an increase in consumption in 2013. This growth resulted between 1% and 3% compared with 2012. Meanwhile, Romania reported an interesting rise in consumption levels in 2013 (+24% compared to in 2012), even if it must be taken into consideration that the absolute value remains fairly low (3.2 Mhl) compared with the other countries. While Switzerland, New Zealand, and Hungary experienced stable levels of consumption between 2012 and 2013, Australia suffered a modest reduction in its internal market after several years of steady growth (MVSWGA 2012, 2013). All of this resulted in a global wine consumption of 238.7 Mhl, showing a reduction of about 2.5 Mhl compared with in 2012.

1.7 Trends in the World Wine Trade

The world wine trade in 2013 decreased in terms of volume by 2.2%, to 98 Mhl, although growing prices allowed for an increase in total revenues of 1.5% up to U.S. $28.5 billion. The relative low harvest in 2012 in the Northern Hemisphere created an impression of a shortage of wine, which pushed prices up to an average of U.S. $2.90 per L.

16   VALORIZATION OF WINE MAKING BY-PRODUCTS

Larger exports from Chile and South Africa could not compensate for the reduced availability of wine particularly from Italy and Spain (Wines of Chile 2010, 2011, 2013). Bottled wine took the biggest hit in terms of the largest reduction in wine trade for the whole years, which accounted for 2 Mhl out of the total decrease of 2.4 million. On the other hand, sparkling wines were the only ones to increase in volume terms by 3.4%. Wines in bulk and in containers above 2 L (representing the cheapest products) were the most dramatically affected by the increase in prices, which declined for sparkling wines. Nevertheless, bottled wines still account for almost 71% of the total sales, while sparkling makes 20% of the total revenues (although it only represents 6.6% of the total volume), and wines in bulk and in containers account for 11.7%, even though they are 38% of the total volume. The lower average price of sparkling wine did not result from the price reduction of the principal suppliers (Italy, France, and Spain increased them by 3.9%, 0.9%, and 1.2%, respectively), but from the fact that Italian sparkling (particularly Prosecco) grew much faster. Italian sparkling wine, at relatively lower prices than those of the French, increased its consumption by 27%.

1.8 Top Wine Exporters

At the beginning of the year, wine export from Spain and Italy was particularly affected by both the lack of wine in the Northern Hemisphere and the correspondent increase in prices. This reduction accounted for 4.3 Mhl and was not entirely compensated by the increase in wine sales in Chile and South Africa, which together grew by 2.7 Mhl more than the previous year.

Since the beginning of the century, all the top 10 exporters increased their value. Among the top wine exporters, the first five account for more than 70% of the total exports in value and volume (Figure 1.7). France remains clearly the leader. After the 2009 world crisis in wine consumption, Italy and (partially) Spain showed a notable performance during the past 4 years. All the major European countries have been steady regarding the level of wine traded but recording an increase with respect to the export value trends. In volume, Italy and Spain are the leaders even though they were affected by a low crop in 2012 in the Northern Hemisphere. In 2013, Chile surpassed

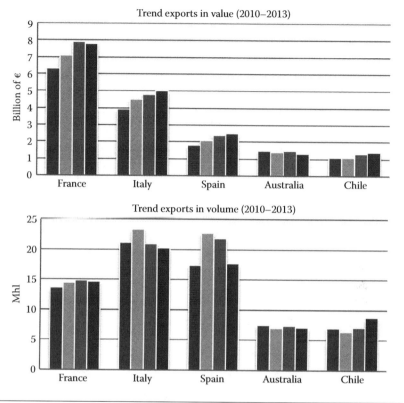

Figure 1.7 Trends of top world exporters: France, Italy, Spain, Australia, and Chile (2010–2013). (Created using information from OIV, State of the Vitiviniculture World Market, 2014, http://www.oiv.int/oiv/info/en_vins_effervescents_OIV_2014, accessed on December 2014.)

Australia, and South Africa took advantage of lower inventories in the North. Chile, South Africa, and New Zealand registered the biggest increase, both in volume and value. The United States decreased in volume, but increased in value.

1.9 Top Wine Importers

Among the top world importers, in terms of volume France had the biggest fall, although its cost increased by 3.6% up to U.S. $719 million in 2013. The United States also imported less but more expensive wine, keeping its first place in the ranking of world wine importers, even further ahead from the second rank, the United Kingdom, which decreased the value of its purchase by 5.2% in 2013. Germany stands first in terms of the volume imported with more than 15 Mhl (–1.7%)

18 VALORIZATION OF WINE MAKING BY-PRODUCTS

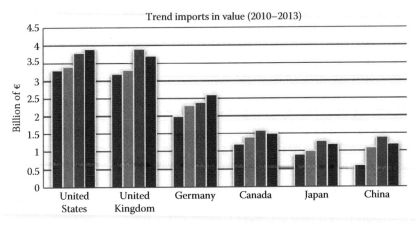

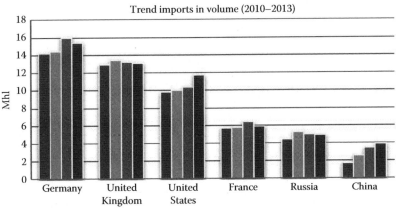

Figure 1.8 Trends of top world importers: the United States, the United Kingdom, Germany, Canada, Japan, China, France, and Russia (2010–2013). (Created using information from OIV, State of the Vitiviniculture World Market, 2014, http://www.oiv.int/oiv/info/en_vins_effervescents_OIV_2014, accessed on December 2014.)

of more expensive wines (Figure 1.8). In Russia, ranking fifth in terms of volume behind France, there was also a huge price increase, leading to an 11.9% growth up to U.S. $1,009 million. Six countries import more than U.S. $1.1 billion of wine each. The United States surpassed the United Kingdom as the top world importer, Germany and Canada grew slowly, and China reached Japan's level although there was a decline in 2013. In terms of volume, France and Russia were among the top six importers, while Germany, the United Kingdom, and the United States import above 10 Mhl of wine. Among the top world markets for wine, the first five in account for more than half the total import in euros. The United States, the United Kingdom, Germany,

Canada, and China reached about U.S. $14.4 billion. Since 2008, both groups of importers recorded an increase in terms of value (China showed the largest increase) and volume (France and the United States). By observing each market, it was noticed that the biggest markets (the United Kingdom, the United States, Germany, and Canada) decreased their import of wine in 2013 in volume, with Canada also reducing its purchase in euros. On the other hand, four large European markets, including two Scandinavian countries (Denmark and Sweden) led to an increase in both volume and value terms.

1.10 The Five Largest Wine Markets

Among the top world markets, the first five account for more than half of the total imports in euros and 49% of the total liters imported in 2013. The United States, the United Kingdom, Germany, Canada, and China reached a total amount of U.S. $14.3 billion and 4.6 billion L (California Sustainable Winegrowing Alliance 2004; CSWA 2013). The United States, as the top global wine importer, bought 1096 million liters (ML) of foreign wine in 2013 for U.S. $5.2 billion. This shows a substantial decrease in volume (–6%). This fall was in the purchase of wines in bulk and containers of more than 2 L (–24%), which was not compensated by the increase in sparkling and bottled wines (8.5% and 2.9%, respectively, in volume). Due to the scarcity in Europe and a large crop in California, there was less quantity of cheap wine. This forced an increase in the average price, related more to the change of mixing the products than to the variation of prices. The severe fall of bulk imports mostly affected Argentina, Spain, and Chile, with Italy and France increasing their market share in volume and especially in value terms. In 2013, imports in the United Kingdom decreased in both value and volume, although in this case it was because of the lesser number of bottled wines. All categories increased their prices, but especially the large volume of cheaper wines forced the decline of the global average price of total imports by 0.2%. Wines introduced in the United Kingdom in bulk and containers showed a 23% increase since the beginning of the century, from 12.4% of total imports in volume in 2000 to 35.2% in 2013. In 2013, all the key suppliers except Chile brought down their sales to the United Kingdom in volume, with only Italy and Spain increasing their revenues as a result of higher prices.

20 VALORIZATION OF WINE MAKING BY-PRODUCTS

In 2013, Germany was a clear example of how less availability of wines and keeping the price of the cheapest wines high affected different markets. The total import of bulk wines (at a 20% higher price) reduced by 2.5% in volume, although increasing the bill by 3.1% up to U.S. $2.8 billion. Meanwhile, huge increase in sales from Chile and South Africa (44.6% and 20.7%, respectively) did not compensate for the decrease of wine coming from Italy, Spain, and France. Also in the case of Canada, the volume of import was less (−1%) although at higher prices (6.8%), which increased the total cost of foreign wines by 5.7%. Particularly affected by these changes was Spain, whose bulk export to Canada decreased by almost 50% and could not be compensated by the sale of other wine categories. Such a huge decrease in bulk wines, compensated only slightly by larger purchases from South Africa and Chile, led to the overall decrease of Canadian import in volume. China, the fifth largest world importer, decreased in both volume and value down to 377 ML. By category, such decrease was caused particularly by lower purchase of wines in bulk (−26% in both volume and value) despite a stable price (−0.3%). Actually, the prices decreased more dramatically for bottled and sparkling wines (−6.5% and −24.7%, respectively), which had different effects, with a slight growth in the import of the former (4.8% in volume) and a much higher growth for the latter (38% in volume), although the amount of sparkling wine bought by China was still very small. Chile took advantage of the situation with an increase of 37% in volume and 10% in value, and South Africa, which had an upper hand in the market at the expense of Spain. However, it must be noted that Chinese wine imports were recently affected by legal debates over allegations against European wine subsidies and potential dumping, which was finally resolved.

1.11 Wine Production–Consumption in China

Asia ranks second only to Europe in coverage (28.7%) of the global vineyards. Although Asia accounts only for 6.9% of the global wine production, since the vast majority of those vineyards are used for table grapes and raisins. Combined with the relatively established wine market of Japan, China is definitely fascinating (considering its rapid growth). For example, Hong Kong is already playing a leading global role as a fine wine hub. According to OIV data, wine consumption

OVERVIEW 21

in China grew 67% by volume from 2000 to 2012. The main driving force includes rising family income, urbanization, and changing lifestyles. Consumers too are evolving their tastes. They have gone from expecting wine to be both sweet and sour (relatively low alcohol) to a largely improved tolerance and acceptance of dryness and tannins. Wine import has seen rapid growth during the last period. In 2012, China Customs recorded about 400 ML of import (272 ML of bottled and 122 ML of bulk wine, respectively). Bottled wine growth results are impressive, up by more than six-fold from 43.5 ML in 2007 and representing a rising share of the total import. The first half of 2013 reported a further 21% year-on-year rise by volume. Nevertheless, China appears more than a wine importer. It also ranks as the fifth largest wine producer according to OIV, and the biggest in Asia, with about 10 times more volume than Japan and India. Some provincial and regional governments strongly support the wine sector; particularly, the northwest areas (Ningxia, Gansu, and Xinjiang) have ambitious wine development plans. The wine industry is increasing not only in the traditional highly populated northeastern coastal areas but also in the more sparsely populated northwest ones, becoming a part of a shift in wine production in China. This development fits into central government plans to maximize the use of farmland and water. Despite this growth, China will not become a solid wine exporter considering present conditions. Economically, it appears hard to produce good-quality winegrapes on a large scale in China because vineyard coverage, even if constantly increasing, is limited because of the limits to irrigation water and relatively high production costs in north China. The domestic wine sector has taken this situation into account as well as the increasingly strong competition from imported wine. The wine industry is reflecting the cooling period of economy, with Chinese wine production for the first half of 2013 seeing a decrease of 7.2% compared with the same period of 2012. Top producers reported a decrease in sales and profits. In addition, although bottled imports have continued to rise, the volume and value of imported bulk wine has oscillated, dropping to 31.4% and 24.2% in the first half of 2013, respectively. After years of unprecedented growth for both China and the wine sector, it shows we have entered a period of slower growth. This evolving market will provide both challenges and opportunities for importers and producers as they pursue to succeed in the world's largest consumer market.

References

ABARES. (2013). Agricultural commodity statistics. Australian Bureau of Agricultural and Resource Economics and Sciences, Canberra, Australian Capital Territory, Australia.

Basalan, M., Gungor, T., Owens, F.N., and Yalcinkaya, I. (2011). Nutrient content and in vitro digestibility of Turkish grape pomaces. *Animal Feed Science and Technology*, 169, 194–198.

Beveridge, T.H.J., Girard, B., Kopp, T., and Drover, J.C.G. (2005). Yield and composition of grape seed oils extracted by supercritical carbon dioxide and petroleum ether: Varietal effects. *Journal of Agricultural Food and Chemistry*, 53, 1799–1804.

Bordiga, M., Coïsson, J.D., Locatelli, M., Arlorio, M., and Travaglia, F. (2013). Pyrogallol: An alternative trapping agent in proanthocyanidins analysis. *Food Analytical Methods*, 6, 148–156.

Bordiga, M., Travaglia, F., Locatelli, M., Coïsson, J.D., and Arlorio, M. (2011). Characterisation of polymeric skin and seed proanthocyanidins during ripening in six *Vitis vinifera* L. cv. *Food Chemistry*, 127, 180–187.

California Sustainable Winegrowing Alliance. (2004). California wine community sustainability report executive summary. http://www.sustainablewinegrowing.org/docs/cswa_2004_report_executive_summary.pdf (accessed on May 2014).

Council Regulation (EC) No. 479/2008 of 29 April 2008 on the common organization of the market in wine, amending Regulations (EC) Nos. 1493/1999, 1782/2003, 1290/2005, and 3/2008 and repealing Regulations (EEC) Nos. 2392/86 and 1493/1999. Official Journal of the European Union.

Council Regulation (EC) No. 491/2009 of 25 May 2009 amending Regulation (EC) No. 1234/2007 establishing a common organization of agricultural markets and on specific provisions for certain agricultural products (Single CMO Regulation). Official Journal of the European Union.

Council Regulation (EC) No. 607/2009 of 14 July 2009 laying down certain detailed rules for the implementation of Council Regulation (EC) No. 479/2008 as regards protected designations of origin and geographical indications, traditional terms, labelling and presentation of certain wine sector products. Official Journal of the European Union.

Council Regulation (EEC) No. 822/87 of 16 March 1987 on the common organization of the market in wine. Official Journal of the European Communities.

Council Regulation (EEC) No. 823/87 of 16 March 1987 laying down special provisions relating to quality wines produced in specified regions. Official Journal of the European Communities.

Council Regulation (EC) No. 1493/1999 of 17 May 1999 on the common organization of the market in wine. Official Journal of the European Union.

CSWA. (2013). Sustainable winegrowing program. http://www.sustainable winegrowing.org/ (accessed on May 2014).

Deng, Q., Penner, M.H., and Zhao, Y. (2011). Chemical composition of dietary fiber and polyphenols of five different varieties of wine grape pomace skins. *Food Research International*, 44, 2712–2720.

European Commission. (2006a). Impact assessment—Annex to the communication from the Commission to the Council and the European Parliament "Towards a Sustainable European Wine Sector." http://ec.europa.eu/agriculture/capreform/wine/fullimpact_en.pdf (accessed on June 2014).

European Commission. (2006b). WINE—Economy of the sector. European Commission, Official Journal of the European Communities. http://ec.europa.eu/agriculture/markets/wine/studies/rep_econ2006_en.pdf (accessed on May 2014).

European Commission. (2007a). Accompanying document to the Proposal for a Council Regulation on the common organisation of the market in wine and amending certain regulations—Impact assessment. http://ec.europa.eu/agriculture/capreform/wine/impact072007/full_en.pdf (accessed on June 2014).

European Commission. (2007b). Fact sheet—Towards a sustainable European wine sector. http://ec.europa.eu/agriculture/publi/fact/wine/072007_en.pdf (accessed on June 2014).

European Commission. (2008). Reform of the EU wine market. http://ec.europa.eu/agriculture/capreform/wine/index_en.htm (accessed on February 2014).

European Commission. (2009–2012). Wine CMO: Financial execution of the national support program. Directorate-General for Agriculture and Rural Development. http://ec.europa.eu/agriculture/markets/wine/facts/index_en.htm (accessed on February 2014).

European Court of Auditors. (2012). The reform of the common organization of the market in wine: Progress to date. Special Report No. 7. http://eca.europa.eu/portal/pls/portal/docs/1/14824739.pdf (accessed on May 2014).

European Landfill Directive (1999/31/EC). The landfill of waste. http://eurlex.europa.eu/legalcontent/EN/TXT/?uri=CELEX:31999L0031 (accessed on April 2014).

Eurostat. (2013). Eurostat statistics database. Luxembourg: Eurostat. http://epp.eurostat.ec.europa.eu/portal/page/portal/education/data/database/ (accessed on February 2014).

FAO. (2012). FAO Statistical Yearbook 2012. FAOSTAT. http://faostat.fao.org (accessed on June 2014).

FAO. (2014). FAOSTAT. http://faostat.fao.org/ (accessed on November 8, 2014).

Fontana, A.R., Antonilli, A., and Bottini, R. (2013). Grape pomace as a sustainable source of bioactive compounds: Extraction, characterization, and biotechnological applications of phenolics. *Journal of Agricultural and Food Chemistry*, 61, 8989–9003.

24 VALORIZATION OF WINE MAKING BY-PRODUCTS

Gonzalez-Paramas, A., Esteban-Ruano, S., Santos-Buelga, C., Pascual-Teresa, S., and Rivas-Gonzalo, J. (2004). Flavanol content and antioxidant activity in winery byproducts. *Journal of Agricultural Food and Chemistry*, 52, 234–238.

International Organization for Standardization. (2004). ISO 14000—Environmental management. http://www.iso.org/iso/iso14000 (accessed on March 2014).

Kim, S.Y., Jeong, S.M., Park, W.P., Nam, K.C., Ahn, D.U., and Lee, S.C. (2006). Effect of heating conditions of grape seeds on the antioxidant activity of grape seed extracts. *Food Chemistry*, 97, 472–479.

Laufenberg, G., Kunz, B., and Nystroem, M. (2003). Transformation of vegetable waste into value added products: (A) the upgrading concept, (B) practical implementations. *Bioresource Technology*, 87, 167–198.

Molina-Alcalde, E., Moumen, A., and Martín García, A. (2008). By-products from viticulture and the wine industry: Potential as sources of nutrients for ruminants. *Journal of the Science of Food and Agriculture*, 4, 597–604.

MVSWGA. (2012). Sustainability report McLaren Vale 2012. http://www.mclarenvale.info (accessed on June 2014).

MVSWGA. (2013). McLaren Vale sustainable winegrowing Australia system (MVSWGA). http://www.sustainablewinegrowing.com.au (accessed on June 2014).

New Zealand Wine (NZW). 2009. New Zealand winegrowers statistical annual 2009. http://wineinf.nzwine.com/statistics_outputs.asp?id=89&cid=6&type=n (accessed on January 2014).

New Zealand Wines. (2014). Sustainability. http://www.nzwine.com/sustainability/ (accessed on May 2014).

OIV. (2004). Resolution CST 1/2004. Development of sustainable vitiviniculture. http://www.oiv.int/oiv/info/enresolution (accessed on January 2014).

OIV. (2008). Resolution CST 1/2008. OIV guidelines for sustainable vitiviniculture: Production, processing and packaging of products. http://www.oiv.int/oiv/info/enresolution (accessed on January 2014).

OIV. (2012). Statistical report on world vitiviniculture. http://www.oiv.int/oiv/info/enizmiroivreport/ (accessed on May 2014).

OIV. (2012–2014). Strategic plan. http://www.oiv.int/oiv/info/enplanstrategique (accessed on May 2014).

OIV. (2013). State of the Vitiviniculture World Market. http://www.oiv.int/oiv/info/enconjoncture/ (accessed on May 2014).

OIV. (2014). State of the Vitiviniculture World Market. http://www.oiv.int/oiv/info/en_vins_effervescents_OIV_2014 (accessed on December 2014).

Pastor del Rio, J. and Kennedy, J.A. (2006). Development of proanthocyanidin in *Vitis vinifera* L. cv. Pinot Noir grapes and extraction into wine. *American Journal of Enology and Viticulture*, 57, 125–132.

Patras, A., Brunton, N.P., Gormely, T.R., and Butler, F. (2009). Impact of high pressure processing on antioxidant activity, ascorbic acid, anthocyanins and instrumental colour of blackberry and strawberry puree. *Innovative Food Science and Emerging Technologies*, 10(3), 308–313.

Rockenbach, I.I., Rodrigues, E., Gonzaga, L.V., Caliari, V., Genovese, M.I., Gonçalves, A.E.S.S., and Fett, R. (2011). Phenolic compounds content and antioxidant activity in pomace from selected red grapes (*Vitis vinifera* L. and *Vitis labrusca* L.) widely produced in Brazil. *Food Chemistry*, 127, 174–179.

Rondeau, P., Gambier, F., Jolibert, F., and Brosse, N. (2013). Compositions and chemical variability of grape pomaces from French vineyards. *Industrial Crops and Products*, 43, 251–254.

SAWIS. (2014). South African wine industry information and systems. http://www.sawis.co.za (accessed on June 2014).

Torres, J.L., Varela, B., García, M.T., Carilla, J., Matito, C., Centelles, J.J., Cascante, M., Sort, X., and Bobet, R. (2002). Valorization of grape (*Vitis vinifera*) by-products. Antioxidant and biological properties of polyphenolic fractions differing in procyanidin composition and flavonol content. *Journal of Agricultural and Food Chemistry*, 50, 7548–7555.

Wine Australia. (2013). How we market Australian wine. http://www.wine australia.com/en/Market%20Development.aspx (accessed on February 2014).

Wines of Argentina. (2011). Wines of Argentina today. http://www.winesofargentina.org/en/wofa/ (accessed on February 2014).

Wines of Chile. (2010). Wines of Chile sustainability program. http://www.winesofchile.org (accessed on May 2014).

Wines of Chile. (2011). Strategic plan 2020. http://www.winesofchile.org/news-press/wines-of-chiles-strategic-plan-2020/ (accessed on February 2014).

Wines of Chile. (2013). Wines of Chile's largest online marketing campaign aimed at China. http://www.winesofchile.org/2012/10/wines-of-chiles-largest-online-marketing-campaign-aimed-at-chinese-market/ (accessed on February 2014).

Yu, J. and Ahmedna, M. (2013). Functional components of grape pomace: Their composition, biological properties and potential applications. *International Journal of Food Science and Technology*, 48, 221–237.

2

WINE MAKING PROCESS

HEND LETAIEF

Contents

2.1	Overview	28
2.2	Vines	29
	2.2.1 General Considerations	29
2.3	Grapes	30
	2.3.1 Grape Growing	30
	2.3.2 Terroir	31
	2.3.3 Soil	32
	2.3.4 Climate	33
	2.3.5 Slope	33
	2.3.6 Good Viticulture Practices	33
	2.3.7 Grape Berry Development and Composition	35
	2.3.8 Vineyard and Grape Waste	37
	2.3.8.1 Pruning Waste	37
	2.3.8.2 Grape By-Products	38
2.4	Wine	39
	2.4.1 Grape Harvest and Documentation	40
	2.4.2 Pre-Fermentation Operations	41
	2.4.2.1 General Considerations	41
	2.4.2.2 Pre-Fermentation By-Products	42
	2.4.3 Pressing	43
	2.4.3.1 Pressing Methods	43
	2.4.3.2 Pressing By-Products	45
	2.4.4 Alcoholic Fermentation	47
	2.4.4.1 Yeast	47
	2.4.4.2 Control of Alcoholic Fermentation	48
	2.4.4.3 Management of Alcoholic Fermentation	49
	2.4.5 Maceration	51
	2.4.6 Malolactic Fermentation	52
	2.4.7 Fermentation By-Products	54

28 VALORIZATION OF WINE MAKING BY-PRODUCTS

2.4.8	Post-Fermentation Operations	55
2.4.9	Wine Aging	56
2.4.10	Wood	56
2.4.11	Reactions	57
2.4.12	Stabilization	58
2.4.13	Fining	59
2.4.14	Tartaric Stabilization	60
2.4.15	Filtration	61
2.4.16	Winery Wastewater	62
References		63

2.1 Overview

Industrial wine production and its complementary products are accompanied by the generation of large quantities of waste streams, namely, the organic waste (solids, skins, pips, marc, etc.), wastewater, emission of greenhouse gases (CO_2, volatile organic compounds, etc.), and inorganic waste (diatomaceous earth, bentonite clay, and perlite). Uncontrollable variables due to both human and physical infrastructures offer an explanation as to why waste management is practiced as end-of-pipe (additive) technologies in numerous wineries, notably wastewater treatment and landfilling of solid waste. Due to the rapidly growing global demand on manufacturing processes to produce more wine with minimal or no environmental footprints, the wine industry is under tremendous legislative pressure to become more efficient (Massette 1994). Thus, with the increasing demand for greening industrial production processes and products, both from customers and legislative authorities, coupled with rising operational and waste treatment costs, the wine industry has started to move toward the adoption of integrated waste preventative approaches, as opposed to the traditional reparatory environmental engineering practices. The final winery waste matrix is usually found to be a combination of interactive factors. Examples of such factors are the type of technology used, reuse and recovery of useful by-products, and the operating practices within a given winery. On the other hand, different production scenarios may have a critical influence on the consumption of raw materials and effluent quantity and quality. The standard vinification process consists of de-stemming, crushing, cooling (storage), screening, fermentation, clarification (maturation), stabilization, and bottling. However, various wine companies use different process routes, which significantly impact waste management for both intrinsic and extrinsic waste. To help frame the problem of waste minimization in wine production, it is important to establish an understanding of the product route from raw materials (grapes) to the final product (bottled wine). This chapter acquaints the reader with two important aspects of enology: grape culture and wine production. It begins with a basic outline of grapevine and viticulture practices followed by an exploration of pre-fermentation practices, fermentation and post-fermentation operations, such as aging, fining, tartaric stabilization, and filtration.

2.2 Vines

2.2.1 General Considerations

The grapevine (*Vitis vinifera* L.) is an ancient crop with a complex taxonomic structure because of its vegetative propagation and wide distribution. It is represented by many cultivars with numerous synonyms and homonyms (Vargas et al. 2007). The grapevine (*V. vinifera*) belongs to the family Vitaceae, which comprises about 60 interfertile wild *Vitis* species distributed in Asia, North America, and Europe under subtropical, Mediterranean, and continental–temperate climatic conditions. More than 60 species are recognized in the grape genus *Vitis*, but globally, almost all wine is made from *V. vinifera*, which is native to the area south of the Caucasus Mountains and the Caspian Sea (Winkler et al. 1974). It is the single *Vitis* species that has acquired significant economic interest over time; some other species, such as the North American *Vitis rupestris*, *Vitis riparia*, or *Vitis berlandieri*, are used as breeding rootstock due to their resistance against grapevine pathogens, such as *Phylloxera*, *Oidium*, and mildew. Indeed, a great majority of cultivars widely cultivated for fruit, juice, and mainly for wine, classified as *Vitis vinifera* L. subsp. *vinifera* (or *sativa*), derive from wild forms (*Vitis vinifera* L. subsp. *sylvestris* [Gmelin] Hegi) (Rossetto et al. 2002; Sefc et al. 2003; Crespan 2004; This et al. 2004).

2.3 Grapes

2.3.1 Grape Growing

The sequence of vineyard development processes from initial planning through to picking, starts with the selection of the vineyard site, choice of the rootstock, vine variety, and clone (Figure 2.1). Soil testing and the preparation and choice of vine density and trellis system follow (the trellis system is the support structures for the vine framework, required for a given training system). Once all of these aspects have been decided, vine planting may start, followed by training and pruning of the vines. Vine pests, vine diseases, and weeds should be controlled throughout the development of a vineyard. The last phase includes sampling the fruits and harvesting them when they reach maturation (Robinson 1999). The duties of the viticulturist during this final phase may include monitoring and controlling pests and diseases,

Figure 2.1 Early summer vineyard (Northern Hemisphere).

fertilizing, irrigating, canopy management, monitoring fruit development and characteristics, deciding when to harvest, and vine pruning during the winter months. With these tasks in mind, viticulture practices may be divided roughly into two classes. The first is terroir decisions, which may be defined as the "total natural environment of any viticultural site" (Robinson 1999). The second is best practices concerning viticulture. These two classes may once again be subdivided into different topics and are briefly discussed in this section.

2.3.2 Terroir

The French term *terroir* defines the geographical and environmental origin where grapes were grown (Laville 1990). This word includes characteristics such as soil composition (minerals, soil density), sunlight, climate (temperature, precipitations), and topography. It can also take into account strains of microorganisms usually found on the berry skin, which participate in wine elaboration. The soil type depends on how the geological parent material has been altered and shaped by physical, chemical, and biological processes. In general, suitable soils for viticulture are those that are not particularly fertile or deep. A *terroir* offering good conditions to a specific grapevine cultivar would then help the plant to produce high-quality grapes (Dominé 2004). Terroirs may be subdivided into three categories, namely, climate, soil, and slope (Figure 2.2).

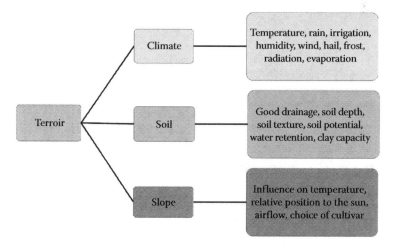

Figure 2.2 Components of a *terroir*. (Created using information from Dominé, A, ed., Vineyard and winery, in: *Wine*, Könemann, Koln, Germany, 2004.)

32 VALORIZATION OF WINE MAKING BY-PRODUCTS

2.3.3 Soil

Many vineyards are inherently heterogeneous in terms of soil properties because of the seemingly haphazard nature of soil-forming factors. On alluvial plains, soil variability can be high, with important properties varying over a few meters. This results in considerable variability for commercial vineyards in terms of vine vegetative growth (vigor), grape yield, ripening, maturity, fruit, and wine composition. Variation in vineyard vigor and soil properties can be associated with spatial variability in grape composition with implications for wine composition, style, and quality. A considerable body of research has reported patterns of spatial variation in vineyard attributes related to soil variability, such as vine vigor, yield, fruit ripeness, and wine color and phenolic composition (Johnson et al. 2001; Lamb et al. 2002; Johnson 2003; Bramley and Hamilton 2004; Bramley 2005; Bramley et al. 2011; Trought and Bramley 2011). Grapevines perform best when they have healthy, well-developed root systems and a wide range of soils suitable for their existence. It has also been shown that it is the physical characteristics of soil that affect the characteristic of the wine, not the chemical characteristics. Physical characteristics include good aeration, loose texture and moderate fertility, a good surface, and internal drainage capability (Vine et al. 1997). Proper soil drainage is a necessity for successful vine culture. Root growth in poorly drained soils is usually limited to the top 2 ft or less, compared to well-drained soils where roots may penetrate six feet or more. This restriction of the root system may cause vine growth and fruit yields to decrease, resulting in plant survival being limited to only a few years (Vine et al. 1997). Too much damp and nourishing reserve in the soil may also stimulate excessive growth of the grapevine that is unfavorable for the quality of the wine. Deeper soil has a higher sustainability and enhances the ability to guarantee stable growing conditions. The soil should also be of medium and open texture and have medium potential that will not induce excessive growth and cause a distorted balance between growth and accumulation of essential components in the grapes. Water retention is determined by the depth, texture, and structure of the soil as well as the quantity of organic material and rocks present. The ideal water retention is medium to high (15–25 mm/m). In dry land conditions, soil with a clay capacity of 10%–30% may improve water retention.

2.3.4 Climate

A glance at the distribution of winegrowing regions on our planet shows at once that wines will grow only in certain conditions. The most important factor for viticulture is climate and, above all, temperature. There is a striking difference between the Northern and Southern Hemispheres. The world map shows that vines clearly prefer moderate conditions. They seldom thrive where temperatures rise above 25°C in the summer months of July and January in the Northern and Southern Hemispheres, respectively. In a large part of Western Europe, which is the zone where the majority of Europe's classic viticultural regions lie, average July temperatures vary between 15°C and 25°C. Rainfall and drought also play an important part. It is almost impossible to grow wine with less than 200 mm of rain a year; on the other hand, too much rain also makes it difficult to cultivate grapes. A moderate climate, with adequate to relatively high rainfall, provides ideal conditions for producing both fragrant white wines with a good structure and acidity and well-balanced red wines with good potential for maturing (Dominé 2004).

2.3.5 Slope

Positioning a vineyard relative to the sun or on hills and valleys plays an important role in the vineyard outcome. The most significant influence of a hill is on the temperature of the area. The slope of a hill causes a 0.6°C decline in temperature with an increase in height of 100 m. Hills also provide warmer temperatures during the night and no extreme temperatures during the day. Southern and southeastern facing hills are cooler than northern and northwestern facing hills. Hills also provide better airflow with cooler air coming down the hill. The gradient of the agricultural site may also have an effect on which cultivar is best suited for the land.

2.3.6 Good Viticulture Practices

The selection of a viticultural site and decisions regarding soil preparation actions and the suitability of the soil for a specific cultivar should be based on proper profile studies. Deciding on the right

34 VALORIZATION OF WINE MAKING BY-PRODUCTS

choice of plant material is crucial to the well-being of the vineyard. The type of soil dictates the choice of rootstock, and the right type of clone must also be selected, since different clones can improve the complexity of a specific cultivar. Certified material should be used whenever available. Certain rootstocks are more resistant to pests and diseases in particular environments than others; therefore, the most pest- and disease-resistant rootstock should be used. The preparation of the soil should only be done once the type of soil has been identified through profile studies. Soil that is deeply cultivated has a better resistance against unfavorable weather. The planting distances also influence the growth of the vines and thus the quality of the wine. The rows should be a standardized 2.2 m apart (based on the potential of the soil), and the distances between the vines should be between 1 and 2 m. The development of the vines should be done in a way that ensures consistent vigor and quality in the vineyard. Enough leaf roof for optimal ripeness is ensured by a large enough canopy system, which should also suit the potential of the soil and the cultivar. The vineyard layout is influenced by the choice of the direction, which is determined by the climate. The direction should also be chosen to prevent soil erosion, allow maximum airflow, and reduce the occurrence of diseases. Decisions should be made on the type of irrigation practice that will be followed. The irrigation system relies on the resource, cultivar, pruning practices, and canopy system; it should also suit the soil type and climate. The right pruning and trellising system must be used, since it has a direct influence on the harvest and is also a means of manipulating the vineyard. The system should accommodate the vivacity of the vines to avoid forming canopies that are too dense, rather allowing maximum airflow in order to reduce the risk of disease occurrence. The foliage management also has an impact on the growth and the strength of the vines. A proper nutrition system should be derived to ensure moderate use of fertilizers. This is of the utmost importance, since it should guarantee moderately growing vineyards. In order to produce the optimal grape and wine quality, each individual situation requires the right fertilizing practice. Fertilizers should only be applied in agreement with soil analyses, since excessive fertilization causes water pollution and induces excessive growth and leaf density. This is an unfavorable situation for the vineyard in the long run. With cultivation practice, mechanical cultivation should be limited to a minimum. A system of minimum cultivation by the use of cover crops is advised. For disease and pest management, chemical control should be eliminated as much as possible in management practices and control measures.

2.3.7 Grape Berry Development and Composition

The development and maturation of grape berries has received considerable scientific scrutiny because of both the uniqueness of such processes to plant biology and the importance of these fruits as a significant component of the human diet and wine industry. For the winemaker, an outstanding attribute of *V. vinifera* is its ability to store enormous quantities of sugar in its berries. From the plant point of view, the ripe phenotype is the summation of biochemical and physiological changes that occur during fruit development and make the organ edible and desirable to seed-dispersing animals (Giovannoni 2001). Control of the ripening timing, berry size, pigmentation, acidity, texture, pathogen susceptibility, and the relative assortment of volatile and non-volatile aroma and flavor compounds in winegrape cultivars are the major concerns to viticulturists. Molecular and biochemical studies of grape berry development and ripening have resulted in significant gains in knowledge over the past years. Understanding how and when various components accumulate in the berry is of critical importance to adjust grape growing practices and thus modify wine typology (Figure 2.3). Grape berries exhibit a double sigmoid growth pattern (Coombe 1992). Growth first occurs mostly by cell division and later by cell expansion. From flowering to approximately 60 days afterward, a first rapid growth phase occurs during which the berry is formed and the seed embryos are produced. Several solutes are accumulated in the berry during the first growth period, contributing in some extent to the expansion of the berry, and reach an apparent maximum around 60 days after flowering (Possner and Kliewer 1985). The most prevalent compounds among all the others are by far tartaric and malic acids. Tartaric acid is accumulated during the initial stages of berry development, and its concentration is highest at the periphery of the developing berry. By contrast, malic acid is accumulated in the flesh cells at the end of the first growth phase. These acids confer the acidity to the wine and are therefore critical to its quality. Hydroxycinnamic

Figure 2.3 Harvested Nebbiolo grapes.

acids are also accumulated during the initial growth period. They are distributed in the flesh and skin of the berry and are important because they are involved in browning reactions and they are precursors of volatile phenols (Romeyer et al. 1983). Tannins, including the monomeric catechins, are present in the skin and seed tissues but nearly absent in the flesh, and they also accumulate during the first growth period (Kennedy et al. 2000a,b, 2001). Several other compounds like minerals, amino acids, micronutrients, and aroma compounds also accumulate during the first phase of berry growth and affect grape berry quality and ultimately wine quality.

In most cultivars, the first growth phase is followed by a lag phase. The duration of this phase is specific to the cultivar considered, and its end corresponds to the end of the herbaceous phase of the fruit. After this period of absence of growth, a second growth phase takes place, which coincides with the onset of ripening. The French word *véraison* used to describe the change in berry skin color, which indicates the beginning of ripening, has been adopted to describe the onset of ripening. The most dramatic changes in grape berry composition occur during this second growth or ripening phase.

Berries switch from a stage where they are small, hard, and acidic with little sugar to a status where they are larger, softer, sweeter, less acidic, and strongly flavored and colored. The flavor that builds in grapes is mostly the result of the acid/sugar balance and the synthesis of flavor and aromatic compounds, or precursors, taking place at this time. The development of these characteristics will largely determine the quality of the final product (Boss and Davies 2001). This ripening phase begins in August in the Northern Hemisphere and lasts about 45 days, depending on the environmental conditions. Overall, the berry approximately doubles in size between véraison and harvest. Many of the solutes accumulated in the grape berry during the first period of development remain at harvest; yet due to the increase in berry volume, their concentration is significantly reduced. However, some compounds produced during the first growth period are indeed reduced (and not simply diluted) on a per-berry basis during the second period of berry growth. Among these is malic acid, which is metabolized and used as an energy source during the ripening phase. Tannins also decline considerably on a per-berry basis after véraison. Aromatic compounds produced during the first growth period also decline (again, on a per-berry basis) during fruit ripening. These include several of the methoxypyrazine compounds that contribute to the vegetal characters of some wines (such as Cabernet Sauvignon and Sauvignon Blanc) (Hashizume and Samuta 1999). The decline in pyrazines is thought to be linked to sunlight levels in the cluster. Therefore, if these compounds are deemed to be undesirable (the current prevailing opinion), then canopy management can be used to reduce them. Although the first growth period contributes to the final quality of the berry, the most important event occurring during the second growth period is a massive increase in compounds, the major ones being glucose and fructose, as a result of a total biochemical shift into fruit ripening mode.

2.3.8 Vineyard and Grape Waste

2.3.8.1 Pruning Waste The net amount of vineyard pruning waste has been recently estimated in about one oven-dry ton per hectare with minor differences between grape varietals and harvesting technologies (Spinelli et al. 2012). This waste has usually low-average

38 VALORIZATION OF WINE MAKING BY-PRODUCTS

moisture content and a high cellulose and lignin content. The residue generated by the annual pruning activity is generally disposed of by mulching it into the vineyard or by on-site combustion, in both cases representing an additional management cost. Mulching the windrowed pruning residues into the soil by means of a mechanical mulcher contributes to maintaining organic matter content, but as a drawback, it may represent a dangerous mean of inoculation in areas infected by diseases as *Phomopsis viticola* and *Botrytis cinerea* (Jacometti et al. 2007). Most of this waste is also burnt and in much smaller proportion is used as fuel. An open air combustion of residues concentrated at the field hedge requires no specific equipment, hence implying a lower cost, but this practice is generally restricted by the environmental laws because of its polluting emissions and negative consequences on air quality (Estrellan and Iino 2010). Furthermore, woody residues from agriculture, similarly to forest residues, may be collected and used as a fuel with an excellent energy balance, a net reduction of polluting emissions, and a clear benefit in terms of global warming control (Kimming et al. 2011). Vineyard pruning residues could partly replace traditional wood assortments for energy and industrial use, and they may play an important subsidiary role in supplying bio-energy plants with renewable fuel, especially in rural areas and where the forest resource is scarce (Moldes et al. 2007). On the other hand, vineyard pruning residues have peculiar quality characteristics compared to other lignocellulosic feedstock, which may affect the choice and the performance of the conversion technology, as well as the potential for cofiring (Molcan et al. 2009). Previous studies have shown that many types of lignocellulosic waste, such as vineyard pruning waste, can be hydrolyzed to extract hemicellulosic sugars, which can be fermented easily and cheaply to produce valuable substances, such as lactic acid and biosurfactants. However, the cellulosic fraction of the lignocellulosic material remains as a by-product of this process (Moldes et al. 2007).

2.3.8.2 Grape By-Products The wine making process generates large amounts of solid waste, which might account for over 30% (w/w) of the grapes used (Makris et al. 2007). The two main by-products obtained from wine cellars are grape pomaces (10%–20% of processed material) and stems (2%–8%). Grape pomace consists of fruit skins, remnants from the fruit pulp, and in certain cases, some stems, with the skins and seeds making up the major part. Several authors have reported the important physicochemical properties of grape pomaces and stems as they are good sources of dietary fibers and antioxidants (Llobera and Canellas 2007). Grape skins, seeds, and stems are not hazardous waste, but the high content of organic matter and their seasonal production can contribute to potential pollution problems, especially regarding the chemical and biological oxygen demand of groundwater (Spigno et al. 2008).

2.4 Wine

Wine making generates different residues characterized by high contents of biodegradable compounds and suspended solids (Figure 2.4). In summary, the residues consist of plant remains derived from the de-stemmed grapes, the sediments obtained during clarification, bagasse from pressing, and lees, which are obtained after different decanting steps (Navarro et al. 2005). The wastewater generated from wine making contains grape pulp, skins, seeds, and dead yeast used in alcoholic fermentation. Waste such as lees and grape marc may exert phytotoxic effects if applied to crops or wetlands (Moldes et al. 2008). Contamination problems related to winery wastewater treatment and disposal have raised concern about subsurface flow contamination and the possibilities of establishing wastewater treatment systems. Wines are generally categorized by experts in five classes: table wines, sparkling wines, dessert wines, aperitif wines, and pop wines. These wines differ in the type of grape variety used and also in the wine making method. The production lines in use in wine cellars for the grape extraction must range from grape picking by hand, gravity tipping into the press, and the gentle extraction of juice for later static settling and racking, to more mechanical extraction including crushing, dynamic drainers, use of pumps, augers, and presses. In any case, the production line should be managed in such a way as to enhance quality. According to various studies, careful handling of the vintage, use of gravity, or conveyor belts and appropriate well-programmed pressing methods all favor the final wine quality (Boulton et al. 1996). Figure 2.4 shows a scheme of both standard white and red wine making technology; the steps in which waste is generated have been noted.

40 VALORIZATION OF WINE MAKING BY-PRODUCTS

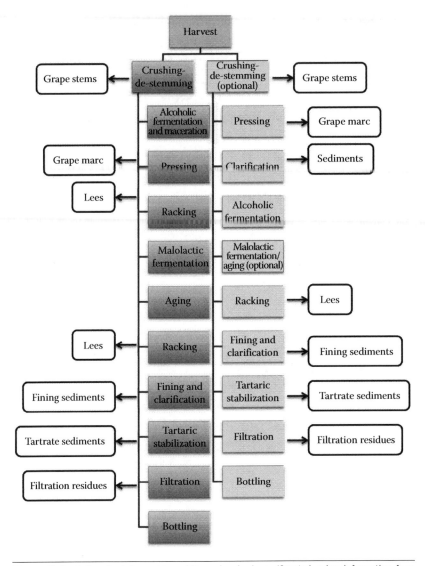

Figure 2.4 A flowchart for making white and red wines. (Created using information from Navarro, P. et al., *Water Sci. Technol.*, 51, 113, 2005.)

2.4.1 Grape Harvest and Documentation

The quality of a wine is greatly affected by the quality of the grapes used. A harvesting decision for "optimal ripeness" requires adequate knowledge of grape compositional factors relevant to achieving a targeted wine style, taking into consideration the grape cultivar, climate, topography, seasonal weather conditions, and vineyard management practices. Traditional measures used to determine grape ripeness include the assay of total soluble solids in juice as an estimate of grape sugar accumulation or an estimate of grape acidity decline as titratable acids or pH (Jackson and Lombard 1993). However, there is general recognition that these measures alone are not sufficient to accurately predict wine composition, notably as many key grape–derived compounds do not track with sugar accumulation, and are also highly dependent upon the grapevine genotype and its environment (Jackson and Lombard 1993). The composition and soundness of the grapes used for making wine is of major importance and dictates the quality of the resulting wine. The degree of ripeness of the grapes at harvest depends on the type of wine to be made with normal maturity being between 18 and 22.5 Brix for dry white wine. However, other varieties such as Chardonnay may benefit from more mature grapes. The grapes used for red wine making usually range from 18 to 25.2 Brix (Rankine 2004). The riper the grapes, the heavier the wine is in body and (generally) in flavor, as well as the higher in alcohol. In optimal conditions, grapes harvested should be cool (between 8°C and 16°C). In warmer areas, this may be achieved by machine harvesting at night. If this is not possible, must cooling may be applied. Sulfur dioxide and ascorbic acid may be added to the grapes directly after harvesting (Rankine 2004).

2.4.2 Pre-Fermentation Operations

2.4.2.1 General Considerations Once all of the requirements are met, the grapes can be harvested and transported to the cellar by means of a truck or tractor depending on the yield. Upon arriving at the cellar, the grape quality is properly analyzed and documented. Grading techniques may include the visual inspection of grapes, batch sampling, or any other means. Grapes are then de-stemmed and crushed, after which the stems are discarded (Rankine 2004). Crushing occurs in order to break the grape skin and release the juice. The grape bunches may be wholly or partially crushed. Stems may also be included in the fermentation if the wine maker wishes to do so, but it is not usual (Rankine 2004).

The thick liquid that forms as result from the crusher de-stemmer, called must, is neither grape juice nor wine but a mixture of grape

42 VALORIZATION OF WINE MAKING BY-PRODUCTS

juice, stem fragments, grape skins, seeds, and pulp (Robinson 1999). Tartaric acid may be added, if required, to ensure that the pH of the resulting juice is within a range of 3.0–3.4. Certain enzymes may also be added at this stage to hasten the yield of the drained juice and should be added as soon as possible after crushing to allow the longest period of time for them to act (Rankine 2004). The activity of these enzymes is strongly dependent on temperature. At 10°C, they have between 15% and 25% of their activity, with optimum activity at a temperature of 45°C–50°C. At 60°C, their activity rapidly starts to decrease until they become completely inactive at 80°C. Consequently, a conflict exists between the desire for cool temperatures for grape-handling operations and the need for efficient activity of the enzymes. In order to ensure that the must is the correct temperature, a mash cooler may be used, which pertains to a heat exchange device and cools the incoming pulp or grapes before fermentation. It is considered essential to maintain as low a temperature as possible during the early stages of the process in hot climate areas. A considerable loss in flavor may occur with higher temperatures in the beginning phase of the wine making process, especially with white wine. It may not always be possible to chill the grapes on arrival at the cellar or during de-stemming processes, in which case it should be done as soon as possible after. The type of mash cooler used depends on the cellar environment. The wine maker may soak the grape skin in the juice if he wishes to extract the flavors and aromas trapped in the skins (Arkell 2003). Certain varieties benefit more from skin contact, and the length of the contact depends on the temperature. Depending on the variety, skin contact may be for up to 18 h at 5°C–10°C. Skin contact is of greatest value with quality grapes from cool regions. With certain production methods, such as premium white wines and red wines produced by carbonic maceration and whole cluster fermentation, the grapes may be pressed without being de-stemmed and crushed.

2.4.2.2 Pre-Fermentation By-Products Stems are usually generated during wine making after grape de-stemming. They are currently disposed of by distilleries, landfill, or in rural areas. Many alternatives to these traditional disposals have been investigated, such as composting, removal of heavy metals from aqueous solutions, recovery of natural antioxidants, and addition to culture medium of the white-rot fungi *Trametes versicolor* to enhance laccase production (Lorenzo et al. 2002; Bertran et al. 2004; Martínez et al. 2006). White-rot fungi, in fact, are capable of preferentially degrading lignin when grown on lignocellulosics, because they are equipped with extracellular oxidative enzymes (such as laccase, manganese-dependent peroxidase, and the hydrogen-producing enzyme aryl alcohol oxidase), which cooperate for oxidation and depolymerization of lignin and several lignin-derived compounds (Curreli et al. 2004). Since stalks, like many other agricultural waste, are of a lignocellulosic nature (with cellulose, hemicelluloses, and lignin as structural components), their chemical processing could be undertaken according to the "biomass refining" philosophy, which is based on the separation of "fractions" according to their chemical properties and employment of these fractions as resources for large-scale commodity products such as fuels, bulk chemicals, and materials.

2.4.3 Pressing

2.4.3.1 Pressing Methods The pressing may be achieved using different types of presses (Arkell and George 2003). Vertical presses exert pressure downward or upward, imitating to a certain extent the traditional treading of the grapes by man. In these presses, the skins tend to accumulate parallel to the surface of pressing, and the liquid is then filtered through the solid materials. In continuous belt presses, the grapes or pomace are tipped into a preliminary draining area from where they pass on to a "sandwich" compression between two cloths and are flattened progressively between rollers. This process is rapid, but the liquid extracted is highly turbid since there is practically no self-filtering. In horizontal membrane presses, the pressure is exerted by filling a membrane with compressed air (pneumatic press) or water under pressure. Horizontal plate presses contain plates that move around in a pressing circuit within the basket of the press. They allow greater control over the process, and by blocking the liquid outlets, pomace contact can continue in the press cavity itself. Due to this flexibility in its use for different quantities of vintage, its suitability for both red and

44 VALORIZATION OF WINE MAKING BY-PRODUCTS

white wines, and the quality of the results obtained from the press fractions, the use of this press has become generalized. The internal structure of the grape is heterogeneous. Three zones can be considered: central (pulp around the seeds and the fibrovascular bundles leading to the peciole), intermediate, and peripheral zone (nearest to the skin). Their main characteristics are described in Table 2.1. In general, the malic acid content increases toward the center of the grape. This heterogeneous distribution has technological repercussions in the pressing. When grapes are subjected to pressure or maceration, the first juice released is from the intermediate area and the last from the peripheral zone. If we consider a pressing program that permits progressive extraction of must with stepped pressure increases, the first fractions will contain juice from the intermediate zone, and these would be richer in sugars and tartaric acid. The later fractions would be richer in malic acid and the last in phenolic compounds, salts, and oxidative enzymes. The seeds contain essential oils and a large part of the catechins responsible for bitterness. Excessive pressing might squash the seeds and stems, releasing fatty substances responsible for unpleasant, excessively bitter, and herbal tastes. In recent years, the use of membrane presses has been progressively introduced into white wine production. The appropriate use of these wine presses demands knowledge of the conditions affecting must yield and quality, plan for maximum production of high-quality must, minimum extraction of phenols (when this is required for the particular type of wine), minimum turbidity, and in many cases, minimum pressing time (Price 2008). High-quality pressing should allow musts with minimum polyphenols content and oxidation levels to be selected and obtained. The guidelines normally set for this are progressive must extraction with a slow increase in pressure, avoiding crushing the

Table 2.1 Grape Pulp Structure and Main Components

INTERNAL STRUCTURE OF GRAPES	JUICE QUALITY	MAIN COMPONENT
Central zone	Average	Sugar and malic acid
Intermediate zone	High	Sugar and tartaric acid
Peripheral zone	Low	Phenolics, minerals, aromas, oxidases

Source: Created using information from Arkell, J. and George, R., *Wine*, New Holland Publishers (UK) Limited, London, UK, 2003.

solid parts of the clusters, limiting enzyme activity especially where it leads to oxidation, and in many cases, permitting self-filtering of juices in the press itself (Benavente and García 1993) (Figure 2.5). Quality is conceived of as a pure must with turbidity between 50 and 150 nephelometric turbidity units (NTU) and generally a low optical density at 420 nm, and also a low total polyphenol index of 280 nm. The pH and total acidity must be appropriate for white wine making (Price 2008).

2.4.3.2 Pressing By-Products The grape marc results from pressing, and it represents the main by-product generated by the wine industry. Pomace obtained as a winery by-product constitutes 10%–20% of the weight of the grapes, and the seed content, on a wet basis, ranges from 20% to 30%. The grape marc cell wall is a complex network composed of 30% neutral polysaccharides, 20% acidic pectic substances, 15% insoluble proanthocyanidins, lignin, and structural proteins and phenols, these last two are cross-linked to the lignin–carbohydrate framework (Pinelo et al. 2006).

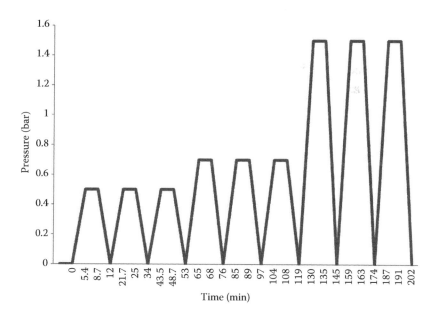

Figure 2.5 Direct pressing with three fractions: (first) up to 0.5 bar, (second) up to 0.7 bar, and (third) up to 1.5 bar. (Created using information from Benavente, J.L.A. and García, M.J., *Viticultura Enología Profesional*, 24, 31, 1993.)

46 VALORIZATION OF WINE MAKING BY-PRODUCTS

Although it has a considerable economic and environmental impact, it provides an opportunity for the development of value-added products through innovative technologies (Llobera and Canellas 2007). Grape pomace from wine processing is already used for the extraction of anthocyanins on an industrial scale. Increased efforts are now directed toward more extensive valorization of the residues from wine processing to obtain high-value co-products such as natural health remedies, food supplements, and novel nutrifunctional food ingredients or to use the material for enzyme production by solid-state fungal fermentation (Arvanitoyannis et al. 2006). This by-product is rich in phenolic compounds and dietary fibers, making it an excellent candidate for nutraceutical, medical, and food applications (Deng et al. 2011). Dietary fibers may be used by the food industry to offer physiological function-alities. Potential health benefits have been attributed to lignins (Mitjans and Vinardell 2005), and the high Klason lignin content of winegrape marc skins makes it suitable for several industrial applications (Stewart 2008). Grape seeds and skins are rich in phe-nolic compounds (Winkler et al. 1974). Approximately, 60% of the polyphenols present in grape are found in the seed, and these contain a relatively high concentration of flavan-3-ols, catechin, and epicatechin being the majority compounds in the grape seed, which account for 50%–70% of the total of flavanols according to the variety (Monagas et al. 2006). In the wine making process, the proanthocyanidins that remain in the grape seeds after the fermentation stage are bioactive compounds that can be extracted in order to obtain valuable products that could be employed by the pharmaceutical or food industries. Different adhesive formu-lations have been prepared from the phenolic compounds from grape marcs, and many positive effects of the phenolic extract from grapes on human health have been described. Grape-seed oil, on the other hand, is considered to be a dietary oil of high quality with a high concentration of unsaturated linoleic acid, vitamin E, and phytosterols. The extracts of procyanidins from grape seeds have an antioxidant activity *in vivo* and could be as important as vitamin E in the prevention of oxidative damage in tissues by the reduction of lipidic oxidation and/or blocking the production of free radicals (Tebib et al. 1997).

2.4.4 Alcoholic Fermentation

2.4.4.1 Yeast During alcoholic fermentation, yeast converts sugars into alcohol. The formed alcohol assists with the extraction of pigments and tannins from the skins of the grapes during maceration. In the last 30 years, the wine industry has tended to move away from spontaneous fermentations toward controlled fermentations initiated by inoculation, which are more reliable (Figure 2.6). In spontaneous fermentations, other yeast species may affect the fermentation process and wine characteristics, but these effects are variable and difficult to predict. *Saccharomyces cerevisiae* is the main yeast used in wine making, due to its high fermentation capacity. More than 200 different *S. cerevisiae* strains are currently available commercially, with highly diverse fermentation properties (Colombié et al. 2005). Studies have shown that differences in assimilable nitrogen and oxygen demands account for most of the differences between strains. The amounts of nitrogen required to regulate the fermentation are highly strain dependent and ranges between 48 and 110 mg/L of ammoniacal nitrogen (Julien et al. 2000). Temperature has been found to

Figure 2.6 Stainless steel tanks, at a controlled temperature.

affect yeast nitrogen requirements in terms of both quantity and quality (Beltran et al. 2007). The choice of strain used by the winemaker is increasingly motivated by the potential impact of that strain on the wine characteristics. Specific strains are now widely recognized to be useful for increasing the fruity character, for improving some varietal characters, and for limiting the production of organic acids or increasing the production of glycerol and volatile phenols (Scanes et al. 1998; Shinohara et al. 2000; Dubourdieu et al. 2006). Some strains produce mannoproteins and improve the color of red wines through their interactions with polyphenolic compounds (Medina et al. 2005). These mannoproteins will have an important effect on the mouthfeel of wines aged on lees.

2.4.4.2 Control of Alcoholic Fermentation Optimizing the control of alcoholic fermentations for wine making is a difficult challenge (Figure 2.7). The control of technological parameters, such as sugar exhaustion, the duration of the fermentation, and the amount of energy required to regulate the fermentation temperature, is important. Several methods for fermentation monitoring have been

Figure 2.7 Stainless steel fermentation vessels.

proposed as follows: density measurement, ethanol concentration, and measurement of the CO_2 produced (El Haloui et al. 1988; Warriner et al. 2002). CO_2 measurement is particularly promising because it provides online estimations of density, sugar, or ethanol concentration and calculates the instantaneous fermentation rate. This rate is of prime importance both technologically (proportional to the amount of energy produced by the fermenting tank) and microbiologically (proportional to the activity of yeasts). Some researchers have tried to develop online methods for measuring specific fermentation by-products using biosensors. Others have investigated the feasibility of simultaneous monitoring of several products by Fourier transform infrared spectroscopy or electronic noses (Pinheiro et al. 2002; Zeaiter et al. 2006).

2.4.4.3 Management of Alcoholic Fermentation Many works have shown that fast fermentations may be detrimental to wine quality, especially for white wines. On the contrary, too long a fermentation process both the subsequent processes and increases the risks of wine damage. The control of fermentation kinetics is generally considered a prerequisite for controlling the characteristics of the wine. Despite progress in fermentation management, incomplete fermentations remain a problem in several regions. Indeed, in recent years, many winemakers have focused on flavor development in ripening grapes. This has resulted in high-sugar musts becoming much more prevalent, and stuck or sluggish fermentations may have a major economic impact. Many factors can affect wine fermentation. Some practices such as acidification (or deacidification), clarification by physical or enzymatic treatments, enrichment with sugar, and addition of activators can impact the fermentation process. In white wine making, fermentation occurs at temperatures cooler than room temperature from 10°C to 18°C (Ough and Amerine 1961; Robinson 1999). Temperature may have a direct impact on aromatic characteristics, favoring either fermentative or varietal aromas. Low temperatures increase the production of volatile compounds (esters, acetates, medium-chain fatty acids) by the yeast during the alcoholic fermentation (Mc Lellan 1986; Torija et al. 2003). Enhancing the production of these volatile compounds may improve the aromatic profile of some wines. However, low temperatures may also result

50 VALORIZATION OF WINE MAKING BY-PRODUCTS

in sluggish or stuck fermentations, and the choice of the yeast strain and an increase in temperature at the end of fermentation are essential. In Sauvignon musts, temperature has been shown to affect the concentrations of varietal aromas, with higher fermentation temperatures (20°C) resulting in significantly higher concentrations of thiols than lower fermentation temperatures (13°C), regardless of the yeast strain used (Howell et al. 2004). In conventional red wine making, the aroma compounds produced by yeast fermentation have a much lesser impact than in white wine making. The fermentation temperature is mostly regulated to favor the transfer of polyphenol compounds from the solid to the liquid phase. A combination of high temperatures (up to 30°C) and an increase in ethanol concentration are generally considered to favor this extraction (Sacchi et al. 2005). Oxygen is also essential, mostly for maintaining cell viability at the end of fermentation. The amount of oxygen required during fermentation has been estimated at almost 10 mg/L, and the best time for oxygenation has been shown to be at the end of the growth phase (Sablayrolles et al. 1996). The addition of nutrients may be useful, particularly to avoid stuck fermentations. This practice is widespread, and the number of commercial nutrient products available is increasing. The addition of ammonium salts (diammonium phosphate or ammonium sulfate) is a highly efficient way to increase the fermentation rate and lower the fermentation duration (Cramer et al. 2002), but the timing of this addition is crucial. If nitrogen is added at the time of inoculation, it is metabolized and used for additional yeast growth. If added at the start of the stationary phase, it is mostly used to reactivate the existing yeasts. The maintenance of sufficiently high levels of turbidity (50–150 NTU) is essential for white wine production. Solid particles act as a source of lipid compounds and provide nucleation sites for decreasing the concentration of dissolved CO_2. Unsaturated fatty acids or sterols may be added to compensate, at least partly, for oxygen deficiencies. The addition of synthetic solutions of lipids is not permitted, but some permitted additives, such as yeast hulls and inactivated yeasts, contain lipid fractions. Most commercial products are combinations of different nutrients, including at least ammoniacal nitrogen, thiamine, and inactivated yeasts. They have a strong impact on fermentation kinetics, particularly if combined with oxygenation. New activators containing inactivated dry yeasts (IDY) for addition during the rehydration phase have recently become available. IDY facilitates the rehydration of active dry yeasts (ADY), due to the transfer of sterols from IDY to ADY, improving fermentation in cases of sterol deficiency (Soubeyrand et al. 2005). Certain metal ions in grape must have been shown to be an important factor in governing fermentation performance by wine yeasts (Birch et al. 2003). Magnesium plays an important role in protecting cells from stress factors. During fermentation, winemakers may add oak chips or shavings to create an oak complexity. It is important to note that these additions only impart oak complexity and are never a satisfactory replacement for oak cask maturation (Rankine 2004).

2.4.5 Maceration

The extraction of phenolics from grapes is a key aspect of red wine making. Phenolic compounds are primarily located in the skins and seeds of *V. vinifera* grapes. Leaving the partially fermented wine on the skins to draw out more tannin, color, and flavor is referred to as *maceration* (Arkell and George 2003). Without maceration, wine made from dark-skinned grapes is merely pink (Robinson 1999). The combined process of fermentation and maceration may take anywhere between 24 h and 3 weeks depending on the color of the final product required. Agitation of the must prior to, during, and after primary fermentation has long been known to increase the extraction of phenolics (Ough and Amerine 1961). Pre-fermentation maceration, or "cold soaking," is used by winemakers to improve the wine color intensity through increased anthocyanin extraction (Parenti et al. 2004), but without increased tannin extraction from seeds (Peyrot des Gachons and Kennedy 2003). Scientists suggest that these methods of extraction will have different effects (Boulton et al. 1996). Winemakers are able to implement various maceration techniques during red wine fermentation, such as plunging or pump overs at regular intervals during the fermentation so that they mix with the fermenting juice. The grape skins may also be submerged with headboards so that the skins are in constant contact with the juice; there exist specifically designed red wine fermenters to assist in this process. Namely, pump overs may extract less color due to uneven leaching of the cap but may extract rough, bitter tannins due to the harsh treatment of the must

52   VALORIZATION OF WINE MAKING BY-PRODUCTS

(which increases the level of fine solids in the wine), while plunging results in a more gentle, even extraction of phenolic compounds (González-Manzano et al. 2004). Importantly, the rate of extraction and the types of phenolic compounds extracted by maceration techniques can change depending on the temperature and percentage of alcohol in the must (González-Manzano et al. 2004; Sacchi et al. 2005). It has been suggested that higher temperatures increase the permeability of hypodermal cells of the skin and seed tissues resulting in greater release of phenolics, while certain phenolic compounds become soluble at different ethanol concentrations, particularly those from the seeds (González-Manzano et al. 2004; Sacchi et al. 2005). However, one of the most important factors in phenolic extraction has been suggested to be the length of contact time between the phenolic-containing tissues and the juice, must, or wine (Casassa and Harbertson 2014). It is on this principle that winemakers base the use of extended maceration on post-fermentation (Sacchi et al. 2005).

2.4.6 Malolactic Fermentation

Red wines usually undergo malolactic fermentation (MLF), also referred to as secondary fermentation since it almost never precedes alcoholic fermentation. It is practically a biological process of wine deacidification in which the dicarboxylic L-malic acid (malate) is converted to the monocarboxylic L-lactic acid (lactate) and carbon dioxide (Davis et al. 1985). Deacidification is particularly desirable for high-acid wine produced in cool climate regions. This process is normally carried out by lactic acid bacteria (LAB) isolated from wine, including *Oenococcus oeni* (formerly *Leuconostoc oenos*) (Dicks et al. 1995), *Lactobacillus* spp., and *Pediococcus* spp. (Davis et al. 1985). *Oenococcus oeni* is the preferred species used to conduct MLF due to its acid tolerance and flavor profile produced. This happens naturally but may also be artificially induced via the injection of lactic bacteria (Arkell and George 2003). During this stage, the wine may be given a light fining of bentonite or egg white to settle the suspended material, and if necessary, the acidity may be adjusted with the addition of tartaric acid to between pH 3.3 and 3.6 (Rankine 2004). One of the main reasons advanced in favor of MLF is that wines that have undergone MLF are, in a microbiological sense, more stable than those that have not sustained MLF. Wines without MLF may undergo this reaction in the bottle; as a consequence, the sediment, haze, and gassiness produced would be considered as spoilage. It is now accepted that the role of MLF is more than just a deacidification process, although deacidification via MLF is still a primary objective in wine fermentation in cool climate regions. The complexity and diversity of the metabolic activity of LAB suggest that MLF may affect wine quality both positively and negatively. Despite the significant influence of MLF on wine aroma, only certain wine attributes modified during MLF can be related to the production or utilization of specific chemical compounds by wine LAB. According to Henick-Kling (1995), MLF enhances the fruity aroma and buttery note but reduces the vegetative, green/grassy aroma. The enhanced fruitiness may be the result of the formation of esters by wine LAB. Esters, such as ethyl acetate and C4 to C10 fatty acid ethyl esters, are largely, if not exclusively, responsible for the fruity aroma of wine (Ebeler 2001). The increased buttery note is known to arise from diacetyl produced from citrate fermentation by wine LAB. The final level of diacetyl in wine is affected by a number of factors, such as bacterial strain, wine type, sulfur dioxide, and oxygen (Henick-Kling 1995; Nielsen and Richelieu 1999). It should be pointed out that diacetyl is formed chemically from the oxidative decarboxylation of α-acetolactate, an unstable intermediary compound produced during citrate metabolism (Ramos et al. 1995). The reduction in vegetative, green/grassy aroma may be due to the catabolism of aldehydes by wine LAB. Hexanal, *cis*-hexen-3-al, and *trans*-hexen-2-al cause the green, grassy, and vegetative aroma off in wine (Ferreira et al. 1995). Presumably, wine LAB (especially oenococci) can also metabolize these aldehydes (Keenan 1968). Besides aroma, MLF is believed to increase the body and mouthfeel of wine and give a longer aftertaste (Henick-Kling 1995). This may be ascribable to the production of polyols and polysaccharides by wine LAB. The production of glycerol, erythritol, and other polyols by oenococci and other wine LAB has been observed (Liu et al. 1995). In addition to the fruity and buttery notes, other flavor characteristics associated with MLF are described as floral, nutty, yeasty, oaky, sweaty, spicy, roasted, toasty, vanilla, smoky, earthy, bitter, ropy, and honey (Henick-Kling 1995). Some LAB possess decarboxylases that decarboxylate amino acids to form corresponding amines and carbon

54 VALORIZATION OF WINE MAKING BY-PRODUCTS

dioxide. Amines are toxic substances that have deleterious effects on human health (Shalaby 1996). The wine LABs vary in their ability to produce amines from amino acids. In wine, it appears that the lactobacilli and pediococci are the main producers of amines, although some oenococci can also produce amines (Leitão et al. 2000). MLF may also be applied to select a few white wine styles. In certain white wines, it is definitely not desired.

2.4.7 Fermentation By-Products

Lees are one of the most important winery by-products. They are also known as dregs, and they are defined as the residue formed at the bottom of recipients containing wine after fermentation, during storage, or after authorized treatments, as well as the residue obtained following filtration or centrifugation. Wine lees are mainly composed of microorganisms (mainly dead yeasts), tartaric acid, inorganic matter, and phenolic compounds. Literature has also reported the presence of anthocyanins and other phenolics in wine lees (Morata et al. 2003). Autolysis phenomena undergone by yeast lees during wine aging produce breakdown of cell membranes, release of intracellular constituents, liberation of hydrolytic enzymes, and hydrolysis of intracellular biopolymers into products of low molecular weights; thus, yeast lees autolysis is of paramount importance in lees composition and in their influence on wine aging (Guilloux-Benatier and Chassagne 2003). The most common microorganisms present in lees are yeast, which are responsible for alcoholic fermentation, but bacteria from MLF may also be present (Salmon et al. 2002). Thus, lees can be responsible for the presence in wines of amino acids, decarboxylase-positive microorganisms, and decarboxylase enzymes (which can be released during yeast lees autolysis), which, under favorable environmental conditions, can lead to biogenic amines formation. The widely reported importance of lees for natural removal of undesirable compounds from wine makes advisable the exploitation of this phenomenon in other fields such as water detoxification and filters production, in which lees could play a key role as natural and cheap decontaminant. Lees could be used as a supplement in animal feed, but when recovered by centrifugation after column distillation, they have a very poor nutritional value, making them unsuitable for this purpose (Maugenet 1973). This is probably due to the high amounts of polyphenols associated with proteins, which render the proteins nonassimilable, or to the presence of toxic elements (generated by treatment of the residues), which then accumulate in yeast lipids (Maugenet 1973). Some of the studies on wine lees have been focused on the role they play in the evolution of toxic compound such as mycotoxins, volatile phenols, pesticides, and defoaming agents during the vinification process. Some authors have proposed the utilization of alcoholic fermentation lees as nutritional media for lactic acid production. For example, wine making lees were employed as nutritional media for *Lactobacillus rhamnosus*, and it was demonstrated that lees from white wine making technology could be employed as general nutritional media for LAB.

2.4.8 Post-Fermentation Operations

Solid waste obtained after de-stemming and grape pressing (skin, pulp, seeds, lees, etc.) is not the only kind of residue obtained in a wine cellar. After clarification and stabilizing, some other by-products are also obtained. Post-fermentation treatment may begin almost immediately after fermentation. These may involve any necessary adjustments to the wine's physicochemical composition, as well as procedures such as aging on lees. Subsequent modifications involve various forms of clarification and fining and chemical and biologic stabilization including oxidation control. Both before and after bottling, further spontaneous chemical changes occur (aging). These affect the wine's visual, gustatory, and olfactory attributes, with potential beneficial or detrimental consequences. Post-fermentation treatments are also necessary to prevent haze formation in finished wines. Undesirable suspended particles and substances can induce turbidity, which is not usually appreciated by consumers. Among all the techniques employed to achieve a limpid wine, tartaric stabilization, fining, and filtration are most favored by winemakers. Precipitation of unstable tartaric acid crystals and colloidal substances is induced, together with the carrying over of metal–polyphenol complexes, polysaccharides, proteins, and ferric phosphate (Ribéreau-Gayon et al. 2000). Tartaric acid is the only by-product generally processed for commercial uses. If we keep in mind that coloring matter is also present, polyphenols are also expected to be available in this residual product.

56 VALORIZATION OF WINE MAKING BY-PRODUCTS

2.4.9 Wine Aging

Wine aging is an important process for high-quality wines. The aging process can be divided into two phases. The first one is called maturation, which refers to changes in wines after fermentation and before bottling, and the second phase of aging is bottling. During wine aging, many reactions occur, which tend to improve the taste and flavor of wine over time. These reactions may occur in either inert or wood containers. Because of the significance of maturing premium wines in barrels before bottling, the taxonomy, distribution, and structural and chemical attributes of oak are noted, as well as barrel construction, conditioning, and care, as well as their alternatives. Only a small percentage of white wine comes into contact with wood; therefore, barrel fermentation and aging is sometimes present in the production of certain white wines, especially wines made from the Chardonnay grape. In addition to improving mouthfeel, body, and aromatic persistence of white wines (Vidal et al. 2004; Pati et al. 2012), aging on fine lees reduces protein haze, protects from tartaric acid precipitation (Lomolino and Curioni 2007), and favors the growth of LAB (Guilloux-Benatier et al. 1993). Polysaccharides and mainly mannoproteins released during the yeast autolysis process seem to be responsible for these enological benefits (Pati et al. 2012).

2.4.10 Wood

Traditional oak barrel aging technology is the oldest and widely accepted technology. The application of wood fragments and physical methods is also promising in accelerating aging process artificially, while the application of micro-oxygenation and lees is reliable to improve the wine quality (Del Barrio-Galán et al. 2011; Guerrero et al. 2011). Wine aging in barrels is one of the most common methods in the wine making process. Oak barrels are beneficial to wines in two different aspects. On one hand, astringency-related phenolic compounds and oak-responsible aromatic compounds are transferred to wine during aging. The volatile compounds extracted from wood are mainly furfural compounds, guaiacol, oak or whisky lactone, eugenol, vanillin, syringaldehyde, and volatile phenols (Ortega-Heras et al. 2004; Matejícek et al. 2005). On the other hand, atmospheric oxygen permeation through the barrel wall allows certain compounds to be oxidized gently, which results in a reduction of astringency and changes in color (Gambuti et al. 2010). The extradimensions of flavor and oxygen ingress the wood provides depend on the origin of the oak, the size and age of the barrels, the number of times they have been used, and aging time. Oak species and their geographical origin play an important role in defining oak compositional differences and oxygen diffusion rates (Sauvageot and Feuillat 1999). There are three different wood-toasting intensities: light, medium, and heavy. Light toasting produces few pyrolytic by-products that result in less aromatic compounds but more tannin. Medium toasting produces many phenolic and furanilic aldehydes, which provides woods a vanillin and roasted character. Heavy toasting destroys or limits the synthesis of phenolic and furanilic aldehydes and simultaneously generates volatile phenols, which cause a smoky and spicy character (Fernández de Simón et al. 2010). Several oak-related volatile compounds extracted from oak gradually become exhausted with barrel reuse. Thus, the initial extraction rate of these compounds in new barrels is higher than that in used barrels, and more compounds related to toasting can be extracted from new barrels (Gómez-Plaza et al. 2004). Although the oak barrel aging technology has been extensively applied for centuries, several disadvantages of this traditional technology exist. Aging in barrel is time, space, and money consuming and normally takes from 3–5 months to 3–5 years or even longer. Moreover, as barrels become older, undesirable microorganisms such as the yeast genera *Brettanomyces* and *Dekkera* may contaminate them. These yeasts can produce significant concentrations of undesirable aromas in wines (Suárez et al. 2007). In addition, during barrel aging, there is a great loss of wine due to evaporation that causes a negative financial impact (Ruiz de Adana et al. 2005).

2.4.11 Reactions

The composition of wine is complex and changes continuously during aging. It has been concluded that high-quality red wine storage conditions (long period of time, permanent contact with wood, low pH values, low temperature, and aqueous environment) are all in favor of good molecular associations between the wine components.

58 VALORIZATION OF WINE MAKING BY-PRODUCTS

Small quantities of oxygen are usually present during oak maturation. A barrel microoxygenation profile can be assumed to require a long aging time during which the wine consumes practically all the oxygen it absorbs (Nevares and Del Alamo 2008). Anthocyanins tend to form new stable compounds during aging. These compounds include the products resulting from direct and acetaldehyde-mediated anthocyanin–tannin condensation reactions (Atanasova et al. 2002; Fulcrand et al. 2006) as well as the products originated from the C4/C5 cycloaddition reaction of anthocyanins with other molecules bearing a polarizable double bond, including pyruvic acid, 4-vinylphenols, vinylflavanols, and acetaldehyde, which conform the so-called pyranoanthocyanins (Rentzsch et al. 2007). Some of these reactions may be favored by the presence of small quantities of oxygen or by the acetaldehyde produced by the effect of oxygen on ethanol. In this way, oxygen or reactive species seems to be involved in the formation of A-type vitisins, while acetaldehyde seems to be involved in the formation of ethyl-linked anthocyanin and tannin adducts, B-type vitisins, vinyl-flavanols, and vinyl-pyranoanthocyanis (Lee et al. 2004; Morata et al. 2007). All these reactions may influence the color and color stability of wine, as well as gustatory qualities related to the structure of the tannins. Their relative importance as well as the structure of the end product depends not only on the initial wine composition but also on the presence of yeast metabolites, which have an effect on the reactivity toward the oxygen of wine polyphenolic compounds (Mazauric and Salmon 2005).

2.4.12 Stabilization

Stabilization can be divided into physical, chemical, and microbiological stabilization. Physical and chemical stabilizations (not insured by filtration) prevent the formation of organic and inorganic hazes and deposits after bottling. Microbiological stabilization by microfiltration is guaranteed by eliminating yeasts and bacteria that can destroy or modify a wine's taste. Wine stability is defined as a state or condition where the wine will not, for some definite period of time, exhibit undesirable physical, chemical, or organoleptic changes. These undesirable changes that denote wine instability are browning or other color deterioration, haziness or very slight cloudiness, cloudiness, deposits, and undesirable taste or odor. Many examples of colloidal instability have been identified in wines, for example, protein and polyphenol instability, iron and copper cloudiness, and pectin and yeast polysaccharide haziness (Ribéreau-Gayon et al. 2000).

After aging, red wines are blended and tasted again, analyzed, and may be stabilized, then filtered and lightly sulfited if necessary. Red wines may need two or three rackings before they can be filtered. They are transferred between vats or barrels to rack off their lees and to allow limited aeration (Arkell and George 2003). After fermentation or aging, white wines as allowed to settle, then racked under carbon dioxide off gross lees and protected against bacteria (Robinson 1999; Arkell and George 2003). At racking, sulfur dioxide and ascorbic acid may be added (Rankine 2004). Racking may be delayed for weeks with some varieties, such as Chardonnay, during which time MLF may occur. Cool temperatures and minimized exposure to air as well as the minimum handling of wine are of renowned importance in white wine making (Rankine 2004). Once the various batches of wine are blended to form a uniform bulk, the wines may be fined and stabilized.

2.4.13 Fining

Fining agents are commonly used in wine production to clarify, to control browning, and to improve stability and/or organoleptic characteristics. In white wine, fining is frequently employed for clarification or stabilization. It involves the formation of a floccular precipitate in wine, which will absorb or entrain the natural haze-forming constituents and colloidal particles while settling. Both the organic and inorganic fining agents (bentonite) are commonly used to clarify and stabilize wines, thus avoiding the appearance of haze in the bottle. Bentonite fining is a low cost and effective method for removing proteins from wine or grape juice. However, it has some negative attributes including dilution of the wine by the bentonite slurry, removal of positive flavor attributes, high labor costs, and handling and disposal problems as well as loss of quality for wine recovered from lees (Waters et al. 2005). Proteins have been used in wines as fining agents for a long time. Protein-based fining agents can determine some declines in astringency and bitterness of young red wines due to their

60 VALORIZATION OF WINE MAKING BY-PRODUCTS

interaction with tannins (Oberholster et al. 2013). Nevertheless, some fining agents (like albumin and gelatin) can affect wine color due to the precipitation of pigments. The various protein fining agents can behave differently, depending on their composition, their origin, and their preparation condition. Nowadays, a wide range of protein fining agents are used, including gelatin, casein, potassium caseinate, egg albumin or isinglass, and more recently, some proteins of vegetable origin. The fining process must be reasonably rapid, and the loss of saleable product in the sediment or lees should be minimal. Moreover, it should not have any undesirable effects, such as the removal of a desired flavor or the addition of undesired flavor components.

2.4.14 Tartaric Stabilization

Tartrate crystals (potassium hydrogen tartrate and calcium tartrate) develop naturally in wine and are the major cause of sediment in bottled wines. Alcoholic fermentation during wine making leads to a decrease in tartrate salt (potassium hydrogen tartrate [KHT]) solubility due to the presence of ethanol. As a consequence, at normal storage temperatures, an untreated wine is supersaturated in KHT, and undesirable precipitation can occur in the bottles. To overcome this problem, the excess of this salt is traditionally removed by cooling the wine to $-4°C$ over several days to induce KHT precipitation prior to bottling. Potassium bitartrate crystals may be added if rapid stabilization is required (Vine et al. 1997; Rankine 2004). The complexity of the cold stabilization process does not allow a precise control of the final KHT concentration achieved by this technique. Besides, this operation can affect wine quality due to the simultaneous precipitation of polysaccharides and polyphenols together with the KHT crystals (Vernhet et al. 1999). These limitations led to the development of other techniques like ion exchange resins and electrodialysis. The treatment by ion exchange resins consists of equilibrating the wine with a cation exchange resin that replaces the wine potassium ions by hydrogen or sodium ions. The treatment by electrodialysis is based on ion electrical migration and is a single-stage operation. This is in contrast with ion exchange, which requires an additional stage of operation for resins regeneration. Other methods include mannoproteins addition. Obtained from hydrolysis of the yeast cell wall, mannoproteins inhibit the crystallization of tartrate salts by lowering the crystallization temperature preventing the occurrence of precipitates in wine. Arabic gum is a protective colloid and also has some effectiveness to prevent tartaric stability (Ribéreau-Gayon et al. 2000). The addition of carboxymethylcellulose and metatartaric acid can also be used to prevent tartaric stability (Gerbaud et al. 2010).

2.4.15 Filtration

The key roles of filtration are to provide limpidity and also microbiological stabilization of wines. Three groups of compounds have been identified in crude wines according to the compound sizes as shown in Figure 2.8. Measuring turbidity, which is expressed by NTU, assesses wine limpidity. Table 2.2 summarizes the wines turbidity and its correlation with the visual aspect of the wines. After filtration, the turbidity of wines must be less than 2 NTU. In order to have a limpid wine before bottling, winemakers implement successive solid–liquid separations using traditional technologies such as centrifugation, filtration on sheets, diatomaceous earth filtration, and the

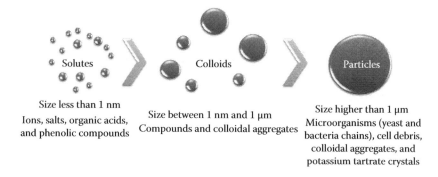

Figure 2.8 Crude wine compounds classified according to their size.

Table 2.2 Correlation between Wine Turbidity (Nephelometric Turbidity Unit) and Visual Aspects of Wine

TYPE	BRILLIANT (NTU)	HAZINESS (NTU)
White wine	<1.1	>4.4
Rose wine	<1.4	>5.8
Red wine	<2.0	>8.0

Source: Created using information from Cook, A.G. et al., *Biol. Trace Elem. Res.*, 103(1), 1, 2005.

62 VALORIZATION OF WINE MAKING BY-PRODUCTS

use of exogenic additives. Although diatomaceous earth filtration is the most used technique to clarify wines, it is nowadays classified as dangerous substances due to the presence of crystalline silica. Massive exposure may cause eye and airways irritations (Cook et al. 2005). Diatomaceous earth has also a negative impact on the environment. After use, it cannot be disposed of but must be transported to waste disposal sites to be treated. So, restrictions for environment and health force the enology sector to search for alternative techniques to traditional filtrations, and cross-flow microfiltration could represent this alternative. Indeed, this one-step technology can substitute the conventional processes, which imply several filtration steps on diatomaceous earth previous to the final microbial stabilization obtained by dead-end filtration on membranes. In addition to a great simplification of the wine processing line, cross-flow microfiltration offers a number of additional advantages such as elimination of earth use and its associated environmental problems as well as the combination of clarification, stabilization, and sterile filtration in one single continuous operation.

2.4.16 Winery Wastewater

The wine industry generates large volumes of wastewater, 1 and 4 L of effluent for each liter of wine produced (Airoldi et al. 2004). The bulk of winery wastewater emanates from cleaning equipment, vats, and floors of wine cellars during seasonal activities associated with wine making. Winery wastewater is typically composed of suspended solids, polyphenols, organic acids, alcohols, sugars (maltose, glucose, and fructose), aldehydes, soaps and detergents, nitrogen compounds, and inorganics, including some traces of heavy metals. It is characterized by a widely variable high organic strength (800–25,000 mg/L chemical oxygen demand [COD]), high salinity (3,000–4,000 µS/cm), high sodicity sodium adsorption ratio (sodium adsorption ratio [SAR] = 8–9), and a high COD associated with its large biodegradable organic fraction (Mosse et al. 2011). The volume, COD range, and organic composition are directly related to cellar activities, including must production, fermentation processes, maturation/stabilization processes, and decanting, and are thus prone to seasonal variation. Each grape varietal also has a unique organic fingerprint, so that, for example,

WINE MAKING PROCESS   63

wastewater generated during crushing of late harvest grape varietals has a comparatively high sugar and low phenolic content, while the converse is true for wastewater generated during the crushing of early harvest red grape varietals (Devesa-Rey et al. 2011). The wine industry recognizes the priority of the environmental and economical sustainable management of its wastewater. There are numerous treatment options (biological, physicochemical, and advanced), all of which aim to achieve a significant reduction in the concentration of organic matter and the solids present, and some may also reduce inorganic load. Biological treatment is particularly well suited to the treatment of winery wastewater, because the majority of the organic components in the waste stream are readily biodegradable. In terms of biodegradability, there is considerable variation in the ratios of the different fractions of winery effluent, including readily biodegradable sugars, moderately biodegradable alcohols, and slowly biodegradable/recalcitrant phenolics. Biological treatment processes include an activated sludge process, a sequencing batch reactor, a fixed-bed biofilm reactor, an air-bubble column bioreactor, a jet-loop-activated sludge reactor, and anaerobic digestion (Mosse et al. 2011). Advanced oxidation processes are currently being proposed as an alternative to biological treatment for this wastewater (Oller et al. 2011). Sand bioreactors have also been shown to be effective for the treatment of winery wastewater (Welz et al. 2012).

References

Airoldi, G., P. Balsari, and F. Gioelli. 2004. Results of a survey carried out in Piedmont region winery on slurry characteristics and disposal methods. Paper presented at the *Third International Specialised Conference on Sustainable Viticulture and Winery Wastes Management*, Barcelona, Spain.

Arkell, J. 2003. *Wine*. London, UK: New Holland Publishers (UK) Limited.

Arvanitoyannis, I.S., D. Ladas, and A. Mavromatis. 2006. Potential uses and applications of treated wine waste: A review. *International Journal of Food Science and Technology* 41(5):475–487.

Atanasova, V., H. Fulcrand, C.L. Guernevé, V. Cheynier, and M. Moutounet. 2002. Structure of a new dimeric acetaldehyde malvidin 3-glucoside condensation product. *Tetrahedron Letters* 43(35):6151–6153.

Beltran, G., N. Rozes, A. Mas, and J.M. Guillamon. 2007. Effect of low-temperature fermentation on yeast nitrogen metabolism. *World Journal of Microbiology and Biotechnology* 23(6):809–815.

Benavente, J.L.A. and M.J. García. 1993. Influencia del prensado en la calidad del vino. *Viticultura Enología Profesional* 24:31–37.

64 **VALORIZATION OF WINE MAKING BY-PRODUCTS**

Bertran, E., X. Sort, M. Soliva, and I. Trillas. 2004. Composting winery waste: Sludges and grape stalks. *Bioresource Technology* 95(2):203–208.

Birch, R.M., M. Ciani, and G.M. Walker. 2003. Magnesium, calcium and fermentative metabolism in wine yeasts. *Journal of Wine Research* 14(1):3–15.

Boss, P.K. and C. Davies. 2001. Molecular biology of sugar and anthocyanin accumulation in grape berries. In *Molecular Biology and Biotechnology of the Grapevine*, K. Roubelakis-Angelakis (ed.). Dordrecht, the Netherlands: Springer.

Boulton, R.B., V.L. Singleton, L.F. Bisson, and R.E. Kunkee. 1996. *Principles and Practices of Winemaking*. New York: Chapman & Hall.

Bramley, R.G.V. 2005. Understanding variability in winegrape production systems 2. Within vineyard variation in quality over several vintages. *Australian Journal of Grape and Wine Research* 11(1):33–42.

Bramley, R.G.V. and R.P. Hamilton. 2004. Understanding variability in winegrape production systems. 1. Within vineyard variation in yield over several vintages. *Australian Journal Grape and Wine Research* 10:32–45.

Bramley, R.G.V., J. Ouzman, and P.K. Boss. 2011. Variation in vine vigour, grape yield and vineyard soils and topography as indicators of variation in the chemical composition of grapes, wine and wine sensory attributes. *Australian Journal of Grape and Wine Research* 17(2):217–229.

Casassa, L.F. and J.F. Harbertson. 2014. Extraction, evolution, and sensory impact of phenolic compounds during red wine maceration. *Annual Review of Food Science and Technology* 5(1):83–109.

Colombié, S., S. Malherbe, and J.-M. Sablayrolles. 2005. Modeling alcoholic fermentation in enological conditions: Feasibility and interest. *American Journal of Enology and Viticulture* 56(3):238–245.

Cook, A.G., P. Weinstein, and J.A. Centeno. 2005. Health effects of natural dust. *Biological Trace Element Research* 103(1):1–15.

Coombe, B.G. 1992. Research on development and ripening of the grape berry. *American Journal of Enology and Viticulture* 43(1):101–110.

Cramer, A.C., S. Vlassides, and D.E. Block. 2002. Kinetic model for nitrogen-limited wine fermentations. *Biotechnology and Bioengineering* 77(1):49–60.

Crespan, M. 2004. Evidence on the evolution of polymorphism of microsatellite markers in varieties of *Vitis vinifera* L. *Theoretical and Applied Genetics* 108(2):231–237.

Curreli, N., A. Rescigno, A. Rinaldi, B. Pisu, F. Sollai, and E. Sanjust. 2004. Degradation of juglone by Pleurotus sajor-caju. *Mycological Research* 108(08):913–918.

Davis, C.R., D. Wibowo, R. Eschenbruch, T.H. Lee, and G.H. Fleet. 1985. Practical implications of malolactic fermentation: A Review. *American Journal of Enology and Viticulture* 36(4):290–301.

Del Barrio-Galán, R., S. Pérez-Magariño, and M. Ortega-Heras. 2011. Techniques for improving or replacing ageing on lees of oak aged red wines: The effects on polysaccharides and the phenolic composition. *Food Chemistry* 127(2):528–540.

Deng, Q., M.H. Penner, and Y. Zhao. 2011. Chemical composition of dietary fiber and polyphenols of five different varieties of wine grape pomace skins. *Food Research International* 44(9):2712–2720.

Devesa-Rey, R., G. Bustos, J.M. Cruz, and A.B. Moldes. 2011. Optimisation of entrapped activated carbon conditions to remove coloured compounds from winery wastewaters. *Bioresource Technology* 102(11):6437–6442.

Dicks, L.M.T., F. Dellaglio, and M.D. Collins. 1995. Proposal to reclassify *Leuconostoc oenos* as *Oenococcus oeni* [corrig.] gen. nov., comb. nov. *International Journal of Systematic Bacteriology* 45(2):395–397.

Dominé, A. (ed.) 2004. Vineyard and winery. In *Wine*. Koln, Germany: Könemann.

Dubourdieu, D., T. Tominaga, I. Masneuf, C.P. des Gachons, and M.L. Murat. 2006. The role of yeasts in grape flavor development during fermentation: The example of Sauvignon Blanc. *American Journal of Enology and Viticulture* 57(1):81–88.

Ebeler, S.E. 2001. Analytical chemistry: Unlocking the secrets of wine flavor. *Food Reviews International* 17(1):45–64.

El Haloui, N., D. Picque, and G. Corrieu. 1988. Alcoholic fermentation in winemaking: On-line measurement of density and carbon dioxide evolution. *Journal of Food Engineering* 8(1):17–30.

Estrellan, C.R. and F. Iino. 2010. Toxic emissions from open burning. *Chemosphere* 80(3):193–207.

Fernández de Simón, B., E. Cadahía, M. Del Álamo, and I. Nevares. 2010. Effect of size, seasoning and toasting in the volatile compounds in toasted oak wood and in a red wine treated with them. *Analytica Chimica Acta* 660(1):211–220.

Ferreira, B., C. Hory, M.H. Bard, C. Taisant, A. Olsson, and Y. Le Fur. 1995. Effects of skin contact and settling on the level of the C18: 2, C18: 3 fatty acids and C6 compounds in Burgundy Chardonnay musts and wines. *Food Quality and Preference* 6(1):35–41.

Fulcrand, H., M. Dueñas, E. Salas, and V. Cheynier. 2006. Phenolic reactions during winemaking and aging. *American Journal of Enology and Viticulture* 57(3):289–297.

Gambuti, A., R. Capuano, M.T. Lisanti, D. Strollo, and L. Moio. 2010. Effect of aging in new oak, one-year-used oak, chestnut barrels and bottle on color, phenolics and gustative profile of three monovarietal red wines. *European Food Research and Technology* 231(3):455–465.

Gerbaud, V., N. Gabas, J. Blouin, and J.-C. Crachereau. 2010. Study of wine tartaric acid salt stabilization by addition of carboxymethylcellulose (cmc): Comparison with the "protective colloids" effect. *International Journal of Vine and Wine Sciences* 44(4):231–242.

Giovannoni, J. 2001. Molecular biology of fruit maturation and ripening. *Annual Review of Plant Physiology and Plant Molecular Biology* 52(1):725–749.

Gómez-Plaza, E., L.J. Pérez-Prieto, J.I. Fernández-Fernández, and J.M. López-Roca. 2004. The effect of successive uses of oak barrels on the extraction of oak-related volatile compounds from wine. *International Journal of Food Science and Technology* 39(10):1069–1078.

66 VALORIZATION OF WINE MAKING BY-PRODUCTS

González-Manzano, S., J.C. Rivas-Gonzalo, and C. Santos-Buelga. 2004. Extraction of flavan-3-ols from grape seed and skin into wine using simulated maceration. *Analytica Chimica Acta* 513(1):283–289.

Guerrero, E.D., R.C. Mejías, R.N. Marín, M.J.R. Bejarano, M.C.R. Dodero, and C.G. Barroso. 2011. Accelerated aging of a Sherry wine vinegar on an industrial scale employing microoxygenation and oak chips. *European Food Research and Technology* 232(2):241–254.

Guilloux-Benatier, M. and D. Chassagne. 2003. Comparison of components released by fermented or active dried yeasts after aging on lees in a model wine. *Journal of Agricultural and Food Chemistry* 51(3):746–751.

Guilloux-Benatier, M., H.S. Son, and M. Feuillat, and M. Bouhier. 1993. Activités enzymatiques: Glycosidases et peptidases chez Leuconostoc oenos au cours de la croissance bactérienne. Influence des macro-molecules des levures. *Vitis* 32.51–57.

Hashizume, K. and T. Samuta. 1999. Grape maturity and light exposure affect berry methoxypyrazine concentration. *American Journal of Enology and Viticulture* 50(2):194–198.

Henick-Kling, T. 1995. Control of malo-lactic fermentation in wine: Energetics, flavour modification and methods of starter culture preparation. *Journal of Applied Bacteriology* 79:29S–37S.

Howell, K.S., J.H. Swiegers, G.M. Elsey, T.E. Siebert, E.J. Bartowsky, G.H. Fleet, I.S. Pretorius, and M.A. Barros Lopes. 2004. Variation in 4-mercapto-4-methyl-pentan-2-one release by *Saccharomyces cerevisiae* commercial wine strains. *FEMS Microbiology Letters* 240(2):125–129.

Jackson, D.I. and P.B. Lombard. 1993. Environmental and management practices affecting grape composition and wine quality—A review. *American Journal of Enology and Viticulture* 44(4):409–430.

Jacometti, M.A., S.D. Wratten, and M. Walter. 2007. Management of under-storey to reduce the primary inoculum of *Botrytis cinerea*: Enhancing ecosystem services in vineyards. *Biological Control* 40(1):57–64.

Johnson, L.E. 2003. Temporal stability of an NDVI-LAI relationship in a Napa Valley vineyard. *Australian Journal of Grape and Wine Research* 9:96–101.

Johnson, L.F., D.F. Bosch, D.C. Williams, and B.M. Lobitz. 2001. Remote sensing of vineyard management zones: Implications for wine quality. *Applied Engineering in Agriculture* 17:557–5650.

Julien, A., J.-L. Roustan, L. Dulau, and J.-M. Sablayrolles. 2000. Comparison of nitrogen and oxygen demands of enological yeasts: Technological consequences. *American Journal of Enology and Viticulture* 51(3):215–222.

Keenan, T.W. 1968. Production of acetic acid and other volatile compounds by *Leuconostoc citrovorum* and *Leuconostoc dextranicum*. *Applied Microbiology* 16(12):1881–1885.

Kennedy, J.A., Y. Hayasaka, S. Vidal, E.J. Waters, and G.P. Jones. 2001. Composition of grape skin proanthocyanidins at different stages of berry development. *Journal of Agricultural and Food Chemistry* 49(11):5348–5355.

Kennedy, J.A., M.A. Matthews, and A.L. Waterhouse. 2000a. Changes in grape seed polyphenols during fruit ripening. *Phytochemistry* 55(1):77–85.

Kennedy, J.A., G.J. Troup, J.R. Pilbrow, D.R. Hutton, D. Hewitt, C.R. Hunter, R. Ristic, P.G. Iland, and G.P. Jones. 2000b. Development of seed polyphenols in berries from *Vitis vinifera* L. cv. Shiraz. *Australian Journal of Grape and Wine Research* 6(3):244–254.

Kimming, M., C. Sundberg, Å. Nordberg, A. Baky, S. Bernesson, O. Norén, and P.-A. Hansson. 2011. Biomass from agriculture in small-scale combined heat and power plants—A comparative life cycle assessment. *Biomass and Bioenergy* 35(4):1572–1581.

Lamb, D.W., R.G.V. Bramley, and A. Hall. August 2002. Precision viticulture-an Australian perspective. Paper read at *XXVI International Horticultural Congress: Viticulture-Living with Limitations*, Adelaide, South Australia, Australia.

Laville, P. 1990. Le terroir, un concept indispensable a l'elaboration et a la protection des appellations d'origine comme a la gestation des vignobles: le cas de la France. *Le Bulletin de l'OIV* 63(709–710):217–241.

Lee, D.F., E.E. Swinny, R.E. Asenstorfer, and G.P. Jones. 2004. Factors affecting the formation of red wine pigments. Red Wine Color. *American Chemical Society* 886:125–142.

Leitão, M.C., H.C. Teixeira, M.T. Barreto Crespo, and M.V. San Romão. 2000. Biogenic amines occurrence in wine. Amino acid decarboxylase and proteolytic activities expression by *Oenococcus oeni*. *Journal of Agricultural and Food Chemistry* 48(7):2780–2784.

Liu, S.Q., C.R. Davis, and J.D. Brooks. 1995. Growth and metabolism of selected lactic acid bacteria in synthetic wine. *American Journal of Enology and Viticulture* 46(2):166–174.

Llobera, A. and J. Canellas. 2007. Dietary fibre content and antioxidant activity of Manto Negro red grape (*Vitis vinifera*): Pomace and stem. *Food Chemistry* 101(2):659–666.

Lomolino, G. and A. Curioni. 2007. Protein haze formation in white wines: Effect of *Saccharomyces cerevisiae* cell wall components prepared with different procedures. *Journal of Agricultural and Food Chemistry* 55(21):8737–8744.

Lorenzo, M., D. Moldes, S. Rodrıguez Couto, and A. Sanroman. 2002. Improving laccase production by employing different lignocellulosic wastes in submerged cultures of *Trametes versicolor*. *Bioresource Technology* 82(2):109–113.

Makris, D.P., G. Boskou, and N.K. Andrikopoulos. 2007. Polyphenolic content and in vitro antioxidant characteristics of wine industry and other agri-food solid waste extracts. *Journal of Food Composition and Analysis* 20(2):125–132.

Martínez, M., N. Miralles, S. Hidalgo, N. Fiol, I. Villaescusa, and J. Poch. 2006. Removal of lead (II) and cadmium (II) from aqueous solutions using grape stalk waste. *Journal of Hazardous Materials* 133(1):203–211.

Massette, M. 1994. Wineries facing regulation. Paper read at *Proceedings of International Specialized Conference on Winery Wastewaters*. June 20–22, Narbonne, France, pp. 13–18.

68   VALORIZATION OF WINE MAKING BY-PRODUCTS

Matejícek, D., O. Mikes, B. Klejdus, D. Sterbová, and V. Kubán. 2005. Changes in contents of phenolic compounds during maturing of barrique red wines. *Food Chemistry* 90(4):791–800.

Maugenet, J. 1973. Evaluation of the by-products of wine distilleries. II. Possibility of recovery of proteins in the vinasse of wine distilleries. *Comptes Rendus des Séances de l'Académie d'Agriculture de France* 59:481–487.

Mazauric, J.-P. and J.-M. Salmon. 2005. Interactions between yeast lees and wine polyphenols during simulation of wine aging: I. analysis of remnant polyphenolic compounds in the resulting wines. *Journal of Agricultural and Food Chemistry* 53(14):5647–5653.

Mc Lellan, M.R. 1986. The effect of fermentation temperature on chemical and sensory characteristics of wines from seven white grape cultivars grown in New York State. *American Journal of Enology and Viticulture* 37(3):190–194.

Medina, K., E. Boido, E. Dellacassa, and F. Carrau. 2005. Yeast interactions with anthocyanins during red wine fermentation. *American Journal of Enology and Viticulture* 56(2):104–109.

Mitjans, M. and M.P. Vinardell. 2005. Biological activity and health benefits of lignans and lignins. *Trends in Comparative Biochemistry and Physiology* 11:55–62.

Molcan, P., L. Gang, T.L. Bris, Y. Yan, B. Taupin, and S. Caillat. 2009. Characterisation of biomass and coal co-firing on a 3MWth combustion test facility using flame imaging and gas/ash sampling techniques. *Fuel* 88(12):2328–2334.

Moldes, A.B., G. Bustos, A. Torrado, and J.M. Domínguez. 2007. Comparison between different hydrolysis processes of vine-trimming waste to obtain hemicellulosic sugars for further lactic acid conversion. *Applied Biochemistry and Biotechnology* 143(3):244–256.

Moldes, A.B., M. Vázquez, J.M. Domínguez, F. Díaz-Fierros, and M.T. Barral. 2008. Negative effect of discharging vinification lees on soils. *Bioresource Technology* 99(13):5991–5996.

Monagas, M., C. Gómez-Cordovés, and B. Bartolomé. 2006. Evolution of the phenolic content of red wines from *Vitis vinifera* L. during ageing in bottle. *Food Chemistry* 95(3):405–412.

Morata, A., F. Calderón, M.C. González, M.C. Gómez-Cordovés, and J.A. Suárez. 2007. Formation of the highly stable pyranoanthocyanins (vitisins A and B) in red wines by the addition of pyruvic acid and acetaldehyde. *Food Chemistry* 100(3):1144–1152.

Morata, A., M.C. Gómez-Cordovés, J. Suberviola, B. Bartolomé, B. Colomo, and J.A. Suárez. 2003. Adsorption of anthocyanins by yeast cell walls during the fermentation of red wines. *Journal of Agricultural and Food Chemistry* 51(14):4084–4088.

Mosse, K.P.M., A.F. Patti, E.W. Christen, and T.R. Cavagnaro. 2011. Review: Winery wastewater quality and treatment options in Australia. *Australian Journal of Grape and Wine Research* 17(2):111–122.

Navarro, P., J. Sarasa, D. Sierra, S. Esteban, and J.L. Ovelleiro. 2005. Degradation of wine industry wastewaters by photocatalytic advanced oxidation. *Water Science and Technology* 51(1):113–120.

Nevares, I. and M. Del Alamo. 2008. Measurement of dissolved oxygen during red wines tank aging with chips and micro-oxygenation. *Analytica Chimica Acta* 621(1):68–78.

Nielsen, J.C. and M. Richelieu. 1999. Control of flavor development in wine during and after malolactic fermentation by *Oenococcus oeni*. *Applied and Environmental Microbiology* 65(2):740–745.

Oberholster, A., L.M. Carstens, and W.J. du Toit. 2013. Investigation of the effect of gelatine, egg albumin and cross-flow microfiltration on the phenolic composition of Pinotage wine. *Food Chemistry* 138(2):1275–1281.

Oller, I., S. Malato, and J.A. Sánchez Pérez. 2011. Combination of advanced oxidation processes and biological treatments for wastewater decontamination—A review. *Science of the Total Environment* 409(20):4141–4166.

Ortega-Heras, M., C. González-Huerta, P. Herrera, and M.L. González-Sanjosé. 2004. Changes in wine volatile compounds of varietal wines during ageing in wood barrels. *Analytica Chimica Acta* 513(1):341–350.

Ough, C.S. and M.A. Amerine. 1961. Studies with controlled fermentation. VI. Effects of temperature and handling on rates, composition, and quality of wines. *American Journal of Enology and Viticulture* 12(3):117–128.

Parenti, A., P. Spugnoli, L. Calamai, S. Ferrari, and C. Gori. 2004. Effects of cold maceration on red wine quality from Tuscan Sangiovese grapes. *European Food Research and Technology* 218(4):360–366.

Pati, S., M. Esti, A. Leoni, M.T. Liberatore, and E. La Notte. 2012. Polysaccharide and volatile composition of Cabernet wine affected by different over-lees ageing. *European Food Research and Technology* 235(3):537–543.

Peyrot des Gachons, C. and J.A. Kennedy. 2003. Direct method for determining seed and skin proanthocyanidin extraction into red wine. *Journal of Agricultural and Food Chemistry* 51(20):5877–5881.

Pinelo, M., A. Arnous, and A.S. Meyer. 2006. Upgrading of grape skins: Significance of plant cell-wall structural components and extraction techniques for phenol release. *Trends in Food Science and Technology* 17(11):579–590.

Pinheiro, C., C.M. Rodrigues, T. Schäfer, and J.G. Crespo. 2002. Monitoring the aroma production during wine–must fermentation with an electronic nose. *Biotechnology and Bioengineering* 77(6):632–640.

Possner, D.R.E. and W.M. Kliewer. 1985. The localization of acids, sugars, potassium and calcium in developing grape berries. *Vitis* 24:229–240.

Price, E.P.P. 2008. Aplicación de la tecnología de membranas en el proceso de vinificación. PhD dissertation, Universidad de Concepción, Concepción, Chile.

70 VALORIZATION OF WINE MAKING BY-PRODUCTS

Ramos, A., J.S. Lolkema, W.N. Konings, and H. Santos. 1995. Enzyme basis for pH regulation of citrate and pyruvate metabolism by *Leuconostoc oenos. Applied and Environmental Microbiology* 61(4):1303–10.

Rankine, B.C. 2004. *Making Good Wine.* Sydney, New South Wales, Australia: Pan Macmillan.

Rentzsch, M., M. Schwarz, and P. Winterhalter. 2007. Pyranoanthocyanins–An overview on structures, occurrence, and pathways of formation. *Trends in Food Science and Technology* 18(10):526–534.

Ribéreau-Gayon, P., Y. Glories, A. Maujean, and D. Dubourdieu. 2000. *Handbook of Enology. The Chemistry of Wine and Stabilisation and Treatments*, vol. 2. Chichester, UK: Wiley.

Robinson, J. 1999. *The Oxford Companion to Wine*, 3rd edn. New York: Oxford University Press.

Romeyer, F.M., J.J. Macheix, J.J. Guiffon, C.C. Reminiac, and J.C. Sapis 1983. Browning capacity of grapes. 3. Changes and importance of hydroxycinnamic acid-tartaric acid esters during development and maturation of the fruit. *Journal of Agricultural and Food Chemistry* 31(2):346–349.

Rossetto, M., J. McNally, and R.J. Henry. 2002. Evaluating the potential of SSR flanking regions for examining taxonomic relationships in the Vitaceae. *Theoretical and Applied Genetics* 104(1):61–6.

Ruiz de Adana, M., L.M. Lopez, and J.M. Sala. 2005. A Fickian model for calculating wine losses from oak casks depending on conditions in ageing facilities. *Applied Thermal Engineering* 25(5):709–718.

Sablayrolles, J.-M., C. Dubois, C. Manginot, J.-L. Roustan, and P. Barre. 1996. Effectiveness of combined ammoniacal nitrogen and oxygen additions for completion of sluggish and stuck wine fermentations. *Journal of Fermentation and Bioengineering* 82(4):377–381.

Sacchi, K.L., L.F. Bisson, and D.O. Adams. 2005. A review of the effect of winemaking techniques on phenolic extraction in red wines. *American Journal of Enology and Viticulture* 56(3):197–206.

Salmon, J.M., C. Fornairon-Bonnefond, and J.P. Mazauric. 2002. Interactions between wine lees and polyphenols: Influence on oxygen consumption capacity during simulation of wine ageing. *Journal of Food Science* 67:1604–1609.

Sauvageot, F. and F. Feuillat. 1999. The influence of oak wood (*Quercus robur* L., *Q. petraea* Liebl.) on the flavor of Burgundy Pinot Noir. An examination of variation among individual trees. *American Journal of Enology and Viticulture* 50(4):447–455.

Scanes, K.T., S. Hohmann, and B.A. Prior. 1998. Glycerol production by the yeast *Saccharomyces cerevisiae* and its relevance to wine: A review. *South African Journal for Enology and Viticulture* 19:17–24.

Sefc, K.M., H. Steinkellner, F. Lefort, R. Botta, A. da Câmara Machado, J. Borrego, E. Maletić, and J. Glössl. 2003. Evaluation of the genetic contribution of local wild vines to European grapevine cultivars. *American Journal of Enology and Viticulture* 54(1):15–21.

Shalaby, A.R. 1996. Significance of biogenic amines to food safety and human health. *Food Research International* 29(7):675–690.

Shinohara, T., S. Kubodera, and F. Yanagida. 2000. Distribution of phenolic yeasts and production of phenolic off-flavors in wine fermentation. *Journal of Bioscience and Bioengineering* 90(1):90–97.

Soubeyrand, V., V. Luparia, P. Williams, T. Doco, A. Vernhet, A. Ortiz-Julien, and J.-M. Salmon. 2005. Formation of micella containing solubilized sterols during rehydration of active dry yeasts improves their fermenting capacity. *Journal of Agricultural and Food Chemistry* 53(20):8025–8032.

Spigno, G., T. Pizzorno, and D.M. De Faveri. 2008. Cellulose and hemicelluloses recovery from grape stalks. *Bioresource Technology* 99(10):4329–4337.

Spinelli, R., C. Nati, L. Pari, E. Mescalchin, and N. Magagnotti. 2012. Production and quality of biomass fuels from mechanized collection and processing of vineyard pruning residues. *Applied Energy* 89(1):374–379.

Stewart, D. 2008. Lignin as a base material for materials applications: Chemistry, application and economics. *Industrial Crops and Products* 27(2):202–207.

Suárez, R., J.A. Suárez-Lepe, A. Morata, and F. Calderón. 2007. The production of ethylphenols in wine by yeasts of the genera Brettanomyces and Dekkera: A review. *Food Chemistry* 102(1):10–21.

Tebib, K., J.M. Rouanet, and P. Besancon. 1997. Antioxidant effects of dietary polymeric grape seed tannins in tissues of rats fed a high cholesterol-vitamin E-deficient diet. *Food Chemistry* 59(1):135–141.

This, P., A. Jung, P. Boccacci, J. Borrego, R. Botta, L. Costantini, M. Crespan et al. 2004. Development of a standard set of microsatellite reference alleles for identification of grape cultivars. *Theoretical and Applied Genetics* 109(7):1448–1458.

Torija, M.J., G. Beltran, M. Novo, M. Poblet, J. Manuel Guillamón, A. Mas, and N. Rozes. 2003. Effects of fermentation temperature and *Saccharomyces* species on the cell fatty acid composition and presence of volatile compounds in wine. *International Journal of Food Microbiology* 85(1):127–136.

Trought, M.C.T. and R.G.V. Bramley. 2011. Vineyard variability in Marlborough, New Zealand: Characterising spatial and temporal changes in fruit composition and juice quality in the vineyard. *Australian Journal of Grape and Wine Research* 17(1):72–82.

Vargas, A.M., M.D. Vélez, M.T. De Andrés, V. Laucou, T. Lacombe, J.M. Boursiquot, J. Borrego, and J. Ibáñez. 2007. *Corinto bianco*: A seedless mutant of Pedro Ximenes. *American Journal of Enology and Viticulture* 58(4):540.

Vernhet, A., K. Dupre, L. Boulange-Petermann, V. Cheynier, P. Pellerin, and M. Moutounet. 1999. Composition of tartrate precipitates deposited on stainless steel tanks during the cold stabilization of wines. Part I. White wines. *American Journal of Enology and Viticulture* 50(4):391–397.

Vidal, S., P. Courcoux, L. Francis, M. Kwiatkowski, R. Gawel, P. Williams, E. Waters, and V. Cheynier. 2004. Use of an experimental design approach for evaluation of key wine components on mouth-feel perception. *Food Quality and Preference* 15(3):209–217.

72 VALORIZATION OF WINE MAKING BY-PRODUCTS

Vine, R.P., E.M. Harkness, T. Browning, and C. Wagner. 1997. *Winemaking: From Grape Growing to Marketplace*. New York: Chapman & Hall.

Warriner, K., A. Morrissey, J. Alderman, G. King, P. Treloar, and P.M. Vadgama. 2002. Modified microelectrode interfaces for in-line electrochemical monitoring of ethanol in fermentation processes. *Sensors and Actuators B: Chemical* 84(2):200–207.

Waters, E.J., G. Alexander, R. Muhlack, K.F. Pocock, C. Colby, B.K. O'Neill, P.B. Høj, and P. Jones. 2005. Preventing protein haze in bottled white wine. *Australian Journal of Grape and Wine Research* 11(2):215–225.

Welz, P.J., J.-B. Ramond, D.A. Cowan, and S.G. Burton. 2012. Phenolic removal processes in biological sand filters, sand columns and microcosms. *Bioresource Technology* 119:262–269.

Winkler, A.J., J.A. Cook, W.M. Kliewer, and L.A. Lider. 1974. Development and composition of grapes. In *General Viticulture*, L. Cerruti (ed.). Berkeley, CA: University of California Press.

Zeaiter, M., J.M. Roger, and V. Bellon-Maurel. 2006. Dynamic orthogonal projection. A new method to maintain the on-line robustness of multivariate calibrations. Application to NIR-based monitoring of wine fermentations. *Chemometrics and Intelligent Laboratory Systems* 80(2):227–235.

3

WINE MAKING BY-PRODUCTS

ZHIJING YE, ROLAND HARRISON, VERN JOU CHENG, AND ALAA EL-DIN A. BEKHIT

Contents

3.1	Overview	73
	3.1.1 Industrial Economy	75
3.2	Overall Nutritional Composition	77
3.3	Vines	80
	3.3.1 Pruning	80
3.4	Grapes	83
	3.4.1 Stems	83
	3.4.2 Pomace	85
	3.4.3 Seeds and Seed Oil	86
3.5	Wines	91
	3.5.1 Yeast Lees	91
	3.5.1.1 Soluble Components of Lees	92
	3.5.1.2 Interaction between Protein Compounds in Lees and Phenolic Compounds	94
	3.5.2 Tartaric Acid	99
	3.5.3 Carbon Dioxide	101
	3.5.4 Wastewater	103
3.6	Challenges and Opportunities	106
	Acknowledgment	106
	References	107

3.1 Overview

Approximately, 67.1 million tons of grapes were utilized in wine production in 2013 (FAO 2014). This generates a considerable amount of waste because as much as 20% of the weight of processed grapes is not found in the final product (Mazza and Miniati 1993). Figure 3.1 shows the vinification process and waste materials generated at each

74 VALORIZATION OF WINE MAKING BY-PRODUCTS

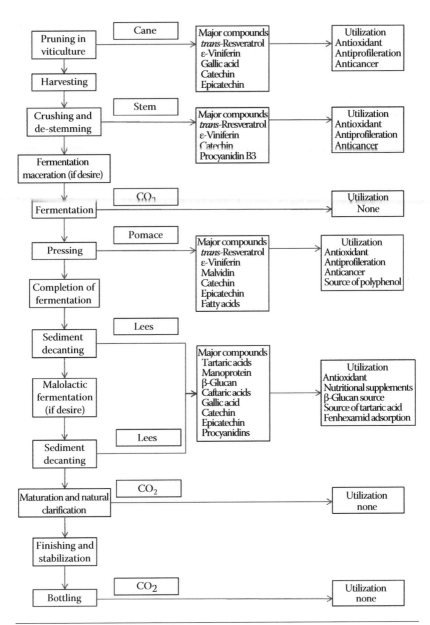

Figure 3.1 The vinification process, composition of by-products and potential utilization. (Modified from Pérez-Serradilla, J.A. and Luque de Castro, M.D., *Food Chem.*, 111, 447, 2008; Gorena, T. et al., *Food Chem.*, 155, 256, 2014; Zhang, A. et al., *Molecules*, 16, 2846, 2011a; Zhang, A. et al., *Molecules*, 16, 10104, 2011b; Ye, Z.J., Characterization of bioactive compounds in lees from New Zealand wines and the effect of enzymatic oxidation on their bioactivity, Masters' of Food Science, University of Otago, Otago, New Zealand, 2014.)

production step. The three main by-products that are generated from wine making are stalks, grape pomace/marc, and wine lees (Bustamante et al. 2008). Stalks and grape pomace (comprising skin, seeds, and pulp) are left over after crushing, draining, and pressing (Jin and Kelley 2009). Wine lees is a sludge (or mud-like) material generated in the wine making process. This material contains dead yeast, yeast residue, or particles precipitated at the bottom of wine tanks or barrels (Hwang et al. 2009). Waste contains significant amounts of organic species (such as sugars, phenolics, polyalcohols, pectins, and lipids) with high chemical and biological oxygen demand and is therefore recognized as an environmental pollutant (Lafka et al. 2007). There is growing interest in the utilization of this waste including its conversion into biofuels and use as nutrient supplements, food ingredients, and animal feeds. For example, Shirikhande (2000) and González-Paramás et al. (2003) reported that extracts from grape seeds and skin contain phenolic compounds that can be used as dietary supplements for better health and well-being. Many studies have investigated the use of waste derived from grapes as a source of natural antioxidants with most of the focus being on the skins, stalks, and seeds (Van Dyk et al. 2013; Wadhwa and Bakshi 2013; Naziri et al. 2014). However, the utilization of waste from wine production, particularly wine lees, is still poorly investigated.

3.1.1 Industrial Economy

Wine making is a well-developed industry using both artisanal skills and scientific knowledge to produce various styles of wines from different varieties of grapes (Jackson 2008). Quality wines are produced from *Vitis vinifera* but other *Vitis* species are also used for wine making (Jackson 2008). Some of the major commercial varieties used are as follows: Cabernet Sauvignon, Merlot, Pint Noir, and Shiraz for red wines, and Chardonnay, Pinot Gris, Riesling, and Sauvignon Blanc for white wines. Wine plays an important role in the world economy. In 2012, approximately 26.4 million tons of wine were produced worldwide, with approximately 80% of this in Europe, the United States, and China (FAOSTAT 2014). A number of so-called "Old World" countries such as France, Italy, and Spain lead world production with 5.29, 4.09, and 3.15 million tons, respectively (Figure 3.2). The United States, Australia, China, Australia, New Zealand, Argentina, Chile, and South Africa are

76   VALORIZATION OF WINE MAKING BY-PRODUCTS

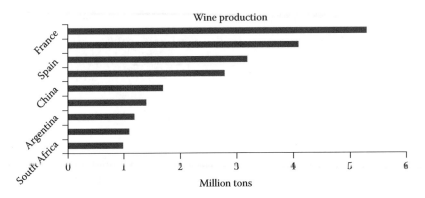

Figure 3.2 Wine production of major countries in 2012. (From FAOSTAT, World grape production. http://faostat3.fao.org/download/Q/QC/E, accessed on December 22, 2014.)

regarded as "New World" countries with the United States being the major producer of wine in this class with a production of 2.82 million tons (about 10% of world production) in 2012. The combined total wine production of Australia, New Zealand, Argentina, and Chile was around 3.67 million tons in 2012. Large-scale production of wines and increased awareness of the potential environmental consequences of inappropriate disposal of wine making by-products have resulted in the territorial authorities of many major wine-producing countries introducing regulations to minimize the negative effects. For example, European Council Regulation (EC) 479/2008 requires that grape marc and lees be sent to alcohol distilleries to extract ethanol leaving behind the exhausted grape marc and liquid waste (vinasse). Similarly, industrial waste must be treated prior to disposal to prevent contamination according to Spanish Law 10/1998. The cost of disposal fees generated when wine companies discard wine making by-products may be significant. Devesa-Rey et al. (2011) reported a higher disposal fee for wine making by-products compared with the basic price for urban waste and industrial wastewater (€0.01202 and €0.03005 m^{-3}, respectively) depending on its nature, characteristics, and degree of contamination. In Spain, heavy penalties are applied to unauthorized discharge of wastewater and vinasse, from a €3,000 fine and temporary withdrawal of license to fines in excess of €40,000 and possibly including a period of imprisonment (Devesa-Rey et al. 2011). Such regulations should encourage wine industries to investigate the novel technologies for waste treatment and to identify other value-added products.

3.2 Overall Nutritional Composition

Vineyard and winery by-products contain valuable chemical compounds not extracted during wine making. Molina-Alcaide et al. (2008) and González-Centeno et al. (2010) have characterized the nutrients available in various wine by-products (Table 3.1). Nutrient concentrations vary depending on the by-product type (e.g., marc versus lees) and wine making (e.g., red versus white versus sparkling). All the by-products were rich in organic matter (ranging from 696 to 888 g/kg DM). The content of polysaccharides was found to be greater in shoots and marcs compared to lees. Lees were found to be rich in protein and had total nitrogen concentrations ranging from 22.6 to 32.5 g/kg DM. Molina-Alcaide et al. (2008) suggested that the combinations of by-products obtained from different vinifications could be used as a valuable source of energy and protein for ruminants. Phenolic compounds in by-products of wine making are currently of particular interest as they are an essential part of the human diet with well-recognized health benefits. Phenolic compounds found in these by-products can be divided into flavonoids and non-flavonoids. Flavonoids consist of anthocyanins, flavonols, and flavan-3-ols, which are collectively characterized by a C6–C3–C6 skeleton with two phenolic rings connected by a pyran ring (Jackson 2008). These compounds show different antioxidant activities based on their structures, especially the number and positions of the hydroxyl groups and the nature of substitutions on the aromatic rings (Balasundram et al. 2006). In addition, shoots, marc, and lees have been found to be rich (17.9–202.6 g/kg DM) in condensed tannins (polymers of flavan-3-ols of various molecular weights) (Molina-Alcaide et al. 2008). Also termed proanthocyanidins, their structures have been described in many studies (Marles et al. 2003; Xie and Dixon 2005; Serrano et al. 2009). The four common flavan-3-ol subunits found in grape berries are as follows: (+)-catechin, (–)-epicatechin, (–)-epicatechin-gallate, and (–)-epigallocatechin (Cheynier et al. 2006). The combination of these four subunits can result in high structural diversity of proanthocyanins and their degree of polymerization (normally referred to as the length of the chain) may vary greatly. Proanthocyanidins have been described with molecular weights of up to 20,000 Da (Watkins 1997). The non-flavonoids have simpler structures

Table 3.1 Chemical Composition of Different Vinification By-Products

COMPONENT	SHOOT	POMACE	STEM	SEED	SKIN	PULP	LEES
Dry matter (g/kg)	506	329–470					144–277
Ash		41–173	55–110	38–58	20–183	80–135	201–304
Organic matter	885	827–959	890–945	942–962	817–980	863–934	696–799
Carbohydrates		290					
Pectin		31			40		
Fructose		81			50		
Glucose		80			11		
Arabinose					12		
Rhamnose					12		
Mannose					72		
Galactose					330		
Xylose					22		
Cellulose					125–200		
Hemicellulose					125		
Holocellulose					465		
Soluble sugar		45–172	55–117				
Neutral detergent fiber	741	569–626		506–542		350–423	122–321

(Continued)

Table 3.1 (Continued) Chemical Composition of Different Vinification By-Products

COMPONENT	SHOOT	POMACE	STEM	SEED	SKIN	PULP	LEES
Acid detergent fiber	518	480–543		467–514		312–401	58–174
Acid detergent lignin	166	370–388		371–416		175–261	28–107
Total dietary fiber		460					
Crude protein		60–106	65–83	119–126	50–190	99–110	
Total nitrogen	7.2	1–17	10–13	19–20		3–18	23–33
Fat							
Dichloromethane		40–140			24–55		
Ether		30–80	9–26	95–145	10–60	59–101	
Hexane							
Tannins							
Free	45.8	16–38					3–98
Fiber-bound	1.25	19–34					5–35
Protein-bound	9.45	56–131			34–138		9–47
Total	57	22–203					18–180

Sources: Molina-Alcaide, E. et al., *J. Sci. Food Agric.*, 4, 597, 2008; González-Centeno, M.R. et al., *LWT—Food Sci. Technol.*, 43, 1580, 2010.
DM, dry matter.

(C6–C1 or C6–C3 skeleton) than flavonoids. In addition to phenolics derived from grape materials, some non-flavonoid phenolics are also produced during yeast metabolism, the most prevalent being tyrosol (Jackson 2008). Some non-flavonoids, such as resveratrol, also show biological activity (Moreno-Arribas and Polo 2009).

3.3 Vines

3.3.1 Pruning

Pruning is an important practice in vineyard management and involves the removal of living shoots, canes, leaves, and other vegetative parts of the vine (Winkler et al. 1974) (Figure 3.3). The goal is to facilitate other vineyard management practices (e.g., those associated with the control of disease) to produce grapes of a desired quality, to produce fruitful shoots, and to regulate the vegetative growth of the vine (Tassie and Freeman 2001). Pruning is carried out in both the dormant and growing stages of the grapevine, normally referred to as winter and summer pruning, respectively. Spur and cane pruning are carried out during the winter. In spur pruning, the distal node

Figure 3.3 Pruning grapevine rows.

WINE MAKING BY-PRODUCTS

and shoot are removed and the spur at the proximal node is retained. In cane pruning, unwanted canes are removed. Summer pruning is often regarded as a part of canopy management which aims to control the amount and position of leaves, shoots, and fruit in space to achieve some desired arrangement (Tassie and Freeman 2001). Therefore, the major waste generated in winter and summer pruning are canes and leaves, respectively. The estimated amounts of material generated during cane pruning varied from 0.56 to 2.01 kg/vine depending on the trellis system and year (Reynolds et al. 1995). Vine spacing varies globally from 500 to 50,000/ha (Coombe and Dry 1992). Therefore, the estimated weight of cane prunings is more than 2.1 million tons from about 7.4 million ha worldwide in 2009 (FAOSTAT 2010). In the vineyard, only moderately vigorous and well-matured canes are used for grapevine propagation (Coombe and Dry 1992). Cane prunings are generally disposed of by composting or burning (Çetin et al. 2011). As an agricultural waste, this material is largely ignored although its phytochemical content could be used for value-added products in the food industry. Çetin et al. (2011) characterized grape canes from 10 different table grape cultivars in Turkey. They found that the total carbohydrate and protein content varied from 35.0 to 44.2 and 12.1 to 28.1 g/100 g, respectively, depending on the genotype. Grape canes are rich in calcium, potassium, iron, magnesium, and zinc with concentrations varying from 6.0 to 10.2, 5.2 to 8.2, 0.26 to 0.68, 1.9 to 11.1, and 0.7 to 9.8 mg/100 g, respectively (Çetin et al. 2011), and could be alternative sources of mineral dietary supplements. Grape canes have also been found to be rich in phenolic compounds ranging from 25.4 to 36.6 mg gallic acid equivalents (GAE)/g. Catechin and epicatechin had highest concentrations (Table 3.2) among phenolic compounds in grape canes (Çetin et al. 2011). Currently, most studies focus on stilbenoids, which are phytoalexins and play an important role in grapevine response to fungal infection, especially *Botrytis cinerea* (Rentzsch et al. 2007). Within this group, resveratrol and viniferin are known to have health-promoting properties including antioxidant and anti-inflammatory activities. Karacabey and Mazza (2008) reported that the content of *trans*-resveratrol and *trans*-ε-viniferin in *Vitis viniferia* L. Pinot Noir was 4.25 and 2.03 mg/g dry matter, respectively.

Table 3.2 Summary of the Potential Bioactive Compounds from Grape Cane

VARIETY	COMPOUNDS FOUND	REFERENCES
V. vinifera	trans-Resveratrol, ε-viniferin	Gorena et al. (2014)
V. vinifera	trans-Resveratrol, ε-viniferin	Rayne et al. (2008)
V. amurensis, V. pentagona, V. vinifera	Polyphenol	Min et al. (2014)
V. vinifera	Stilbenoid	Karacabey and Mazza (2010)
V. amurensis, V. arizonica, V. berlandieri, V. etulifolia, V. cinerea, etc.	E-resveratroloside, E-piceid, E-piceatannol, E-resveratrol, E-ε-viniferin, E-ω-viniferin, E-ampelopsin E, E-vitisin B, E-amurensin B	Pawlus et al. (2013)
V. vinifera	trans-Piceatannol, trans-resveratrol, ε-viniferin	Vergara et al. (2012)
V. vinifera, V. pentagona, V. davidii	resveratrol	Zhang et al. (2011a)
V. vinifera, V. amurensis, V. pentagona, V. davidii	Gallic acid, protocatechuic acid, vanillic acid, syringic acid, catechin, epicatechin, trans-resveratrol	Zhang et al. (2011b)
V. vinifera	trans-Resveratrol, trans-ε-viniferin, ferulic acid	Karacabey et al. (2012)
V. vinifera	trans-Resveratrol, trans-ε-viniferin, ferulic acid	Karacabey and Mazza (2008)

3.4 Grapes

3.4.1 Stems

In traditional wine making, stems were often left with grapes during crushing, pressing, and even during fermentation, especially for the production of red wine (Jackson 2008) (Figure 3.4). This may improve the drainage during pressing and add more tannins in a poor vintage. However, this practice is no longer common because of negative organoleptic effects in the wine, for example, increased astringency. The modern trend is to separate the processes of de-stemming and crushing in order to minimize the excessive uptake of phenols and lipids from vine parts (Jackson 2008). Until recently, grape stems were another poorly characterized wine making by-product (Tables 3.1 and 3.3). The grape stem is rich in dietary fiber which constitutes up to 77% of its dry matter (Llobera and Cañellas 2007). The dietary fiber in the grape stem comprises mainly neutral sugars and lignin (43.4% and 31.6% of the dry matter, respectively) (Llobera and Cañellas 2007). Llobera and Cañellas (2007) found

Figure 3.4 Grape stems.

84 VALORIZATION OF WINE MAKING BY-PRODUCTS

Table 3.3 Composition of Manto Negro Grape By-Products (% Dry Matter)

	% DRY MATTER		% DRY MATTER
Protein	7.29 ± 0.20	Soluble pectins	1.04 ± 0.08
Soluble sugars	1.70 ± 0.06	Condensed tannins	10.3 ± 0.50
Oil	1.65 ± 0.01	Total dietary fiber	77.2 ± 1.97
Ash	5.48 ± 0.16		

Source: Llobera, A. and Cañellas, J., *Food Chem.*, 101, 659, 2007.

that the lignin in the grape stem contains important amounts of condensed tannins which could explain its excellent ability to scavenge free radicals. In a later study of the carbohydrate composition of stems from 10 different grape varieties, González-Centeno et al. (2010) found cellulose was the predominant component followed by pectin, ranging from 40 to 48 and 27 to 37 mol%, respectively. Among pectin polysaccharides, homogalacturonan (about 68 mol% of total pectin in fresh grape) was predominant, followed by rhamnogalactoronan I (26.5 mol% of total pectin) and rhamnogalactoronan II (about 7 mol% of total pectin). Phenolic compounds are the second most abundant chemical component found in the grape stem, and can amount to more than 10.3% of its dry matter (Llobera and Cañellas 2007). Anastasiadi et al. (2012) studied grape stems from six native Greek red and white *Vitis vinifera* cultivars and found the total phenolic content ranged from 367 to 587 and 372 to 574 mg/g, respectively. In agreement with Barros et al. (2014), they found that the phenolic content of stems from red varieties was greater than that in white ones. The predominant phenolic compounds in both red and white grape stems were (+)-catechin, followed by procyanidin B3, ε-viniferin, and *trans*-resveratrol (Anastasiadi et al. 2012). Barros et al. (2014) analyzed the correlation between the concentrations of extracted phenolic compounds and antioxidant activity as measured by the DPPH, ORAC, ABTS, FRAP, and $O_2^{\cdot-}$ radical scavenging capacity; they found that procyanidin dimer B, isorhamnetin-3-O-(6-O-feruloyl)-glycoside, quercetin-3-O-glucoside, and malvidin-3-O-(6-O-caffeoyl)-glucoside were highly correlated to antioxidant activities. In addition, grape stem extracts were also found to inhibit cell proliferation of colon (HT29), breast (MCF-7 and MDA-MB-23), renal (786-0 and Caki-1), and thyroid (K1) cancer cells.

3.4.2 Pomace

Grape pomace is a major by-product of wine making (Figure 3.5). It consists of pressed skins, disrupted cells from grape pulp, seeds, and stems (Figure 3.6). When grapes are processed for red wine, the skin and seeds usually remain in contact with the fermenting grape juice for an extended period of time ranging from a few days to a number of weeks. Thus, the grape skin and seeds are subjected to a prolonged extraction with a weak aqueous ethanol solution (Pinelo et al. 2006). Limited extraction during the wine making process results in a high phenolic content in the grape pomace. The average distribution of polyphenolic compounds is about 1% in the pulp, 5% in the skin, and approximately 62% in the seeds (Thorngate and Singleton 1994). Cheng (2011) investigated the phenolic composition of grape pomace from Pinot Noir and Pinot Meunier. He found that catechin was the major phenolic compound in Pinot Noir grape pomace ranging from 9.0 to 11.8 mg/g, followed by epicatechin (5.0 to 6.7 mg/g), procyanidins (2.88 to 3.98 mg/g), and malvidin-3-glucoside (5.2 to 5.9 mg/g) (Table 3.4). Similar trends have also been found in other grape

Figure 3.5 The solid remains of grapes: stems, seeds, and skins.

Figure 3.6 Grape skins.

varieties (Kammerer et al. 2004; Yilmaz and Toledo 2006; Ruberto et al. 2007). Ye (2014) has also characterized the phenolic content of five grape pomace samples from different locations and vinification backgrounds. The total phenolic content ranged from 17.3 to 41.0 mg GAE/g DM, and the total tannin content ranged from 10.1 to 19.2 mg epicatechin equivalents (ECE)/g DM. Vinification techniques, including "no cap management" which is likely to result in higher fermentation temperatures and extended maceration, were found to result in a lower total phenolic content in the resulting pomace. Sacchi et al. (2005) and Ough and Amerine (1961) have reported that higher fermentation temperatures led to a higher total phenolic content in Pinot Noir wine. Similarly, extended maceration is a common practice to increase the release of phenolic compounds in wine.

3.4.3 Seeds and Seed Oil

As by-products of wine making, grape seeds account for approximately 17% of the weight of fresh grape pomace, and 38%–52% of dry matter (Fernandes et al. 2013; Toscano et al. 2013) (Figure 3.7). A grape berry usually contains two seeds which represent 0%–6% of

Table 3.4 Phenolic Profile of Extracts (mg/g of Extract) Obtained from Grape Pomace

PHENOLIC COMPOUNDS	VARIETIES								
	PN[a]	PM[a]	NA[b]	NM[b]	NC[b]	FR[b]	CS[b]	Me[c]	CM[d]
Non-flavonoids									
Gallic acid	0.28–0.33	0.08–0.12						0.11	
Flavonoids									
Flavanols									
Catechin	8.97–11.81	3.50–4.35							
Epicatechin	5.00–6.74	2.13–3.12							
Epicatechin gallate	0.08–0.13	Trace							
Procyanidin A[b]	3.81–4.39	0.69–0.83							
Procyanidin B[b]	2.88–3.98	0.47–0.75							
Anthocyanins									
Delphinidin-3-glucoside[c]	0.10–0.18	0.05–0.06	0.21	1.17	0.81	0.21	0.16		0.44–1.11
Cyanidin-3-glucoside[c]	Trace	Trace	Trace	0.64	Trace	0.09	—		1.51–3.81

(Continued)

Table 3.4 (*Continued*)   Phenolic Profile of Extracts (mg/g of Extract) Obtained from Grape Pomace

PHENOLIC COMPOUNDS	VARIETIES								
	PN[a]	PM[a]	NA[b]	NM[b]	NC[b]	FF[b]	CS[b]	Me[c]	CM[d]
Petunidin-3-glucoside[c]	0.34–0.53	0.13–0.16	0.43	1.43	1.30	0.28	0 32		0.53–1.34
Peonidin-3-glucoside[c]	0.33–0.46	0.57–0.76	0.20	1.48	0.42	0.29	0 15		0.99–2.49
Malvidin-3-glucoside[c]	5.18–5.94	1.74–2.44	5.61	4.09	10.38	2. 6	5 70		4.12–10.19
Vitisin[c]	0.04–0.07	Trace	—	—	—	—	—		
Total anthocyanin	6.01–7.18	2.55–3.47	28.7	9.10	45.27	3. 5	9 61		
Flavonols									
Quercetin	0.33–0.68	0.15–0.18	1.19	0.88	2.77	0. 8	0 54		
Quercetin methyl glucoside	0.32–0.82	0.39–0.45	—	—	—	—	—		

[a] Cheng (2011).

[b] Ruberto et al. (2007).

[c] Yilmaz and Toledo (2006).

[d] Kammerer et al. (2004).

NA, Nero d'avola; NM, Nerello Mascalese; Me, Merlot; PN, Pinot Noir; PM, Pinot Meunier; NC, Nerello Cappuccino; FR, Frappato; CS, Cabernet Sauvignon; CM, Cabernet Mitos.
   Trace: <0.05.

Figure 3.7 Grape seeds.

berry weight (Rankine 1989; Cadot et al. 2006). However, the number and weight of seeds vary according to berry weight and ripening and there appears to be a quadratic relationship between the average berry size and the average seed number (Cadot et al. 2006). Worldwide, over 3 million tons of grape seeds were discarded annually (Fernandes et al. 2013). Grape seeds contain protein, lipids (fats and oils), carbohydrates, and approximately 5%–8% polyphenolics depending on the variety (Amerine and Joslyn 1967; Shi et al. 2003). Polyphenolics in grape seeds are mainly flavonoids, including monomeric flavan-3-ols (catechin, epicatechin, gallocatechin, epigallocatechin, and epicatechin gallate), procyanidin dimers, trimers, and more highly polymerized procyanidins (Silva et al. 1991; Prieur et al. 1994). Cheng (2011) investigated the phenolic composition of Pinot Noir grape seeds and showed catechin was the predominant flavonoid, followed by epicatechin and procyanidins; the concentration of epicatechin gallate was low in grape seeds. The oil content of grape seeds varies from 8% to 20% on a dry matter basis (Rombaut et al. 2014). Sabir et al. (2012) studied grape seeds from 17 varieties of *Vitis vinifera*, one *Vitis labrusca* variety, one hybrid variety, and two American rootstock varieties.

90   VALORIZATION OF WINE MAKING BY-PRODUCTS

They found that the oil content varied greatly among the genotypes, ranging from 7.3% (Perle de Csaba) to 22.4% (Italy) (Sabir et al. 2012). Similar findings have been reported by Tangolar et al. (2009) in which the oil content varied between cultivars and ranged from 10.5% (Razaki) to 16.7% (Salt creek). The oil content of grape seeds is also affected by the stage of berry development. Rubio et al. (2009) studied seed oil extraction throughout the different stages of berry development. They found that the seed oil content of berries collected in the period between fruit set and cell division was only 4.4% (dry matter basis); seed oil content reached 15% prior to veraison and remained approximately same until harvest (Rubio et al. 2009). The stage of berry development also affects the fatty acid composition in seeds in the early but not later stages of development (Rubio et al. 2009). Fernandes et al. (2013) studied 10 traditional Portuguese grape varieties and found polyunsaturated fatty acids to be predominant (63.6%–73.5%), whereas the proportion of monounsaturated and saturated fatty acids were 14.2%–21.3% and 11.6%–14.9%, respectively. Similar results have also been reported by Lutterodt et al. (2011). In grape-seed oil, linoleic acid (18:2) is the most abundant fatty acid, followed by oleic acid (18:1), stearic acid (18:0), and palmitic acid (C16:0) (Table 3.5) (Lutterodt et al. 2011; Prado et al. 2012; Sabir et al. 2012;

Table 3.5 Effects of Origin and Grape Variety (Red versus White) on the Fatty Acid Profile in Grape-Seed Oil

		FATTY ACID (%)					
VARIETY	ORIGIN	14:0	16:0	18:0	18:1	18:2	18:3
Red	United States	0.04–0.06	5.8–7.1	2.5–3.1	13.7–20.3	69.7–77.8	Traces
	New Zealand	0.01	6.76	3.71	12.38	76.23	0.45
White	United States	0.04–0.07	6.8–8.1	2.7–3.5	15.9–21.5	66.9–73.9	traces
	Spain	ND	8.12	5.6	19.59	66.16	0.37
Various	Turkey	ND	6.5–9.0	4.0–5.4	17.8–25.5	60.1–70.1	0.0–0.9
Unknown	India	0.01	6.76	3.71	12.38	76.23	0.45

Sources: Bail, S. et al., *Food Chem.*, 108, 122, 2008; Bekhit, A.E.D. et al., Manipulating the functionality of grape seeds products through reflective mulching and wine fermentation, *43rd Annual Conference of New Zealand Nutrition Society*, Christchurch, New Zealand, December 9–10, 2008; De Marchi, F. et al., *J. Mass Spectrom.*, 47, 1113, 2012; Oomah, B.D. et al., *J. Agric. Food Chem.*, 46, 4017, 1998; Shivananda Nayak, B. et al., *Phytother. Res.*, 25, 1201, 2011.

Note: The results are presented in % of fatty acids of the extracted oils.
ND, Not detected.

Fernandes et al. 2013). Lutterodt et al. (2011) found that the linoleic acid content ranged from 66.0 g/100 g to 75.3 g/100 g of total fatty acids in Ruby Red and Concord seed oil. In addition, trilinolein was found to be the most abundant triglyceride in different hybrid grape varieties, followed by dilinoleoyl-oleoylglycerol and dilinoleoyl-palmitoylglycerol (De Marchi et al. 2012). The composition of fatty acids in various cultivars can be different; for example, palmitic, oleic, and stearic acids were found to be the predominant fatty acids in Pkatsiteli grape-seed oil. Grape-seed oil is also rich in tocopherols: α-tocopherol or vitamin E (ranged from 85.5 to 260.5 mg/kg oil) is predominant, followed by γ-tocopherol (ranged from 2.5 to 30.2 mg/kg oil) (Sabir et al. 2012; Fernandes et al. 2013).

3.5 Wines

3.5.1 Yeast Lees

EEC Regulation No. 337/79 defines wine lees as "the residue that forms at the bottom of receptacles containing wine, after fermentation, during storage or after authorized treatments, as well as the residue obtained following the filtration or centrifugation of this product." Thus, wine lees is a complex mixture that may contain dead yeast materials, plant cell debris, polyphenols adsorbed to proteins, lipids, tartaric acid crystals, and many other compounds (Table 3.6). Consequently, the composition of lees can vary greatly depending on many factors including the composition of grapes, the conditions of the vinification, including settling and aging periods. Dead yeast cells and yeast debris are normally the major components of lees. Numerous genera and species of yeasts are involved in wine making (Bisson 2009), although most belong to *Saccharomyces cerevisiae* (Reed and Nagodawithana 1991; Benítez et al. 2011). Dead yeast cells start to appear toward the end of the alcoholic fermentation due to increased ethanol content and lack of nutrients. Death of yeast cells is followed by autolysis (destruction of cells by autologous enzymes) of the cell material. Yeast cells contain lipids, amino acids, proteins, nucleotides, β-glucan, and mannoproteins which are released during autolysis and these compounds contribute to the stability and organoleptic properties of wine (Charpentier 2010). Protease A is the major proteolytic enzyme which is responsible for 80% of nitrogen released

92 VALORIZATION OF WINE MAKING BY-PRODUCTS

Table 3.6 Summary of Elemental Composition and Main Physicochemical Characteristics of Wine Lees

ELEMENTAL COMPOSITION	RANGE	PHYSICOCHEMICAL PARAMETERS	RANGE
Ca (g/kg DM)	3.6–15.5	C_{org} (g/kg DM)	226–376
Cu (mg/kg DM)	13–1187	C_{ws} (g/kg DM)	44.3–168.9
Fe (mg/kg DM)	84–1756	EC (dS/m)	4.0–13.8
K (g/kg DM)	17.6–158.1	OM (g/kg DM)	598–936
Mg (g/kg DM)	0.4–3.7	pH	3.6–7.2
Mn (mg/kg DM)	<0.2–21	Pol (g/kg DM)	1.9–16.3
P (g/kg DM)	1.61–10.3	TN (g/kg DM)	17.2–59.7
Zn (mg/kg DM)	14–84	Proteins (%)	14.5–15.7
		Lipids (%)	5.0–5.9
		Sugars (%)	3.5–4.8
		Dietary fiber (%)	21.2–21.9
		Tartaric acid (%)	24.5–24.7
		Ash (%)	10.5–10.6

Sources: Bustamante, M.A. et al., *Waste Manage.*, 28, 372, 2008; Gómez, M.E. et al., *J. Agric. Food Chem.*, 52, 4791, 2004.

EC, electric conductivity; OM, organic matter; C_{org}, oxidizable organic carbon; C_{ws}, water-soluble carbon; TN, total nitrogen; Pol, water-soluble polyphenols.

during autolysis (Charpentier 2010). Polysaccharides (primarily mannoproteins and β-glucan) are the major components of yeast cell walls and account for 20%–30% of the cell dry matter (Alexandre and Guilloux-Benatier 2006). Charpentier and Feuillat (2008) determined the nitrogen and polysaccharide content in lees at different stages of wine making and concluded that the composition of lees is dependent on the settling duration. In a recent study, lees were divided into two groups (heavy lees and light lees) according to the settling duration: heavy lees were defined as that which settles within 24 h corresponding to a particle size range of approximately 100 μm to 2 mm, and light lees which settles after more than 24 h with a particle size from 1 to 50 μm (Charpentier 2010).

3.5.1.1 Soluble Components of Lees During wine making, lees is separated from wine and this process is called racking. Solid materials separated from wine are then pressed for maximum wine yield. However, some wine cannot be extracted by pressing and remains in lees. Rice (1976) reported the moisture content of pressed lees of white and red wine after post-alcoholic fermentation to be 52.9% and 54.0%, respectively. These "wet" lees contained 5.38% (w/w) and 5.41% (w/w) of alcohol. Lees from different stages of wine making may have different moisture contents. For example, lees obtained post-malolactic acid fermentation had a moisture content of about 60% which was slightly higher than that obtained after fermentation (Rice 1976), although alcohol content (5.54% w/w) was similar. Lees also contain lipids extracted from grape seeds during fermentation and from yeast cell walls. Gómez et al. (2004) studied lees from a sherry (a fortified white wine) production and found that the major fatty acids were palmitic (C16) and linolenic acids (C18) (Table 3.7). The degree of unsaturation of lipids extracted from lees is close to that of grape seeds lipids (Prado et al. 2012). The composition of amino acids, proteins, nucleotides, β-glucan, and mannoproteins in lees are unknown due to limited studies in this area. Tartaric acid (TA) can also be a major chemical component of wine lees. Ye (2014) studied 16 wine lees (6 Riesling [RL] and 10 Pinot Noir [PN]); the TA content obtained in RL and PN lees was 1.49–3.23 and 1.08–5.90 g TA/kg fresh lees, respectively. This research also suggested that the TA content might depend on the vinification technique and grape origin. Most RL lees had lower total TA content than PN lees, which might be attributed to cold settling prior to fermentation during which potassium bitartrate precipitates and is removed by racking, and it also reflects the acid content of grapes at harvest required for different styles of wine.

Table 3.7 Fatty Acid Composition of Sherry Wine Lees

FATTY ACID		RELATIVE AMOUNT (%)
Capric acid	C10:0	2.32
Lauric acid	C12:0	4.42
Miristic acid	C13:0	1.98
Palmitic acid	C16:0	33.29
Palmitoleic acid	C16:1	1.80
Margaric acid	C17:0	0.30
Stearic acid	C18:0	10.40
Oleic acid	C18:1	7.82
Linoleic acid	C18:2	21.26
Linolenic acid	C18:3	5.88
Araquidonic acid	C20:0	2.10
Erucic acid	C22:0	6.10
Lignoceric acid	C24:0	2.32

Source: Gómez, M.E. et al., *J. Agric. Food Chem.*, 52, 4791, 2004.

94 VALORIZATION OF WINE MAKING BY-PRODUCTS

3.5.1.2 Interaction between Protein Compounds in Lees and Phenolic Compounds Wine lees is known to adsorb phenolic compounds from wine as well to release phenolic compounds, enzymes, fatty acids, mannoproteins, glucans, amino acids, and polypeptides to wine. This can modify the initial composition and consequently the wine quality (Pérez-Serradilla and Luque de Castro 2008). For example, mannoproteins released during the autolysis of yeast cell can stabilize the wine color and reduce wine astringency by interaction with phenols (Fornairon-Bonnefond et al. 2002; Guadalupe et al. 2010). Bustamante et al. (2008) reported that wine lees contain between 1.9 and 16.3 g of polyphenols/kg depending on the wine type and processing. Therefore, the estimated weight of polyphenols potentially obtained from wine lees of 20,000 tons was determined to be between 38 and 326 tons. Phenolic compounds would be potentially recoverable from wine lees and can add value to the wine industry. Ye (2014) studied the composition of phenolic compounds extracted from lees itself (Tables 3.8 and 3.9). Vasserot et al. (1997) reported that the interaction between lees and anthocyanins was mainly based on weak and reversible hydrogen bonds. However, Mazauric and Salmon (2005) argued that hydrogen bonding could not entirely explain the adsorption of anthocyanins on yeast lees in a complex environment such as wine. For example, anthocyanins were not only adsorbed by lees during oak aging of red wine, but they react with oxygen and generate new compounds (e.g., anthocyanins bound to proanthocyanidins by ethyl bridges) (Rivas-Gonzalo et al. 1995). Moreover, anthocyanins also react with flavanol polymers (King et al. 1980; Rivas-Gonzalo et al. 1995). Nonanthocyanin phenolic compounds are also adsorbed by yeast lees (Mazauric and Salmon 2005, 2006; Pérez-Serradilla and Luque de Castro 2008). Mazauric and Salmon (2005) measured wine polyphenols after contact with yeast lees in wine suspensions, and showed that adsorption followed biphasic kinetics. Other mechanisms of interaction between phenols and yeast lees have been proposed by researchers. For example, Salmon et al. (2000) who are in agreement with earlier researchers (Haslam et al. 1992; Kawamoto and Nakatsubo 1997) hypothesized that yeast can adsorb polyphenols by reacting with the yeast cell membrane possibly by polyphenol colloids interacting with proteins by van der Waal's bonds. In general, the available information suggests that the mechanisms of interaction are very complicated and further investigations are still required for a full understanding.

Table 3.8 Phenolic Profile of PN Lees Extracts (mg/g DM Extracts) Quantified by LC-MS Analysis

PHENOLIC COMPOUNDS	WINE LEES									
	PN1	PN2	PN3	PN4	PN5	PN6	PN7	PN8	PN9	PN10
Flavonoids										
Flavan-3-ols										
Catechin	2.24 ± 0.04	1.88 ± 0.08	3.32 ± 0.20	1.32 ± 0.06	2.70 ± 0.26	3.30 ± 0.82	2.18 ± 0.20	0.42 ± 0.00	7.98 ± 6.56	5.96 ± 0.46
Epicatechin	1.64 ± 0.10	1.24 ± 0.04	3.34 ± 0.04	0.90 ± 0.04	1.74 ± 0.12	1.82 ± 0.12	1.64 ± 0.02	0.90 ± 0.01	5.78 ± 0.10	2.72 ± 0.02
Epicatechin gallate	0.72 ± 0.04	0.38 ± 0.12	0.76 ± 0.08	0.42 ± 0.16	0.54 ± 0.18	0.96 ± 0.04	0.78 ± 0.00	0.56 ± 0.00	1.48 ± 0.64	0.98 ± 0.04
Epigallocatechin	0.02 ± 0.00	0.02 ± 0.00	0.02 ± 0.00	0.02 ± 0.00	0.02 ± 0.00	0.02 ± 0.00	0.02 ± 0.00	0.00 ± 0.00	0.08 ± 0.00	0.04 ± 0.00
Procyanidin A	1.08 ± 0.10	0.70 ± 0.48	1.38 ± 0.04	0.54 ± 0.08	1.08 ± 0.10	0.82 ± 0.04	0.74 ± 0.00	0.32 ± 0.02	3.38 ± 1.68	1.82 ± 0.04
Procyanidin B	1.62 ± 0.10	1.22 ± 0.08	3.30 ± 0.06	1.10 ± 0.04	1.56 ± 0.04	1.90 ± 0.10	1.72 ± 0.04	1.04 ± 0.02	3.52 ± 2.12	2.46 ± 0.10
Procyanidin C	—	0.02 ± 0.02	0.10 ± 0.02	0.00 ± 0.00	0.06 ± 0.00	—	—	—	—	0.14 ± 0.00
Anthocyanins										
Cyanidin-3-glucoside	—	—	—	—	—	—	—	—	—	—
Delphinidin-3-glucoside	—	—	—	—	—	—	—	—	—	—
Malvidin-3-glucoside	2.80 ± 0.02	2.49 ± 0.05	2.71 ± 0.02	3.13 ± 0.16	2.81 ± 0.08	3.03 ± 0.02	0.76 ± 0.02	1.45 ± 0.01	1.27 ± 1.06	2.09 ± 0.03
Peonidin-3-glucoside	—	—	—	—	—	—	—	—	—	—
Petunidin-3-glucoside	—	—	—	—	—	—	—	—	—	—

(*Continued*)

Table 3.8 (*Continued*)   Phenolic Profile of PN Lees Extracts (mg/g DM Extracts) Quantified by LC-MS Analysis

PHENOLIC COMPOUNDS	WINE LEES									
	PN1	PN2	PN3	PN4	PN5	PN6	PN7	PN8	PN9	PN10
Flavonols										
Quercetin	1.49 ± 0.13	1.14 ± 0.09	2.43 ± 0.13	2.05 ± 0.11	2.24 ± 0.08	2.99 ± 0.03	0.61 ± 0.03	0.72 ± 0.03	3.24 ± 1.94	0.95 ± 0.02
Quercetin methyl-glucoside	0.55 ± 0.06	0.45 ± 0.05	0.53 ± 0.05	0.38 ± 0.04	0.44 ± 0.04	0.55 ± 0.07	0.49 ± 0.00	0.49 ± 0.00	0.93 ± 0.23	0.55 ± 0.01
Non-flavonoids										
Gallic acid	1.80 ± 0.02	1.88 ± 0.06	2.24 ± 0.04	1.57 ± 0.08	1.98 ± 0.04	1.76 ± 0.03	0.89 ± 0.02	0.53 ± 0.00	1.66 ± 0.82	1.57 ± 0.04
Hydroxycinnamic acids										
Caftaric acid	1.22 ± 0.18	1.14 ± 0.32	1.36 ± 0.07	1.43 ± 0.16	1.27 ± 0.32	1.50 ± 0.20	0.89 ± 0.18	1.98 ± 0.12	3.27 ± 1.41	2.15 ± 0.05
Cinnamic acid	—	—	—	0.01 ± 0.00	—	—			—	—
p-Coumaric acid	0.02 ± 0.00	0.02 ± 0.01	0.03 ± 0.02	0.04 ± 0.00	0.06 ± 0.04	0.05 ± 0.02	—	—	0.02 ± 0.01	0.01 ± 0.00
Hydrobenzoric acid	0.05 ± 0.01	0.12 ± 0.01	0.00 ± 0.00	0.02 ± 0.00	—	—	0.02 ± 0.02	—	0.04 ± 0.04	0.03 ± 0.03
Stilbenoids										
Resveratrol	0.07 ± 0.01	0.03 ± 0.01	0.04 ± 0.01	0.03 ± 0.00	0.06 ± 0.00	0.07 ± 0.01	0.03 ± 0.00	0.17 ± 0.00	0.41 ± 0.16	0.15 ± 0.00

Source: Ye, Z.J., Characterization of bioactive compounds in lees from New Zealand wines and the effect of enzymatic oxidation on their bioactivity, Masters' of Food Science, University of Otago, Otago, New Zealand, 2014.

Table 3.9 Phenolic Profile of RL Lees Extracts (mg/g DM Extracts) Quantified by LC-MS Analysis

PHENOLIC COMPOUNDS	WINE LEES					
	RL1	RL2	RL3	RL4	RL5	RL6
Flavonoids						
Flavan-3-ols						
Catechin	0.66 ± 0.34	0.78 ± 0.02	0.50 ± 0.04	7.94 ± 0.60	3.36 ± 0.54	0.22 ± 0.02
Epicatechin	0.02 ± 0.02	0.14 ± 0.02	0.06 ± 0.10	1.20 ± 0.14	1.04 ± 0.32	0.28 ± 0.04
Epicatechin gallate	0.10 ± 0.06	0.16 ± 0.00	0.02 ± 0.00	3.56 ± 0.06	7.12 ± 0.74	3.36 ± 0.18
Epigallocatechin	—	—	—	0.06 ± 0.00	0.02 ± 0.00	—
Procyanidin A	0.28 ± 0.12	0.78 ± 0.10	0.38 ± 0.08	0.00 ± 0.00	10.22 ± 1.22	0.72 ± 0.02
Procyanidin B	0.40 ± 0.22	0.00 ± 0.00	0.02 ± 0.02	0.62 ± 0.28	0.80 ± 0.34	0.00 ± 0.00
Procyanidin C	0.30 ± 0.45	2.32 ± 0.02	2.44 ± 0.14	7.42 ± 0.36	7.34 ± 0.12	3.58 ± 0.66
Anthocyanins						
Cyanidin-3-glucoside	—	—	—	—	—	—
Delphinidin-3-glucoside	—	—	—	—	—	—
Malvidin-3-glucoside	—	—	—	—	—	—
Peonidin-3-glucoside	—	—	—	—	—	—
Petunidin-3-glucoside	—	—	—	—	—	—

(*Continued*)

98   VALORIZATION OF WINE MAKING BY-PRODUCTS

Table 3.9 (*Continued*)   Phenolic Profile of RL Lees Extracts (mg/g DM Extracts) Quantified by LC-MS Analysis

PHENOLIC COMPOUNDS	WINE LEES					
	RL1	RL2	RL3	RL4	RL5	RL6
Flavonols						
Quercetin	6.78 ± 0.79	0.65 ± 0.02	1.52 ± 0.23	5.02 ± 0.14	5.21 ± 0.73	0.77 ± 0.31
Quercetin methyl-glucoside	0.55 ± 0.03	0.22 ± 0.03	0.61 ± 0.14	0.75 ± 0.00	4.15 ± 0.13	0.95 ± 0.04
Non-flavonoids						
Gallic acid	0.76 ± 0.03	0.00 ± 0.00	0.09 ± 0.15	0.00 ± 0.00	0.24 ± 0.02	0.00 ± 0.00
Hydroxycinnamic acids						
Caftaric acid	0.84 ± 0.00	—	0.05 ± 0.01	0.25 ± 0.06	8.81 ± 0.57	3.20 ± 0.14
Cinnamic acid	1.74 ± 0.21	0.01 ± 0.00	0.17 ± 0.02	—	0.04 ± 0.04	0.02 ± 0.02
p-Coumaric acid	2.37 ± 0.29	0.12 ± 0.01	0.31 ± 0.07	0.35 ± 0.02	0.07 ± 0.12	0.01 ± 0.01
Hydrobenzoric acid	5.99 ± 1.11	0.14 ± 0.01	0.59 ± 0.03	1.80 ± 0.04	0.97 ± 0.03	0.10 ± 0.01
Stilbenoid						
Resveratrol	0.49 ± 0.09	0.02 ± 0.01	0.04 ± 0.01	—	—	—

Source: Ye, Z.J., Characterization of bioactive compounds in lees from New Zealand wines and the effect of enzymatic oxidation on their bioactivity, Masters' of Food Science, University of Otago, Otago, New Zealand, 2014.

3.5.2 Tartaric Acid

Tartaric acid is one of the major acids synthesized by plants. In grapes, tartaric acid is mainly located in the cells of the skin and outer flesh, with smaller amounts distributed throughout the flesh. Although it can exist in three stereoisomeric forms, only L-(+) tartaric acid is found to be naturally occurring. The other forms are D-(–) tartaric acid and the achiral meso form (Dega-Szafran et al. 2009). During wine making, a considerable amount of tartaric acid from cell fluid is extracted as a result of crushing and pressing of the berries. Two different tartrates may be precipitated during wine making, namely potassium bitartrate and calcium tartrate. Juice is often supersaturated with respect to potassium bitartrate. As alcoholic fermentation progresses, the solubility of potassium bitartrate decreases with the increase in alcohol concentration. This triggers a slow precipitation (Amerine and Singleton 1977; Jackson 2008). Cold treatments are commonly used to remove the excess potassium bitartrate. This treatment speeds up the crystallization of potassium bitartrate and its precipitation rate compared to normal storage temperatures. The precipitate is then filtered while cold to prevent re-dissolving of the crystals (Amerine and Singleton 1977). Calcium tartrate is the other salt of tartaric acid that potentially can precipitate from wine. This can occur when amounts of calcium carbonate are used in deacidification of wine. Moreover, the use of cement cooperage, filter pads, and fining agents can increase the calcium concentration in wine. Calcium tartrate crystals are more difficult to remove during wine making than potassium bitartrate crystals since their rate of precipitation is not increased by chilling. Fortunately, the formation of calcium tartrate crystals in wine is less common (Jackson 2008). Recently, the recovery of tartaric acid from winery waste has attracted the attention of researchers (Andrés et al. 1997; Versari et al. 2001; Yalcin et al. 2008; Kaya et al. 2014). In general, four different processes are used to recover tartaric acid from waste including adsorption, extraction, ion exchange, and electrodialysis (Table 3.10) (Kaya et al. 2014). Alumina and amberlite IRA-67 were used as adsorbents to recover tartaric acid from winery wastewater, the latter showed the maximum adsorption efficiency of 97% (Uslu and Inci 2009; Uslu et al. 2009) under laboratory conditions. In other work, a tri-iso-octylamine and

100 VALORIZATION OF WINE MAKING BY-PRODUCTS

Table 3.10 Summary of Some Processes Applied in the Recovery of Tartaric Acid

PROCESS	RELATIVE ISSUES	REFERENCES
Adsorption	Alumina was used as an adsorbent to recover 22% (maximum % at 298 K) of L-(+)-tartaric acid from winery wastewater. Adsorption is dependent on the acid concentration and the amount of alumina.	Uslu and Inci (2009)
	Amberlite IRA-67 was used as adsorbent which achieved the maximum adsorption efficiency of 97.18%.	Uslu et al. (2009)
Extraction	Tri *iso* octylamino and *iso* docanol mixture was used for the extraction of tartaric acid solution at 25°C.	Poposka et al. (2000)
	Aqueous Amberlite LA-2 (amine) solution used for extraction. The extraction efficiency was 83.06%.	Marchitan et al. (2010)
	MIBK, hexane, 1-octanol, 2-octanol, 1-decanol, 1-hexanol, and tri-n-butyphosphate (TBP) was used for the extraction of model waste (aqueous tartaric acid solution). The highest extraction of tartaric acid was found in aqueous TBP solution (0.25 mol/L).	Upadhyay and Keshav (2012)
	50% acidified ethanol (contains 1% HCl) was used to extract tartaric acid from lees of Pinot Noir (PN) and Riesling (RL) wine. 1.49–3.23 and 1.08–5.90 g tartaric acid/kg fresh lees were obtained from lees of PN and RL, respectively.	Ye (2014)
Ion exchange	Aqueous KOH solution (at 80°C) was used to dissolve the waste material, followed by adding saturated pure tartaric acid to precipitate potassium tartrate. K^+ and $[SO_4]^{2-}$ ions were removed using ion exchange.	Yalcin et al. (2008)
Electrodialysis	Tartaric acid from elute of ion exchange after grape juice treatment was recovered by using electrodialysis. Obtained tartaric acid has purity about 60%. The maximum concentration of 200 kg/m^3 can be achieved.	Andrés et al. (1997)
	Bipolar membrane electrodialysis with three-compartment configuration provides the membrane with high selectivity which helps increase the purity of tartaric acid. Strong acid possess lower electrical resistance which helps reduction of energy consumption.	Zhang et al. (2009)

iso-decanol mixture, Amberlite LA-2 (amine), hexane, 1-octanol, 2-octanol, 1-decanol, 1-hexanol, and tri-n-butyphosphate (TBP) were used to extract tartaric acid from aqueous solutions (Poposka et al. 2000; Marchitan et al. 2010; Upadhyay and Keshav 2012); Ye (2014) extracted the tartaric acid from Pinot Noir and Riesling lees. The weaknesses of solvent extraction are the use of a batch process and no unique solvent is able to cover all different winery waste materials.

WINE MAKING BY-PRODUCTS

Ion exchange can only recover tartaric acid from aqueous solution. Electrodialysis is a novel technology, which can recover the tartaric acid from an eluate post ion exchange. Zhang et al. (2009) applied bipolar membrane electrodialysis with three-compartment configuration which improved the selectivity of membrane and the purity of tartaric acid. In addition, this configuration also showed a potential in the reduction of energy consumption.

3.5.3 Carbon Dioxide

In wine making, carbon dioxide is generated and released during both alcoholic and malolactic fermentations. During alcoholic fermentation, large volumes of carbon dioxide (approximately 260 mL carbon dioxide per g of glucose) accompany the conversion of sugars to ethanol by yeasts (Jackson 2008). Also, small amounts of carbon dioxide are generated by lactic acid bacteria and the breakdown of amino acids and phenols (Jackson 2008). Ciani and Maccarelli (1998) studied both Saccharomyces and non-Saccharomyces strains and found that Saccharomyces strains generally release more carbon dioxide (>1.0 g CO_2/day) than non-Saccharomyces strains (<0.6 g CO_2/day). The rate of carbon dioxide release was different and even within species depended on the yeast strain (Ciani and Maccarelli 1998). Most of the carbon dioxide produced by yeasts during fermentation is lost to the atmosphere, while a proportion remains dissolved in the wine; further loss of carbon dioxide occurs during maturation with only 2 g/L remaining in finished wine. In addition, carbon dioxide produced during wine making stage is not pure but is mixed with various volatile compounds including ethanol, higher alcohols, monoterpenes, ethyl, and acetate esters (Jackson 2008). CO_2 capture technologies are required to recover CO_2 lost during fermentation. These have been well investigated with major applications in gas production and flue reduction. CO_2 can be captured and separated from gas mixtures by using different processes including absorption, adsorption, membrane, and cryogenic distillation (Table 3.11). The most well-known commercialized process is absorption with amine solutions (such as MEA) which are commonly used to strip CO_2 from gas mixtures. However, this process is also regarded as having a low CO_2 loading capacity, requires intensive energy inputs, and is associated with poor

102 VALORIZATION OF WINE MAKING BY-PRODUCTS

Table 3.11 Summary of Some Technologies Used for CO_2 Capture or Separation

PROCESS	CHEMICAL COMPOUNDS USED	COMMENTS
Absorption	MEA[a]	Low CO_2 loading capacity, intensive energy input, degradation of amine. Capacity: 0.5 mol CO_2/mol MEA.
	MEA/MDEA	Mixture of MEA and MDEA, Lower energy consumption than MEA.
	Aqueous ammonium	Sorbent used in wet scrubber, the energy consumption is only 60% of MEA; by-products are $(NH_4)NO_3$, $(NH_4)HCO_3$, $(NH_4)_2SO_4$.
Adsorption	PEI[b]	May be a cost-effective technology with adsorption capacity of 246 mg/g PEI.
	Activated carbon	Loading capacity can reach 65.7 mg CO_2/g adsorbent.
	Lithium compounds	High-temperature CO_2 adsorbent; the reaction between lithium zirconate and CO_2 is reversible at 450°C–590°C; lithium silicate adsorbs CO_2 at 720°C and releases when temperature keeps increasing, along with large capacity (360 mg CO_2/g sorbent), rapid adsorption, and good stability.
Membrane	Polyimide PDP[c] Polysulfone Polyethersulfone Polyetherimide	All polymeric membranes. The costs of membranes are low but cannot achieve high CO_2 purity. High-energy consumption when creating a pressure difference across the membrane.
Cryogenic distillation	N/A	Physical method. Separation of gas based on condensation point. The purity of O_2 separate in this system is no less than 99%.

Source: Yang, H. et al., *J. Environ. Sci.*, 20, 14, 2008.

[a] Methylaminoethanol.
[b] Polyethylenimine.
[c] Polydimethylphenlene oxide.

stability of the amine solutions. Adsorption processes with lithium compounds are able to handle high-temperature flue gases, as well as having a large CO_2 loading capacity (360 mg CO_2/g sorbent) and good stability. However, this process is not suitable for recovering CO_2 at typical wine fermentation temperatures as lithium compounds adsorb CO_2 at 720°C (Yang et al. 2008). Processes using low-cost gas separation membranes are economical. However, further improvements are required in the areas of CO_2 purity and energy consumption. De Assis Filho et al. (2013) have recently demonstrated the feasibility of cryogenic distillation for CO_2 recovery from fermentation. This method is based on the different condensation points of the components of air (at 1.8 MPa, the dew point of O_2, N_2, and CO_2 are $-130°C$, $-154°C$, and $-12°C$, respectively) such that a CO_2 condensate can be collected while O_2 and N_2 remain in the gas phase (De Assis Filho et al. 2013). However, Xu et al. (2010) have suggested that the fact that the feed stream from fermentation is highly concentrated in CO_2 (>90%, v/v) means that dehydration and compression are possibly the only purification processes required to recover CO_2, and these two processes are simple and relatively cheap.

3.5.4 Wastewater

Large amounts of wastewater are generated every year throughout wine production. Kumar and Kookana (2006) reported that a ton of crushed grapes produced 3–5 kL of wastewater. However, the volume of winery wastewater generated varies greatly depending on the size of the winery and the processing stage. These factors also affect the composition of the winery wastewater (Table 3.12) (Chapman et al. 2001). Organic compounds, such as organic acids, sugars and alcohol in wastewater mainly came from juice, wine, and lees lost in transfer operations, while the residues of caustic soda/citric acid arise from the cleaning of pipes (Chapman et al. 2001). Phenolic compounds are a major contaminant in winery wastewater which includes simple

Table 3.12   Chemical Parameters and Composition of Winery Wastewater (mg/L)

	VINTAGE[a]	NONVINTAGE[b]		VINTAGE	NONVINTAGE
Tartaric acid	530	350	BOD	1500–6000	500–3500
Lactic acid	250	120	EC	1.5–3.5	0.9–1.3
Acetic acid	100	50	pH	4–8	6–10
Glucose	2500	230	TOC	2162	1542
Fructose	2500	270	Total Kjeldahl nitrogen	34.1	30.4
Glycerol	190	120	Total phosphorus	7.3	7.2
Ethanol	2400	2900			

Source: Chapman, J. et al., *Winery Wastewater Handbook*, Adelaide, South Australia, Australia, 2001.

[a] The term *vintage* means the period involving major wine making activities, such as harvesting, crushing, and pressing.

[b] The term *nonvintage* means the period not involving major wine making activities.

104 VALORIZATION OF WINE MAKING BY-PRODUCTS

phenolics, phenolic acids and aldehydes, cinnamic acids, coumarins, benzophenones and xanthones, and stilbenes, benzoquinones, anthraquinones, lignans, and flavonoids (Soto et al. 2011). The total organic content of winery wastewater is normally measured as biochemical oxygen demand (BOD), chemical oxygen demand (COD), and total organic carbon (TOC), which provides information on the overall quality of wastewater (Mosse et al. 2011). BOD and TOC of wastewater ranged from 500 to 6000 and 1542 to 2162 mg/L, respectively, through the year (Chapman et al. 2001). The COD of wastewater from wine making in so-called Old World countries ranged from 738 to 296,000 mg/L and was found to be greater than that from the New World which ranged from 320 to 45,000 (Mosse et al. 2011). Moderate quantities of winery waste and wastewater can be used to enhance the fertility of the soil due to its high organic matter (Bustamante et al. 2011). However, continuous application of winery wastewater to soil can increase soil salinity, which can eventually cause soil dispersion (Conradie et al. 2014). Therefore, the management of winery wastewater has attracted the attention of researchers from wine making regions around the world. A number of treatments have been designed to achieve significant reduction in the organic matter content of winery wastewater. These can be divided into biological treatments (e.g., aerobic treatment, anaerobic treatment, wetland), physicochemical treatments (e.g., electrodialysis, ion exchange, and reverse osmosis), and some emerging technological treatments (e.g., advanced oxidation process). Currently, three biological aerobic treatments (aerated lagoons, activated sludge, and sequencing batch reactors) are in widespread use in industries; in contrast, anaerobic treatments, including anaerobic sequencing batches, upflow anaerobic sludge blankets (UASBs), and anaerobic lagoons are only used occasionally. Most of the physicochemical treatments are rarely used, although pilot-scale applications of ion exchange and reverse osmosis have been undertaken (Mosse et al. 2011). The major barrier to the application of more effective treatments are financial, including the cost of equipment/land, equipment maintenance, energy/chemical consumption, and employment of expertise (Mosse et al. 2011). Therefore, there is a demand for the development of affordable winery wastewater treatment using emerging technologies. Although organic carbon can be removed from winery wastewater by both biological and chemical oxidation, the latter can oxidize non-biodegradable substances in wastewater (Mosse et al. 2011). Hydrogen peroxide is an oxidant often used for waste water treatment, such as wet air oxidation (WAO) and advanced oxidation processes (AOP). In WAO, hydrogen peroxide is often applied with the presence of different catalysts (copper and iron ions are most commonly used). In AOP, a high dosage of hydrogen peroxide was used to reduce chemical oxygen demand (COD) instead of using different catalysts. The dosage of hydrogen peroxide required to achieve the desired level of oxidation varies with different techniques and substrates. Several metal ions can be used with hydrogen peroxide acting as catalyst to increase the modification of phenol structure rate. Rivas et al. (1999) reported a combined addition of hydrogen peroxide and a bivalent metal (copper or manganese) to enhance the rate of phenol removal. The use of H_2O_2 can lead to a fast structural changing of phenolic compounds and accelerate the oxidation of phenolic compounds. Horseradish peroxidase (HRP) can also be used to catalyze the oxidation reaction. Ye (2014) studied the oxidation of a model waste system using hydrogen peroxide in the presence of HRP (Figure 3.8). In a period of 24 h, there was no oxidation of tannins in

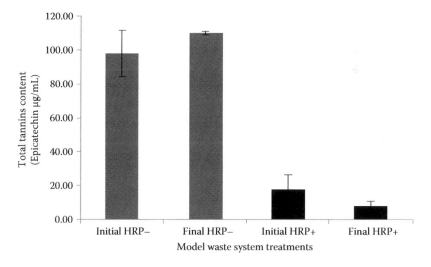

Figure 3.8 The concentration of the total tannins in a model waste system containing catechin: grape-seed extract (50:50 w/w) oxidized with (HRP+) and without (HRP−) horseradish peroxidase. (From Ye, Z.J., Characterization of bioactive compounds in lees from New Zealand wines and the effect of enzymatic oxidation on their bioactivity, Masters' of Food Science, University of Otago, Otago, New Zealand, 2014.)

the control (untreated) sample. In the treated sample, a significant loss of tannin occurred ($P < 0.05$) within a minute. Ye (2014) also studied the free radical scavenging capacity and found that antioxidant activity at pre- and post-oxidation stages were not significantly different. Therefore, the hydrogen peroxide–induced oxidation (in the presence of HRP) has the potential to reduce organic carbon in winery wastewater with the products possessing free radical scavenging capacity.

3.6 Challenges and Opportunities

By-products and wastewater produced during wine making require significant further research both to fully extract their economic value and to minimize the unwanted environmental effects arising from their disposal. Current utilization of wine by-products focuses on grape stems, pomace, and seeds for nutrient supplements, oenological additions, and animal feed. The extraction of bioactive compounds from vine prunings and lees is less well considered. Vine prunings are produced at various times of the year and require chopping and grinding before solvent extraction. Mechanical reduction of particle size is not required for lees, but the amounts produced are very small compared to pomace so that there are difficulties associated with their collection for further processing. Carbon dioxide is mainly released during wine fermentation. Currently, there is no evidence that it is being recovered from production. The financial investment required to set up the wastewater treatment systems is a major barrier, although emerging technologies such as advanced oxidation processes are promising. Most studies have focussed on the reduction of chemical oxygen demand in wastewater, but in future more emphasis on the recovery of useful products is required.

Acknowledgments

The authors acknowledge the funding received from the New Zealand Ministry for Environment (Community Environment Fund & Waste Minimisation Fund, Deed Number 20398) and the Sustainable Farm Fund (Project Number 09/099). This work is part of the New Zealand Grape and Wine Research Program, a joint investment by the Plant and Food Research and NZ Winegrowers.

References

Alexandre, H. and Guilloux-Benatier, M. (2006). Yeast autolysis in sparkling wine—A review. *Australian Journal of Grape and Wine Research*, 12(2), 119–127. doi: 10.1111/j.1755–0238.2006.tb00051.x

Alnaizy, R. and Akgerman, A. (2000). Advanced oxidation of phenolic compounds. *Advance Environmental Research*, 4, 233–244.

Amerine, A. and Joslyn, M.A. (1967). Composition of grapes. In: *Table Wines: The Technology of Their Production*, 2nd edn., Amerine, A. and Joslyn, M.A. (eds.), pp. 234–238. Berkeley, CA: University of California Press.

Amerine, M.A. and Singleton, V.L. (eds.). (1977). *Wine: An Introduction*. Berkeley, CA: University of California Press.

Anastasiadi, M., Pratsinis, H., Kletsas, D., Skaltsounis, A.-L., and Haroutounian, S.A. (2012). Grape stem extracts: Polyphenolic content and assessment of their in vitro antioxidant properties. *LWT—Food Science and Technology*, 48, 316–322.

Andrés, L.J., Riera, F.A., and Alvarez, R. (1997). Recovery and concentration by electrodialysis of tartaric acid from fruit juice industries waste water. *Journal of Chemical Technology & Biotechnology*, 70, 247–252.

Bail, S., Stuebiger, G., Krist, S., Unterweger, H., and Buchbauer, G. (2008). Characterisation of various grape seed oils by volatile compounds, triacylglycerol composition, total phenols and antioxidant capacity. *Food Chemistry*, 108, 1122–1132.

Balasundram, N., Sundram, K., and Samman, S. (2006). Phenolic compounds in plants and agri-industrial by-products: Antioxidant activity, occurrence, and potential uses. *Food Chemistry*, 99(1), 191–203.

Barros, A., Gironés-Vilaplana, A., Teixeira, A., Collado-González, J., Moreno, D.A., Gil-Izquierdo, A., Rosa, E., and Domínguez-Perles, R. (2014). Evaluation of grape (*Vitis vinifera* L.) stems from Portuguese varieties as a resource of (poly) phenolic compounds: A comparative study. *Food Research International*, 65, 375–384. http://dx.doi.org/10.1016/j.foodres.2014.07.021.

Benítez, T., Rincón, A.M., and Codón, A.C. (2011). Chapter 3—Yeasts used in biologically aged Winesw. In: *Molecular Wine Microbiology*, Alfonso, V.C.S., Rosario, M., Ramon Gonzalez Garcia, R.M., Carrascosa Santiago, A.V., and Ramon Gonzalez, G. (eds.), pp. 51–84. San Diego, CA: Academic Press.

Bekhit, A.E.D., Qiao, W., Creasy, G., Hider, R., and Dawson, C. (December 9–10, 2008). Manipulating the functionality of grape seeds products through reflective mulching and wine fermentation. *43rd Annual Conference of New Zealand Nutrition Society*, Christchurch, New Zealand.

Bisson, L.F. and Joseph, C.M.L. (2009). Yeasts. In: *Biology of Microorganisms on Grapes in Must and in Wine*, König, H., Unden, G., and Fröhlich, J. (ed.), pp. 47–60. Berlin, Germany: Springer.

Bustamante, M.A., Moral, R., Paredes, C., Pérez-Espinosa, A., Moreno-Caselles, J., and Pérez-Murcia, M.D. (2008). Agrochemical characterisation of the solid by-products and residues from the winery and distillery industry. *Waste Management*, 28(2), 372–380.

108 VALORIZATION OF WINE MAKING BY-PRODUCTS

Bustamante, M.A., Said-Pullicino, D., Agulló, E., Andreu, J., Paredes, C., and Moral, R. (2011). Application of winery and distillery waste composts to a Jumilla (SE Spain) vineyard: Effects on the characteristics of a calcareous sandy-loam soil. *Agriculture, Ecosystems & Environments*, 140, 80–87.

Cadot, Y., Miñana-Castelló, M.T., and Chevalier, M. (2006). Anatomical, histological, and histochemical changes in grape seeds from *Vitis vinifera* L. cv Cabernet franc during fruit development. *Journal of Agricultural and Food Chemistry*, 54, 9206–9215.

Çetin, E.S., Altinöz, D., Tarçan, E., and Baydar, N.G. (2011). Chemical composition of grape canes. *Industrial Crops and Products*, 34, 994–998.

Chapman, J., Baker, P., and Wills, S. (2001). *Winery Wastewater Handbook: Production, Impacts and Management.* Adelaide, South Australia, Australia: Winetitles.

Charpentier, C. (2010). Ageing on lees (sur lies) and the use of speciality inactive yeasts during wine fermentation. In: *Managing Wine Quality*, Reynolds, A.G. (ed.), Vol. 2, pp. 164–187. Oxford, UK: Woodhead Publishing.

Charpentier, C. and Feuillat, M. (2008). Elevage des vins rouges sur lies. Incidence d l'addition d'une β-glucanase sur lacomposition en polysaccharides et leurs interaction avec polyphenols. *Revue des Enologues*, 129, 31–35.

Cheng, V.J. (2011). Evaluation of the bioactivity of New Zealand wine by-products. Masters' of Food Science, University of Otago, Otago, New Zealand.

Cheynier, V., Dueñas-Paton, M., Salas, E., Maury, C., Souquet, J.M., Sarni-Manchado, P., and Fulcrand, H. (2006). Structure and properties of wine pigments and tannins. *American Journal of Enology and Viticulture*, 57(3), 298–305.

Chowdhury, A.K. and Ross, L.W. (1975). The catalytic wet oxidation of strong wastewaters. *AIChE Symposium Series*, 151, 46–58.

Ciani, M. and Maccarelli, F. (1998). Oenological properties of non-Saccharomyces yeasts associated with wine-making. *World Journal of Microbiology & Biotechnology*, 14, 199–203.

Conradie, A., Sigge, G.O., and Cloete, T.E. (2014). Influence of winemaking practices on the characteristics of winery wastewater and water usage of wineries. *South African Journal of Enology and Viticulture*, 35, 10–19.

Coombe, B.G. and Dry, P.R. (1992). *Viticulture*, Vol. 2: Practice. Adelaide, South Australia, Australia: Winetitiles.

De Assis Filho, R.B., Danielski, L., de Carvalho, F.R., and Stragevitch, L. (2013). Recovery of carbon dioxide from sugarcane fermentation broth in the ethanol industry. *Food and Bioproducts Processing*, 91, 287–291.

De Marchi, F., Seraglia, R., Molin, L., Traldi, P., De Rosso, M., Panighel, A., Dalla Vedova, A., Gardiman, M., Giust, M., and Flamini, R. (2012). Seed oil triglyceride profiling of thirty-two hybrid grape varieties. *Journal of Mass Spectrometry*, 47, 1113–1119.

Dega-Szafran, Z., Katrusiak, A., and Szafran, M. (2009). Structural and spectroscopic studies of the 1:1 complex of meso-tartaric acid with 1,4-dimethylpiperazine di-betaine. *Journal of Molecular Structure*, 920(1–3), 202–207.

Devesa-Rey, R., Vecino, X., Varela-Alende, J.L., Barral, M.T., Cruz, J.M., and Moldes, A.B. (2011). Valorization of winery waste vs. the costs of not recycling. *Waste Management*, 31, 2327–2335.

FAOSTAT. (2010). The total area of grape harvested in 2009. http://faostat3. fao.org/download/Q/QC/E, accessed on October 22, 2014.

FAOSTAT. (2014). World grape production. http://faostat3.fao.org/download/Q/QC/E, accessed on December 22, 2014.

Fernandes, L., Casal, S., Cruz, R., Pereira, J.A., and Ramalhosa, E. (2013). Seed oils of ten traditional Portuguese grape varieties with interesting chemical and antioxidant properties. *Food Research International*, 50, 161–166.

Fornairon-Bonnefond, C., Camarasa, C., Moutonet, M., and Salmon, J.M. (2002). New trends on yeast autolysis and wine aging on lees: A bibliographic review. *Journal of International Science Vigne Vin*, 36, 49–69; Cited from Pérez-Serradilla, J.A. and Luque de Castro, M.D. (2008). Role of lees in wine production: A review. *Food Chemistry*, 111, 447–457.

Gómez, M.E., Igartuburu, J.M., Pando, E., Luis, F.R., and Mourente, G. (2004). Lipid composition of lees from sherry wine. *Journal of Agricultural and Food Chemistry*, 52(15), 4791–4794.

González-Centeno, M.R., Rosselló, C., Simal, S., Garau, M.C., López, F., and Femenia, A. (2010). Physico-chemical properties of cell wall materials obtained from ten grape varieties and their byproducts: Grape pomaces and stems. *LWT—Food Science and Technology*, 43, 1580–1586.

González-Paramás, A.M., Esteban-Ruano, S., Santos-Buelga, C., de Pascual-Teresa, S., and Rivas-Gonzalo, J.C. (2003). Flavanol content and antioxidant activity in winery by-products. *Journal of Agricultural and Food Chemistry*, 52(2), 234–238.

Gorena, T., Saez, V., Mardones, C., Vergara, C., Winterhalter, P., and Baer, D.V. (2014). Influence of post-pruning storage on stilbenoid levels in *Vitis vinifera* L. Canes. *Food Chemistry*, 155, 256–236.

Guadalupe, Z., Martínez, L., and Ayestarán, B. (2010). Yeast mannoproteins in red winemaking: Effect on polysaccharide, polyphenolic and colour composition. *American Journal of Enology & Viticulture*, 61, 191–200.

Haslam, E., Lilley, T.H., Warminski, E., Liao, H., Cai, Y., Martin, R., and Luck, G. (1992). *Polyphenol Complexation*. Washington, DC: American Chemical Society.

Hwang, J.Y., Shyu, Y.S., and Hsu, C.K. (2009). Grape wine lees improves the rheological and adds antioxidant properties to ice cream. *LWT—Food Science and Technology*, 42(1), 312–318.

Jackson, R.S. (2008). *6-Chemical Constitutents of Grapes and Wine: Wine Science*, 3rd edn., pp. 270–331. San Diego, CA: Academic Press.

Jin, B. and Kelly, J. (2009). Biotechnological potential of wine industry residues. In: *Biotechnology for Agro-Industrial Resides Utilization*, Nigam, S. and Poonam, P.A (eds.), Vol. XVIII, pp. 293–326. Amsterdam, the Netherlands: Springer.

Kammerer, D., Claus, A., Carle, R., and Schieber, A. (2004). Polyphenol screening of pomace from red and white grape varieties (*Vitis vinifera* L.) by HPLC-DAD-MS/MS. *Journal of Agricultural and Food Chemistry*, 52: 4360–4367.

Karacabey, E. and Mazza, G. (2008). Optimization of solid-liquid extraction of resveratrol and other phenolic compounds from milled grape canes (*Vitis vinifera*). *Journal of Agricultural & Food Chemistry*, 56, 6318–6325.

Karacabey, E. and Mazza, G. (2010). Optimisation of antioxidant activity of grape cane extracts using response surface methodology. *Food Chemistry*, 119, 343–348.

Karacabey, E., Mazza, G., Bayindirh, L., and Artik, N. (2012). Extraction of bioactive compounds from milled grape canes (*Vitis vinifera*) using a pressurized low-polarity water extractor. *Food Bioprocess and Technology*, 5, 359–371.

Kawamoto, H. and Nakatsubo, F. (1997). Effects of environmental factors on two-stage tannin-protein coprecipitation. *Phytochemistry*, 46, 379–483.

Kaya, C., Şahbaz, A., Arar, Ö, Yüksel, Ü., and Yüksel, M. (2014). Removal of tartaric acid by gel and macroporous ion exchange resins. *Desalination & Water Treatment*, 55, 514–521. doi: 10.1080/19443994.2014.919239.

King, G.A., Swenny, J.C., Radford, T., and Iacobucci, G.A. (1980). The ascorbic/O_2 degradation of anthocyanidins. *Bulletin Liaison du Groupe Polyphenols*, 9, 121–128; Cited from Pérez-Serradilla, J.A. and Luque de Castro, M.D. (2008). Role of lees in wine production: A review. *Food Chemistry*, 111, 447–547.

Kumar, A. and Kookana, R. (2006). Impact of winery wastewater on ecosystem health—An introductory assessment. Final Report, Grape and wine Research and Development Corporation, CSL 02/03.

Lafka, T.I., Sinanoglou, V., and Lazos, E.S. (2007). On the extraction and antioxidant activity of phenolic compounds from winery wastes. *Food Chemistry*, 104(3), 1206–1214.

Llobera, A. and Cañellas, J. (2007). Dietary fibre content and antioxidant activity of Manto Negro red grape (*Vitis vinifera*): Pomace and stem. *Food Chemistry*, 101, 659–666.

Lutterodt, H., Slavin, M., Whent, M., Turner, E., and Yu, L. (2011). Fatty acid composition, oxidative stability, antioxidant and antiproliferative properties of selected cold-pressed grape seed oils and flour. *Food Chemistry*, 128, 391–399.

Marchitan, N., Cojocaru, C., Mereuta, A., Duca, G., Cretescu, I., and Gonta, M. (2010). Modeling and optimization of tartaric acid reactive extraction from aqueous solutions: A comparison between response surface methodology and artificial neural network. *Separation & Purification Technology*, 75, 273–285.

Margalit, Y. (1997). *Concepts in Wine Chemistry*, 1st edn. San Francisco, CA: The Wine Appreciation Guild Ltd.

Marles, M.A.S., Ray, H., and Gruber, M.Y. (2003). New perspectives on proanthocyanidin biochemistry and molecular regulation. *Phytochemistry*, 64, 367–383.

WINE MAKING BY-PRODUCTS **111**

Mazza, G. and Miniati, E. (1993). Grapes. In: *Anthocyanins in Fruits, Vegetables, and Grains,* pp. 149–199. Boca Raton, FL: CRC Press; Cited in: Lafka, T.I., Sinanoglou, V., and Lazos, E.S. (2007). On the extraction and antioxidant activity of phenolic compounds from winery wastes. *Food Chemistry,* 104, 1206–1214.

Mazauric, J.P. and Salmon, J.M. (2005). Interaction between yeast lees and wine polyphenols during simulation of wine aging: I. Analysis of remnant polyphenolic compounds in the resulting wines. *Journal of Agricultural & Food Chemistry,* 53, 5647–5653.

Mazauric, J.P. and Salmon, J.M. (2006). Interaction between yeast lees and wine polyphenols during simulation of wine aging: II. Analysis of desorbed polyphenol compounds from yeast lees. *Journal of Agricultural & Food Chemistry,* 54, 3876–3881.

Min, Z., Guo, Z., Wang, K., Zhang, A., Li, H., and Fang, Y. (2014). Antioxidant effects of grape vine cane extracts from different Chinese grape varieties on edible oils. *Molecules,* 19, 15213–15223.

Molina-Alcaide, E., Moumen, A., and Martín-García, A.L. (2008). By-products from viticulture and the wine industry: Potential as sources of nutrients for ruminants. *Journal of the Science of Food and Agriculture,* 4, 597–604.

Moreno-Arribas, M.V. and Polo, M.C. (2009). *Wine Chemistry and Biochemistry.* New York: Springer.

Mosse, K.P.M., Patti, A.F., Christen, E.W., and Cavagnaro, T.R. (2011). Review: Winery wastewater quality and treatment options in Australia. *Australian Journal of Grape and Wine Research,* 17, 111–122.

Navas, P.B. (2009).Chemical composition of the virgin oil obtained by mechanical pressing form several grape seed varieties (*Vitis vinifera* L.) with emphasis on minor constituents. *Archivos Latinoamericanos de Nutrición,* 59, 214–219.

Naziri, E., Nenadis, N., Mantzouridou, F.T., and Tsimidou, M.Z. (2014). Valorization of the major agrifood industrial by-products and waste from Central Macedonia (Greece) for the recovery of compounds for food applications. *Food Research International,* 65, 350–358.

Oomah, B.D., Liang, J., Godfrey, D., and Mazza, G. (1998). Microwave heating of grapeseed: Effect on oil quality. *Journal of Agricultural and Food Chemistry,* 46, 4017–4021.

Ough, S.C. and Amerine, M.A. (1961). Studies on controlled fermentation V. Effects on colour, composition, and quality of red wines. *American Journal of Enology and Viticulture,* 12, 9–19.

Pawlus, A.D., Sahli, R., Bisson, J., Rivière, C., Delaunay, J., Richard, T., Gomès, E., Bordenave, L., Waffo-Téguo, P., and Mérillon, J. (2013). Stilbenoid profiles of canes from *Vitis* and *Muscadinia* species. *Journal of Agricultural & Food Chemistry,* 61, 501–511.

Pérez-Serradilla, J.A. and Luque de Castro, M.D. (2011). Microwave-assisted extraction of phenolic compounds from wine lees and spray-drying of the extract. *Food Chemistry,* 124(4), 1652–1659.

Pinelo, M., Sineiro, J., and Nunez, M.J. (2006). Mass transfer during continuous solid-liquid extraction of antioxidants from grape by-products. *Journal of Food Engineering,* 77, 57–63.

112 VALORIZATION OF WINE MAKING BY-PRODUCTS

Poposka, F.A., Prochazka, J., Tomovska, R., Nikolovski, K., and Grizo, A. (2000). Extraction of tartaric acid from aqueous solutions with tri-iso-octylamine (Hostarex A 324). Equilibrium and kinetics. *Chemical Engineering Science*, 55, 1591–1604.

Prieur, C., Rigaud, J., Cheynier, V., and Moutounet, M. (1994). Oligomeric and polymeric procyanidins from grape seeds. *Phytochemistry*, 36, 781–784.

Prado, J.M., Dalmolin, I., Carareto, N.D.D., Basso, R.C., Meirelles, A.J.A., Vladimir Oliveira, J., Batista, E.A.C., and Meireles, M.A.A. (2012). Supercritical fluid extraction of grape seed: Process scale-up, extract chemical composition and economic evaluation. *Journal of Food Engineering*, 109, 249–257.

Rankine, B.C. (1989). *Making Good Wine: A Manual of Winemaking Practice for Australia and New Zealand*, p. 207, Melbourne, Australia: Sun Books.

Rayne, S., Karacabey, E., and Mazza, G. (2008). Grape cane waste as a source of *trans*-resveratrol and *trans*-viniferin: High-value phytochemicals with medicinal and anti-phytopathogenic applications. *Industrial Crops Production*, 27, 335–340.

Reed, G. and Nagodawithana, T.W. (1991). *Yeast Technology*, 2nd edn. New York: Van Nostrand Reinhold.

Rentzsch, M., Schwarz, M., and Winterhalter, P. (2007). Pyranoanthocyanins— An overview on structures, occurrence, and pathways of formation. *Trends in Food Science and Technology*, 18, 526–534.

Reynolds, A.G., Wardle, D.A., and Naylor, A.P. (1995). Impact of training system and vine spacing on vine performance and berry composition of Chancellor. *American Journal of Enology & Viticulture*, 1, 88–97.

Rice, A.C. (1976). Solid waste generation and by-product recovery potential from winery residues. *American Journal of Enology and Viticulture*, 27, 21–26.

Rivas, F.J., Kolackowski, S.T., Beltran, F.J., and McLurgh, D.B. (1999). Hydrogen peroxide promoted wet air oxidation of phenol: Influence of operating condition and homogeneous metal catalysts. *Journal of Technology & Biotechnology*, 74, 390–398.

Rivas-Gonzalo, J.G., Bravo-Haro, S., and Santos-Buelga, C. (1995). Detection of compounds formed through the reaction of malvidin 3-monogluco-side and catechin in the presence of acetaldehyde. *Journal of Agricultural & Food Chemistry*, 43, 1444–1449.

Rombaut, N., Savoire, R., Thomasset, B., Castello, J., Van Hecke, E., and Lanoisellé, J.L. (2014). Optimization of oil yield and oil total phenolic content during grape seed cold screw pressing. *Industrial and Crops Production* 63, 26–33. http://dx.doi.org/10.1016/j.indcrop. 2014.10.001.

Ruberto, G., Renda, A., Daquino, C., Amico, V., Spatafora, C., Tringali, C., and De Tommasi, N. (2007). Polyphenol constituents and antioxidant activity of grape pomace extracts from five Sicilian red grape cultivars. *Food Chemistry*, 100, 203–210.

Rubio, M., Alvarez-Ortí, M., Alvarruiz, A., Fernández, E., and Pardo, J.E. (2009). Characterization of oil obtained from grape seeds collected during berry development. *Journal of Agricultural and Food Chemistry*, 57, 2812–2815.

Sabir, A., Unver, A., and Kara, Z. (2012). The fatty acid and tocopherol constituents of the seed oil extracted from 21 grape varieties (*Vitis* spp.). *Journal of the Science of Food and Agriculture*, 92, 1982–1987.

Sacchi, K.L., Bisson, L.F., and Adams, D.O. (2005). A review of the effect of winemaking techniques on phenolic extraction in red wines. *American Journal of Enology and Viticulture*, 56, 197–206.

Salmon, J.M., Fornairon-Bonnefond, C., Mazauric, J.P., and Moutounet, M. (2000). Oxygen consumption by wine lees: Impact on lees integrity during wine aging. *Food Chemistry*, 71, 519–528.

Serrano, J., Puupponen-Pimiä, R., Daucr, A., Aura, A., and Saura-Calixto, F. (2009). Review tannins: Current knowledge of food sources, intake, bioavailability and biological effects. *International Journal of Nutrition and Food Research*, 53, 310–329.

Shi, J., Yu, J.M., Pohorly, J.E., Young, J.C., Bryan, M., and Wu, Y. (2003). Optimization of extraction of polyphenols from grape seed meal by aqueous ethanol solution. *Food Agriculture and Environment*, 1(2), 42–47.

Shivananda, N.B., Dan Ramdath, D., Marshall, J.R., Isitor, G., Xue, S., and Sh, J. (2011). Wound-healing properties of the oils of *Vitis vinifera* and *Vaccinium macrocarpon. Phytotherapy Research*, 25, 1201–1208.

Shirikhande, A.J. (2000). Wine by-products with health benefits. *Food Research International*, 336, 469–474.

Silva, J.M., Rigaud, J., Cheynier, V., Chemina, A., and Moutounet, M. (1991). Procyanidin dimmers and trimers from grape seeds. *Phytochemistry*, 30, 1259–1264.

Soto, M.L., Moure, A., Domínguez, H., and Parajó, J.H. (2011). Recovery, concentration and purification of phenolic compounds by adsorption: A review. *Journal of Food Engineering*, 105, 1–27.

Tangolar, S.G., Ozoğul, Y., Tangolar, S., and Torun, A. (2009). Evaluation of fatty acid profiles and mineral content of grape seed oil of some grape genotypes. *International Journal of Food Science and Nutrition*, 60, 32–39.

Tassie, E., Freeman, B.M., Coombe, B.G., and Dry, P.R. (eds.). (2001). *Viticulture*, Vol. 2: Practice. Adelaide, South Australia, Australia: Winetitiles, pp. 66–84.

Thorngate, J.H. and Singleton, V.L. (1994). Localization of procyanidins in grape seeds. *American Journal of Enology and Viticulture*, 45(2), 259–262.

Toscano, G., Riva, G., Foppa Pedretti, E., Corinaldesi, F., and Rossini, G. (2013). Analysis of the characteristics of the residues of the wine production chain finalized to their industrial and energy recovery. *Biomass and Bioenergy*, 55, 260–267.

Upadhyay, B. and Keshav, A. (2012). Modelling of recovery of tartaric acid using various solvents. *International Conference on Chemical, Ecology and Environmental Science*, Bangkok, Thailand.

114 VALORIZATION OF WINE MAKING BY-PRODUCTS

Uslu, H. and Inci, I. (2009). Adsorption equilibria of L-(+)-tartaric acid onto alumina. *Journal of Chemical & Engineering*, 54, 1997–2001.

Uslu, H., Inci, I., Bayazit, S.S., and Demir, G. (2009). Comparison of solid-liquid equilibrium data for the adsorption of propionic acid and tartaric acid from aqueous solution onto Amberlite IRA-67. *Industrial & Engineering Chemistry Research*, 48, 7767–7772.

Van Dyk, J.S., Gama, R., Morrison, D., Swart, S., and Pletschke, B.I. (2013). Food processing waste: Problems, current management and prospects for utilisation of the lignocellulose component through enzyme synergistic degradation. *Renewable and Sustainable Energy Reviews*, 26, 521–531.

Vasserot, Y., Caillet, S., and Maujean, A. (1997). Study of anthocyanin adsorption by yeast lees. Effect of some physicochemical parameters. *American Journal of Enology Viticulture*, 48, 433–437.

Vergara, C., Baer, D., Mardones, C., Wilkens, A., Wernekinck, K., Damm, A., Macke, S., Gorena, T., and Winterhalter, P. (2012). Stilbene levels in grape cane of different cultivars in southern Chile: Determination by HPLC-DAD-MS/MS method. *Journal of Agricultural & Food Chemistry*, 60, 929–933.

Versari, A., Castellari, M., Spinabelli, U., and Galassi, S. (2001). Recovery of tartaric acid from industrial enological wastes. *Journal of Chemical Technology & Biotechnology*, 76, 485–488.

Wadhwa, M. and Bakshi, M.P.S. (2013). *Utilization of Fruit and Vegetable Wastes as Livestock Feed and as Substrates for Generation of Other Value-added Products*, Harinder P.S. Makkar (ed.). Rome, Italy: Food and Agriculture Organization of the United Nations (FAO).

Watkins, T.R. (1997). *Wine Nutritional and Therapeutic Benefits*. Washington, DC: American Chemical Society.

Winkler, A.J., Cook, J.A., Kliewer, W.M., and Lider, L.A. (1974). *General Viticulture*. Berkeley, CA: University of California Press.

Xie, D.Y. and Dixon, R.A. (2005). Proanthocyanidin biosynthesis—Still more questions than answers? *Phytochemistry*, 66, 2127–2144.

Xu, Y., Isom, L., and Hanna, M.A. (2010). Review: Adding value to carbon dioxide from ethanol fermentation. *Bioresource Technology*, 101, 3311–3319.

Yang, H., Xu, Z., Fan, M., Gupta, R., Slimane, R.B., Bland, A.E., and Wright, I. (2008). Progress in carbon dioxide separation and capture: A review. *Journal of Environmental Science*, 20, 14–27.

Yalcin, D., Ocalik, O., Altiok, E., and Bayraktar, O. (2008). Characterization and recovery of tartaric acid from wastes of wine and grape juice industries. *Journal of Thermal Analysis & Calorimetry*, 94, 767–771.

Ye, Z.J. (2014). Characterization of bioactive compounds in lees from New Zealand wines and the effect of enzymatic oxidation on their bioactivity. Masters' of Food Science, University of Otago, Otago, New Zealand.

Yilmaz, Y. and Toledo, R.T. (2006). Oxygen radical absorbance capacities of grape/wine industry byproducts and effect of solvent type on extraction of grape seed polyphenols. *Journal of Food Composition & Analysis*, 19(1), 41–48.

WINE MAKING BY-PRODUCTS

Zhang, A., Fang, Y., Li, X., Meng, J., Wang, H., Li, H., Zhang, Z., and Guo, Z. (2011a). Occurrence and estimation of *trans*-resveratrol in one-year-old canes from seven major Chinese grape production regions. *Molecules*, 16, 2846–2861.

Zhang, A., Fang, Y., Wang, H., Li, H., and Zhang, Z. (2011b). Free-radical scavenging properties and reducing power of grape cane extracts from 11 selected grape cultivars widely grown in China. *Molecules*, 16, 10104–10122.

Zhang, K., Wang, M., and Gao, C. (2009). Ion conductive spacers fro the energy-saving production of the tartaric acid in bipolar membrane electrodialysis. *Journal of Membrane Science*, 341, 246–251.

4

TECHNOLOGICAL ASPECTS OF BY-PRODUCT UTILIZATION

ALAA EL-DIN A. BEKHIT, VERN JOU CHENG, ROLAND HARRISON, ZHIJING YE, ADNAN A. BEKHIT, TZI BUN NG, AND LINGMING KONG

Contents

4.1	Introduction	118
4.2	Animal Feed and Health	124
4.3	Fertilizers	129
4.4	Combustion Process	130
4.5	Biomass for Biofuels	131
4.6	Distillation	133
4.7	Conventional and Nonconventional Extraction Techniques for the Extraction of Phenolics	136
	4.7.1 Extraction Parameters	143
	4.7.1.1 Sample Pretreatment	143
	4.7.1.2 Solvent-to-Sample Ratio	147
	4.7.1.3 Types of Solvents	148
	4.7.1.4 Time and Temperature	149
	4.7.2 Novel Conventional Technologies for the Extraction of Bioactives from Wine By-Products	151
4.8	Food and Non-Food Applications	155
	4.8.1 Phenolics	155
	4.8.1.1 Health Supplements and Extracts	156
	4.8.1.2 Bioactivity of Phenolic Compounds	160
	4.8.1.3 Antioxidant Activity	160
	4.8.1.4 Antibacterial and Antifungal Activities	162
	4.8.1.5 Antiviral Activity	164
	4.8.2 Food Ingredients	166
	4.8.3 Health Applications	167
	4.8.4 Applications in Meat Products	169

118 VALORIZATION OF WINE MAKING BY-PRODUCTS

4.8.5	Grape-Seed Oil	171
4.8.6	Non-Food Applications	173
4.9	Fermentation	174
4.10	Green Material	175
4.10.1	Grape Stalks	175
4.10.2	Grape Leaves	176
4.11	Challenges and Opportunities	177
	Acknowledgments	177
	References	178

4.1 Introduction

Several by-products are generated by de-stemming, pressing, and decantation steps during the wine making process. These materials are very rich in biodegradable organic matter and can support microbial growth and emission of environmentally undesirable odors and compounds. The wine residues, if not treated efficiently, have the potential to initiate environmental hazards ranging from surface and groundwater pollution to foul odors. Indeed, a significant negative environmental impact is seen due to discharge of wine waste (News 2014) and significant fines have been issued for dumping of wine residues (Devesa-Rey et al. 2011; EPA 2014). Inappropriate disposal of grape pomace attracts flies and pests and this can create unwanted hazards (Bustamante et al. 2007). In some cases, where the wine solid residues are used as fertilizers or composts at excessive rates, plant growth is affected due to nitrogen immobilization or an increase of nitrogen leaching in soils treated with wine solids (Flavel et al. 2005; Bustamante et al. 2007). Supplementation of the wine by-product composts with minerals and micronutrients such as phosphorous can lead to problems pertaining to their heavy metal content (Del Castilho et al. 1993; Karaka 2004). Pinamonti et al. (1997) argued that the heavy metal content is a critical factor leading to restricted agriculture use of wine waste compost. Furthermore, tannins and other compounds from pomace can cause oxygen depletion in the soil and infiltrate surface, soil and groundwaters (Arvanitoyannis et al. 2006). The problems mentioned indicate that current wine waste management practices can potentially lead to serious environmental pollution. It is becoming more important and a subject of increasing concern for wine processors and scientists to find a safe and economical use for grape and wine waste. Furthermore, utilization of wine by-products and the production of environmentally sustainable wine can have marketing advantage. Research from New Zealand indicated that consumers support the purchase of wine produced by green production practices and are prepared to pay a higher price for sustainably produced wine (Forbes et al. 2009). By-products generated from wine making can be broadly classified as follows: solid by-products (stalks and grape pomace), highly viscous by-products (wine lees), and low-viscosity by-products (wastewater) (Figure 4.1). The physical nature of the materials can to a large extent affect their economic potential for utilization, influencing the composition and level of compounds of interest as well as the costs associated with bioconversion and maintaining the stability of the by-product (i.e., handling of solids or liquids will need different engineering requirements and storage of the material will depend on their stability against microbial and oxidative processes). Grape pomace (GP) is equivalent to about 20%

Figure 4.1 Grape stems.

of the grapes used in wine making and is responsible for the largest amount of by-product generated from wine making with an estimated 10,930,834 Mt (Van Dyk et al. 2013). Grape pomace contributes to approximately 62% of the organic waste (Naziri et al. 2014) and therefore practical utilization is an urgent issue from an environmental perspective. GP may contain grape skin and pulp (10%–12% of grapes), seeds (3%–6%), and stalks (2.5%–7.5%) (Wadhwa et al. 2013) (Figure 4.2). The composition depends largely on the wine production system (white wine versus red wine), the grape variety and maturity, production size, the use of machinery for the separation of grapes before crushing, separation techniques used, and type of wine variety that is being produced. Several compounds can be recovered from GP such as oil (12%–17% of the grape seeds weight), protein, carbohydrates (up to 15% sugars and 30%–40% fiber), phenolics/pigments (0.9%), and tartrate (0.05%–0.08%), which are suitable for a range of useful products that can be used in food, pharmaceutical, and agricultural industries. Given the massive amounts produced and its organic load, the utilization of GP is seen as an important activity to eliminate the environmental problems associated with

Figure 4.2 Grape skins.

the disposal of GP as well as an opportunity to add value and generate extra income from the by-product. Lees and stalks represent significant sources of organic waste, 14% and 12% of organic by-product generated during wine making, respectively (Naziri et al. 2014). Grape seeds can be separated from GP and used for the production of linolenic-rich oil (12%–20%), carbohydrates (60%–70%), protein (about 11%), and phenolics (5%–8%) (Naziri et al. 2014) (Figure 4.3). These valuable compounds have several useful applications as animal feed, and food, cosmetic, and pharmaceutical ingredients. The conventional uses of solid wine by-products are animal feed (Louli et al. 2004) and fertilizer/compost (Kammerer et al. 2005) without or with little further processing. However, in recent years, there has been a growing interest in the exploitation of wine industry waste in other applications as the awareness of the potential commercial value has become known. For example, it was found that during the wine making process, while soluble phenolic compounds that are present in the vacuoles of the plant cells are extracted, a large amount of phenolic compounds bound to the

Figure 4.3 Grape seeds.

122 VALORIZATION OF WINE MAKING BY-PRODUCTS

cell walls are not (Meyer et al. 1998). Compounds extracted from wine by-products with high phenolic content may be able to exert positive effects on human health, to protect against cardiovascular disease, to produce anti-inflammatory activity and anticarcinogenic effects as well as being used as food antioxidants (Shrikhande 2000; van de Wiel et al. 2001; Khanna et al. 2002). Many researchers have reported that the extracted natural antioxidants and various phenolic compounds from wine by-products, including grape seeds, skin, exhausted pomace, grape stalks, wine lees, and grape pomace, are safer to use than synthetic antioxidants (Alonso et al. 2002; Negro et al. 2003; Arvanitoyannis et al. 2006). The recovery and utilization of winery by-products could not only reduce the waste disposal burden and other environmental issues, but could also lead to the development of new healthier, robust ingredients, and extract compounds for both the pharmaceutical and food industries. Due to the aforementioned beneficial properties, wine by-products are now available as dietary supplements. Products such as GSE, grape extract, and red wine powder are sold commercially. Table 4.1 summarizes the physicochemical properties and potential use of wine by-products. Recently, viable new business opportunities have emerged from within the wine industry. There is an increasing interest from several industries, including the food industry, to utilize antioxidants and other phenolics from wine by-products and this has resulted in several applications being developed. One of these is the extraction of anthocyanins, known as "enocyanin," to be utilized as food colorants (Alonso et al. 2002). Patents have been awarded to researchers related to their work in processing and developing commercial products in cosmetic and pharmaceutical industries from grape seeds, skin, and related flavonoids (Carson et al. 2001; Henry et al. 2001; Pykett et al. 2001; Ray and Bagchi 2001). Other than human-orientated applications, wine waste can also potentially be utilized for alternative applications such as absorption of heavy metals (namely lead and cadmium) (Farinella et al. 2007); and for the production of pullulan, a polysaccharide that has many food and pharmaceutical applications (LeDuy and Boa 1983). An interesting and newly reported approach toward the integrated utilization of grape skins consists of consecutive or simultaneous extraction with a neutral organic solvent and water under reflux (Mendes et al. 2013).

ASPECTS OF BY-PRODUCT UTILIZATION 123

Table 4.1 Summary of Compounds Available in Wine By-Products and Their Use

	COMPOUND OF INTEREST	FUNCTIONS	REFERENCES
Wastewater	Tartaric acid and malic acid	Food and pharmaceutical industries	Smagge et al. (1992)
	Tartaric acid	Additive in medicine and cosmetics, acidulant compounds in soft drinks	Andrés et al. (1997)
Grape pomace	Fiber	Dietary fiber supplement, bread improver	Torre et al. (1995), Valiente et al. (1995), Martin-Carron et al. (1997), and Saura-Calixto (1998)
	Anthocyanins	Antioxidant and colorant	Zeller (1999)
	Oil	Grape-seed oil	Molero et al. (1995a,b) and Bustamante et al. (2007)
	Phenolic	Bioactives	Bonilla et al. (1999) and Louli et al. (2004)
	Total sugar	Yeast production	Lo Curto and Tripodo (2001)
	Total sugar, tartrates and malic acid	Energy production	Rodrigo-Sener and Pascual-Vidal (2001) and Bustamante et al. (2007)
	Solid-state fermentation	Animal feed	Sánchez et al. (2002)
	Dietary fiber, phenolic	Prolong shelf life of fish products	Sánchez-Alonso et al. (2007)
	Flavonol	Source of flavanols	Torres and Bobet (2001)
	Ethanol precipitate	Pullulan production	Israilides et al. (1998)
	Phenolic	Dietary supplements/ bioactives	Shrikhande (2000), Alonso et al. (2002), González-Paramás et al. (2004), and Bekhit et al. (2011a)
Grape seeds	Lignocellulosic	Laccasse production	Moldes et al. (2003)
	Phenolic	Grape-seed snack	Louli et al. (2004)
Grape skin	Organic matter, minerals	Fertilizer for cultivation of strawberries	Manios (2004)
	Phenolic	Tea infusions	Cheng et al. (2010)
Grape waste	Organic residues	Gas production for heating purpose	Di Blasi et al. (1999)
	Organic matter	Fertilizer for corn seed	Ferrer et al. (2001)
Grape stalks	Carboxyl, amino or phenolic groups	Effluents decontamination	Villaescusa et al. (2004)
Grape bagasse		Low-cost alternative biosorbent of metals for effluent treatment	Farinella et al. (2007)
Wine lees	Phenolics	Improves rheological, antioxidants, antiviral, anthelmintic properties	Hwang et al. (2009), Bekhit et al. (2011b), and Ye et al. (2011)

124 VALORIZATION OF WINE MAKING BY-PRODUCTS

4.2 Animal Feed and Health

Wine by-products have been considered as a potential cheap animal feed since limited processing is required. Generally speaking, GP has poor metabolizable energy per kilogram of dry matter compared to commercial animal feed (Table 4.2; Dairy 2014). GP from red grapes has better animal nutrition value compared with white GP due to higher contents of dry matter, crude protein, and fibers (Table 4.2, Basalan et al. 2011). White GP has a lower content of dry matter and crude protein compared to red GP (Basalan et al. 2011), probably due to the presence of the high sugar content and pulp found in white GP that can bind moisture, leading to an apparent lower, proximate composition. The composition of GP varies greatly among different grape varieties and geographical production locations as indicated by the range of the values in Table 4.2. Pinot Noir GP from central Otago has higher ME (7.9 MJ/kg DM, unpublished data) and the nutrients profile is better than red GP of varieties from various countries, including Pinot Noir. This may be related to several factors associated with wine making (e.g., fermentation and pressing conditions) or environmental effects. Therefore, the nutritional value of GP needs to be determined on a case-by-case basis. The nutritional value of seeds and skin plus pulp is higher than that of GP, but the nutritional content of stalks is lower than that of GP (Table 4.2). The digestibility of GP *in situ* was reported to be less than half of that normally found *in vitro* (Basalan et al. 2011). The digestibility of three varieties of GP (Cabernet Sauvignon, Gewürztraminer, and Tinta Maderira) with cow rumen fluid were 38.6%, 35.9%, and 25.9% (Famuyiwa and Ough 1982), which appeared to be related to the polyphenol content of the GP (19.4, 22.9, and 32.8 gallic acid equivalent g/kg GP). These results suggest an inherent digestibility variation among GP from various grape varieties. Therefore, care should be exercised in the evaluation of nutrition based on *in vitro* methods and from different grape varieties. Generally, there is no difference in the digestibility of white GP or red GP (Spanghero et al. 2009); but some reports suggested a lower digestibility of red GP (Baumgartel et al. 2007). The presence of polyphenols and tannins (Table 4.3) is the main contributor to the low digestibility of wine by-products. Polyphenols and tannins complex with proteins in the feed and with digestive enzymes

Table 4.2 Nutrient Content and Composition of Various Wine Waste Used for Animal Feed

	POMACE	SEED (g/kg DM)	SKIN/PULP (g/kg DM)	VINE SHOOT (g/kg DM)	STALKS (g/kg FM)	LEES
Dry matter (g/kg)	243–494 0.8↑	488	188	506		185a, 144b, 277c
Seeds (g/kg DM)	426–567a 1.5↑	—	—			
pH	3.98–4.43 4.2↑	—	—			
Organic matter	827–962 NC	942; 910 (dehulled)	863a; 920–934b 0.4↓	885		799a, 790b, 696c
Crude protein (CP)	62.2–108.4a	125–126; 259.8 (dehulled) 2.5↑	99–110a 3.7↓	45	6.1	149.4a, 203.1b, 141.3c
Ether extract	47.7–56.3a 117–145b	95; 12.4 (dehulled) 12.6↓	59–101a 4.7↓	4.74		15.1a, 8.4b, 23.1c
Neutral detergent fiber (NDF)	409–626a	506; 465.1 (dehulled) 7.8↑	350a; 402–423b 6.6↑	741		291a, 321b, 122c
Acid detergent fiber (ADF)	495–543a	467; 380.1 (dehulled) 11.4↑	312a; 380–401b 12.5↑	518		149a, 174b, 57.5c
CP bound to ADF	130.6–143a	140 29.8↑	183a; 214–275b 63.5↑	9.45		46.9a, 19.4b, 9.29c
Lignin	388–437a	371; 281.7 (dehulled) 10.9↑	17a; 242–261b 19.4↑	166	17.4	97.1a, 107b, 28.4c
Calcium (mg/kg)	4.4–5.9a	5.3; 10.3 (dehulled) 3.6↓	3.9–4.9a 4.8↑		1.5	
Phosphorus (mg/kg)	3.1–3.2a	3.2; 8.6 (dehulled) 6.3↓	2.9a; 2.7–3.0b 7.4↑			
Potassium (mg/kg)	7.0–14.0	15.4; 104.9 (dehulled) 7.1↑	23.1–45.2a 3.8↑		9.0	
Magnesium (mg/kg)	1.2–1.3b	1.2; 29.4 (dehulled) 8.3↑	1.0a; 0.9–1.1b 10.0↑		0.2	
Sulfur (mg/kg)	1.3–1.4a	1.5 NC	1.7–1.8a;bNC			
Zinc (mg/kg)	9.8–14a	17; 42.5 (dehulled) 6.7↓	12–18 NC		0.1	
Manganese (mg/kg)	8.2–23a	17 5.0↓	13a; 17b 6.7↑			
Iron (mg/kg)	58–180.8a	174; 119.5 (dehulled) 6.9↑	111–398a 15↑			
Copper (mg/kg)	40–49b	16 5.9↑	23–124a 1.2↑			
ME (MJ/kg DM)	5.58–6.89	12.05	5.52–12.70	4.47		

Sources: Spanghero, M. et al., *Anim. Feed Sci. Technol.*, 152, 243, 2009; Mirzaei-Aghsaghali, A. and Maheri-Sis, N., *World J. Zool.*, 3, 40, 2008; Cavani, C. et al., *Ann. De Zootech.*, 37, 1, 1988; Basalan, M. et al., *Anim. Feed Sci. Technol.*, 169, 194, 2011; Sousa, E.C. et al., *Food Sci. Technol. Camp.*, 34, 135, 2014; Alipour, D. and Rouzbehan, Y., *Anim. Feed Sci. Technol.*, 137, 138, 2007; Molina-Alcaide, E. et al., *J. Sci. Food Agric.*, 8, 597, 2008; Prozil, S.O. et al., *Ind. Crops Prod.*, 35, 178, 2012. Arrows indicate the change due to ensilage: ↑, increase; ↓, decrease; NC, no change; a, red; b, white; c, sherry.

126 VALORIZATION OF WINE MAKING BY-PRODUCTS

Table 4.3 Contents of Antinutrient Compounds in Wine By-Products

	BY-PRODUCT FRACTION	CONCENTRATION (GAE g/kg DM)	REMARKS	REFERENCES
Total phenolics	Seeds	51–90		Spanghero et al. (2009)
	Pulp	31–62		
	Pomace	19.4–32.8	White > red	Famuyiwa and Ough (1982)
	Seed	671–704 mg GAF/g extract		Baydar et al. (2007)
	Bagasse	24 mg CAE/g extract		
	Skin	20		Saura-Calixto (1998)
	Seed	1.43–22.23		Bartolomé et al. (2004)
	Pomace	24–138 TAE		Larrauri et al. (1996)
		24.1		Louli et al. (2004)
Saponins	Seed	40–48	White > red	Spanghero et al. (2009)
	Pulp	26–47		
Alkaloids	Seed	13–15		Spanghero et al. (2009)
	Pulp	9–11		
Anthocyanins[a]	Seed	3–99	Red > white	Spanghero et al. (2009)
	Pulp	6–44		
Total condensed tannins	White GP	39.7 (CS)		Famuyiwa and Ough (1982)
	White skin	138 g/kg DM		Mendes et al. (2013)
	GP (white and red)	215.6–516.4		Rondeau et al. (2013)
	Vine shoots	56.5		Molina-Alcaide et al. (2008)
	Red GP	90.5		
	White GP	202.6		
	White lees	94.9		
	Red lees	180.3		
	Sherry lees	17.9		
	Mixed GP and lees	107.1		
	Stalks	159		Prozil et al. (2012)
	Pomace (red)	223		Llobera and Cabéellas (2007)
	Stalks	103		

[a] (mg/kg DM).
GAE, gallic acid equivalent; TAE, tannic acid equivalent; CAE, catechin equivalents.

leading to disruption to the digestion process and loss of nutrients. Tannins can interfere with the absorption of important minerals (e.g., iron) and can cause aberration and damage to the mucosal lining of the gastrointestinal tract. Ensiling could be a useful technique to improve the nutrition of wine by-products. For example, ensiling for GP 45 days reduced antinutrients, anthocyanins, and saponins (Spanghero et al. 2009), condensed and free tannins (Alipour and Rouzbehan 2007), but not the bound tannins. The addition of polyethylene glycol, which binds to tannins, can improve the digestibility and the nutrition of GP (Alipour and Rouzbehan 2007). The inclusion of tannase (tannin acyl hydrolases, EC 3.1.1.20) may be useful to improve the digestibility of feed since tannase can catalyze the breakdown of tannins (Lekha and Lonsane 1997). While these techniques can improve the digestibility of wine by-products, the low energy of these materials still remains a hurdle toward the use of wine by-products as 100% feed. The use of wine by-products as part of a feeding regime to support high-energy feed is a more realistic target. Famuyiwa and Ough (1982) concluded that Cabernet Sauvignon may be used as an animal feed due to its relative high digestibility and low polyphenol content. The authors also cited Prokop (1979) who reported the use of GP at 20% supplementation level in beef cattle finishing without any negative impact on the growth rate of the animals. The effect of dried GP, containing 5.4% total tannins and 0.34% total polyphenols, on the growth of Lori-Bakhtiari male lambs was investigated by Bahrami et al. (2010). Standard feed was formulated to contain 5%, 10%, 15%, or 20% dried GP and the diet was fed to the lambs over 84 days. Dried GP at 5% and 10% significantly increased the crude protein, neutral detergent fiber, and lamb growth compared to the control diet. A high percentage of dried GP (15% or 20%) significantly negatively affected the growth rate of lambs. It is worth mentioning that 10% replacement of dried lucerne and soybean meal did not affect the performance of rabbits (Cavani et al. 1988). The inclusion of wine by-products at 6% and 7%, compared to 6% and 7% water, did not affect the daily growth of beef cattle and the meat quality of the treated animals (Moote et al. 2014). Nistor et al. (2014) used dried GP (composition on dry matter basis (88.6%), which was 68% ash, 12.17% crude protein, 49.41% crude fiber, 6.11% lipids, 1.82% soluble sugars, 0.38% calcium, 0.42% phosphorus,

128 VALORIZATION OF WINE MAKING BY-PRODUCTS

0.13% magnesium, and 5.2% tannins) to supplement the feeding of Tsigai lambs over 90 days at 0.1, 0.125, and 0.15 kg/head/day. The authors found no significant negative effect on the growth of lambs up to 0.125 kg/head/day and significant reduction and weight gain at 0.15 kg/head/day. The authors found similar trends for milk production in Romanian Spotted cows with 3 and 4 kg of GP/head/day having no effect on milk production and composition, but these parameters were affected negatively at a supplementation rate of 5 kg/head/day. GP was used as a supplementary feed to high-energy forages to reduce methane emissions (Molina-Alcaide et al. 2008; Triplakou and Zervas 2008; Beef Central 2011). Research from Australia indicates that feeding dairy cattle with GP reduced methane emissions by 20% and improved milk production by 5% as well as improved the profile of fatty acids in milk (Beef Central 2011). An alternative strategy to use the wine by-products for animal feed is to generate single-cell protein, which can be incorporated in animal feed (Nicolini et al. 1993). Zepf and Jin (2013) reported a 280% increase in the protein content of grape marc (from 7% to 27%) by using three fungal strains (*Aspergillus oryzae* DAR 3699, *Aspergillus oryzae* RIB 40, and *Trichoderma reesei* RUT C30). The process can be optimized through controlling inoculum size, steam treatment, temperature, and moisture content to achieve the bioconversion in 5 days. Grape by-products contain significant amounts of tannins and condensed tannins, especially in the stalks. While the presence of tannins in animal feed is undesirable from a nutritional point of view, tannins can play an important role in animal health. Tannins in feeds can increase the growth of the bacterium *Butyrivibrio fibrisolvens* and decrease the numbers of the bacterium *Butyrivibrio proteoclasticus* (Vasta et al. 2010), thus leading to reduction bloating and rumen acidosis. Therefore, tannins from wine by-products can potentially reduce the need to supplement the feed with yeast or antimicrobials such as monensin to prevent bloating in animals. The potential antiparasitic of lees from several wine varieties was investigated with an *in vitro* anthelmintic assay using *Haemonchus contortus* eggs (Bekhit et al. 2011a). Shiraz lees did not exhibit any anthelmintic activity whereas an inhibitory effect on larval development was found with Pinot Noir samples. Geographical effects were found in the larval development of Pinot Noir lees. Pinot Noir lees from the South of

New Zealand was more effective than its counterpart from North Island. The use of Pinot Noir lees as drench for lambs grazed on contaminated pasture resulted in insignificant (7%) increase in cumulative live weight gain relative to Control (P > 0.05). Gastrointestinal parasitism is a major problem for animal growth and causes significant financial losses for farmers. Moreover, its treatment is a significant cost for farmers—with drenching accounting for 57% of farm expenditure on animal health with an average cost of $2 per head. Anthelmintic drugs are commonly used for the treatment of gastrointestinal parasites. However, increasing drug resistance, environmental issues, and heightened consumer awareness and concerns about the use of chemicals in animal production are driving the demand for new, effective, and environmentally sustainable approaches to parasite management. The problem of parasitism is even more severe in organic farming systems where anthelmintic drug use is restricted. Parasitic gastroenteritis is the most challenging aspect of high-value organic livestock production. The use of wine waste as an anthelmintic treatment option could reduce the average drenching cost substantially and enable many more farmers to migrate toward an organic production system—thereby enabling them to capture the market premium organics offer. The international organic farming sector has enjoyed an average growth of 25%–35% pa, with organic produce fetching price premiums of ~25% more than conventional produce. Wine waste could provide the critical means of achieving organic certification—particularly for North American markets where no chemical drenches are permitted for organic products. Moreover, in doing so, farmers will reduce their impact on the environment by avoiding the use of chemical drenches and also help reduce waste streams from the wine industry.

4.3 Fertilizers

Vine prunings (about 5 tons/ha/year) are rich in lignins, cellulose, and several elements such as nitrogen and potassium (Wadhwa et al. 2013), which can be transformed to compost and used in other applications. The use of wine sludge and grape stalks at a ratio of 1:2 resulted in a rich compost suitable for vineyard solids, which commonly have a low organic matter content (Bertran et al. 2004).

The best results were found when ground stalks were used and the composting was carried out at a moisture content around 55%, temperature around 65°C, and an oxygen concentration not less than 5%–10%. This led to conditions suitable for microbial activity and better carbon:nitrogen ratio. Vermicomposting of GP, lees, vinasse, and vine shoots using *Eisenia andrei* for 8 months was investigated by Romero et al. (2007). The process resulted in reductions of total organic C content and C/N ratio, and an increase of total extractable C and humic acid contents. The addition of poultry or cattle manure can improve the physicochemical and the biological characteristics of the compost. Fresh GP, mixed with chicken droppings at a level of 10% w/w, and applied at 1–4 tons/ha, improved the germination of corn in greenhouses and in field trials (Ferrer et al. 2001). The compost was supplemented with triphosphate salt for the field trial and the optimum application rate was found to be 3 tons/ha. Better performance was achieved by GP compost compared to industrial chemical fertilizers as judged by the higher dry matter of corn. A mixture of exhausted GP and stalks was used in the plug seedling production of lettuce, tomatoes, peppers, and melons (Carmona et al. 2012). The use of 100% wine waste had a low water holding capacity and resulted in shorter seedlings (about 30% shorter) and lower dry weight and root neck in the tomatoes and melons compared to commercial compost. The authors suggested the addition of conditioner or mixing commercial compost with GP to decrease any negative effects that whole GP compost may exert on seedlings, but no phytotoxicity or nitrogen immobilization was found due to the use of GP compost. GP composting can include the addition of other vine crushed green waste (Lempereur and Penavayre 2014). The cost of composting GP is estimated to vary from $69.3–$110/ton, depending on the distance and processing of the material (Lempereur and Penavayre 2014).

4.4 Combustion Process

The gross heat of combustion of grape charcoal briquettes and dried pressed GP was estimated to be about 90% and 65% of commercial briquette (Walter and Sherman 1974). Direct combustion of grape and wine dried residue is easy to use but has low efficiency. Air and steam gasification can potentially be used as solutions for improving the thermal efficiency (Fiori and Florio 2010). The production of biofuel pellets from grape waste was investigated by Gil et al. (2010). The pellets had low durability but this property could be enhanced by mixing with other agriculture residues.

4.5 Biomass for Biofuels

High production wine regions have the economic potential of producing biodiesel from grape seed due to the large amounts produced (Ramos et al. 2009; Fernández et al. 2010). The extraction yield of grape-seed oil varies among different varieties and depending on the extraction method used (conventional press, solvent, CO_2 supercritical, microwave-assisted, pulsed electric field). The extraction efficiency can be improved by using a combination of these methods or the addition of extraction aids such as triacetin, a food grade adjuvant used in oil extraction, which can improve the extraction of grape-seed oil by 23% (Fernández et al. 2010). Grape seeds contain a large amount of unsaturated fatty acids (Figure 4.4).

Figure 4.4 Grape seeds separated from pomace.

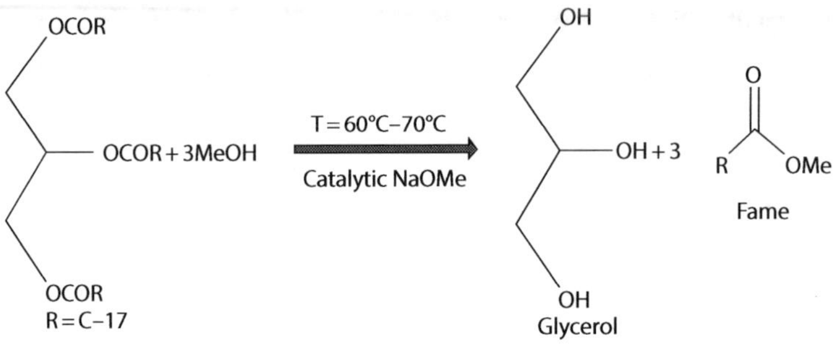

132 VALORIZATION OF WINE MAKING BY-PRODUCTS

Figure 4.5 Conversion of grape-seed oil to biofuel.

This high level of unsaturated fatty acids can improve the low-temperature properties of biodiesel (e.g., flow characteristics, cloud points, and pour points) (Figure 4.5); however, it has the disadvantage of a low oxidation stability (Bhale et al. 2009; Fernández et al. 2010). The conversion of crude oils into biofuels involves refining (removal of water, phospholipids, free fatty acid, and any other impurities) and transesterification (reaction of the refined oil with ethanol or methanol to form methyl or ethyl esters in the presence of catalysts such as NaOH, KOH, or CH_3NaO) steps. The use of ethanol or methanol for transesterification can influence the properties of generated biodiesel. For example, methyl esters have lower melting points and crystallization temperatures compared to ethyl esters (43.1°C versus 56.7°C; and 0°C versus –10°C) (Fernández et al. 2010) and thus ethyl esters offer some advantages in a cold climate. Wine yeast has been an obvious candidate for ethanol production from GP since ethanol is the main end product of their anaerobic glycolysis metabolism. Hang et al. (1986) produced ethanol from GP after solid-state fermentation using wine yeast. Several yeast types (*Saccharomyces cereviseae* Montrachet #52, *Saccharomyces cereviseae* Epernay #2, *Saccharomyces bayanus* California Champagne #505, *Saccharomyces bayanus* Pasture Champagne #595) can be used for ethanol production. Recently, Lempereur and Penavayre (2014) reported the cost associated with distillation of wine by-products in France. The authors estimated the transport and recovery of product cost to range from $23.1–$27.5/ton depending on the distance and wine distillery, which can be in part subsidized by local authorities. The same authors reported the use of wine by-product for the

ASPECTS OF BY-PRODUCT UTILIZATION 133

production of methane through anaerobic digestion. The range of methane produced ranged from 37.5 to 260 mL CH_4/g of wine by-product depending on the grape variety, treatment, geographical location, and storage time. The cost associated with the anaerobic digestion process was estimated to be in the range of $22–$90.8/ton (Lempereur and Penavayre 2014). Interest in using GP to generate butanol has emerged due to its safety (butanol is less volatile than ethanol), technological (butanol has about 25% more energy and less corrosive than ethanol), and environmental advantages. Law (2010) fermented Chardonnay grape pomace using *Clostridium saccharobutylicum* to generate acetone, butanol, and ethanol mixture (ABE). The optimum conditions for ABE generation resulted in the production of 0.21 g/L/h. The use of GP can be very useful to improve the economics of biogas production from dairy manure. Lo and Liao (1986) found the addition (3% wine lees at an addition rate of 1:1 dairy manure and wine lees) of fresh wine lees and aged wine waste improved methane production (liter $CH_4/L/day$) by 11- and 3-fold, respectively. Biogas can also be generated using thermophilic anaerobic digestion of alcohol distillery wastewater. The biogas contained up to 76% methane (Vlissidis and Zouboulis 1993).

4.6 Distillation

Wine GP and lees contain residual ethanol, which can be recovered by distillation. Naziri et al. (2014) estimated the recovery of 4–8 L of ethanol (96% v/v) per 100 kg of wine lees with an estimated extra income of $0.45/% vol/hL for 52% v/v or $1.58/% vol/hL for 92% v/v strength. The nutritional value of the distilled wine by-product is normally lower than undistilled material, but the material can be used for other applications such as recovery of tartaric acid, fermentation for further ethanol/methanol production or organic acids. Zheng et al. (2012) suggested that the best strategy to utilize wine by-products for the generation of ethanol starts by the recovery of the residual ethanol in the by-product and then to use the remaining substrate for cellulosic ethanol production. The high ethanol content in GP obtained after alcoholic fermentation during wine making and the low pH of that GP inhibited the growth of lactic acid bacteria and negatively affected ethanol production.

134 VALORIZATION OF WINE MAKING BY-PRODUCTS

Zheng et al. (2012) found that *Lactobacillus plantarum* B38 generated a high ethanol content (160.7 mg/g DM) after 7 days of fermentation, whereas *Lactobacillus brevis* NRRL B-1836 generated the highest ethanol content (161.1 mg/g DM) after 28 days of fermentation of fresh GP which was obtained from white wine production. In that study, simultaneous saccharification and fermentation of raw fresh GP generated the highest ethanol concentration (about 0.4 g/g DM) compared to ensiled fresh GP (about 0.14 g/g DM) and raw fermented GP (about 0.02 g/g DM). The use of white grape skin water-soluble hexose for the production of ethanol using *Saccharomyces cerevisiae* 4072 yielded 0.42–0.51 g/g of sugar in the skin (Mendes et al. 2013). One of the most famous distillation products of wine by-products is the well-known Italian spirit "grappa." Da Porto (1998) stated that grappa (at least 37.5% v/v alcohol) accounted to about 40% of the Italian spirit production. Data reported by Galletto and Rossetto (2005) show that the production of grappa increased by 7.4% over the period of 1999–2003 and grappa accounted to about 57% of distilled spirits. The product is traditionally made by steam distillation of marc or mixture of wine lees and marc (25 kg/100 kg of marc). The addition of aniseeds or other botanicals is permitted in Italy (Da Porto 1998). The product is very rich in volatile compounds and thus does not require a maturation step, as with other spirits such as whisky, brandy, or cognac. The quality of grappa is dependent on the quality of the materials used (grapes, viticulture practices and factors influencing the berry composition), wine making conditions, and handling of the marc/lees after pressing/racking (Da Porto and Freschet 2003). Similar products are produced in Portugal (bagaceira) and Spain (Orujo). The distillation process generates three fractions. The first (called the *head fraction*) contains components that are more volatile than ethanol (methanol, methyl-butanol/propanol, ethyl acetate, acetal, and acrolein), which have a pungent and strong aroma, as well as several other compounds such as ethyl hexanoate, ethyl octanoate, ethyl decanoate, and acetaldehyde (Silva et al. 2000). The second fraction called *the heart fraction*, contains mostly ethanol but also contains some higher alcohols such as propanol, hexanol, and butanol. The third fraction called the *tail fraction*, contains long-chain alcohols, esters, and lesser volatile acids. Some chemical reactions

ASPECTS OF BY-PRODUCT UTILIZATION   135

(esterification, hydrolysis) occur because of the high distillation temperatures, direct contact with the copper distillers, and the pomace properties (pH, acidity, and so on). Silva et al. (1996, 2000) reported that bagaceira, orujo, and grappa have 45.0%, 58.1%, and 72.9% v/v ethanol and 3389.2, 5169, and 8869.2 mg of methanol/L, respectively. The European Community regulations (EC Regulation No. 1567/89) have set the acceptable methanol content in the spirits above not to exceed 10 g/L of absolute alcohol. The legal limit for methanol in grappa in the United States is set at 7.0 g/L of absolute alcohol or 2.8 g/L of 40% alcohol (Hang and Woodams 2008). Da Porta and Freschet (2003) reported that the concentrations of methanol in Italian grappa to be between 27.61 and 39.39 g/L of absolute alcohol in marc stored for 80 days under various treatments. The highest methanol content (39.39 g/L of absolute alcohol) was found in grappa from marc treated with potassium metabisulfite solution (10%) treated as a rate of 20 g/0.1 ton of marc. Grappa produced in the United States from Niagara, Cayuga White, Gewurztraminer, Chardonnay, and Riesling GP contained 0.44, 0.38, 1.38, 0.55, and 1.12 g of methanol/L of 40% alcohol, which is lower than the legal limit (Hang and Woodams 2008). The use of citric acid (20 g/L) to preserve white GP decreased methanol content in the obtained distillate (Gerogiannaki-Chrisopoulou et al. 2004). There is a close relationship between the pectin methylesterase activity of grape pomaces, the degree of pectin methylation, and the concentration of methanol before distillation. The inhibition of grape pomaces pectin methylesterase can control methanol concentration (Zocca et al. 2008). High-quality spirits are obtained from fresh GP which has not been exposed to air or further secondary fermentation that causes the generation of methanol. Diéguez et al. (2001) reported that the most suitable method to store GP is to keep it in containers that permit little or no contact of the pomace with air (a plastic sack or plastic container), and is small enough to avoid accumulation of heat, and increase in temperature during alcoholic fermentation (Diéguez et al. 2001), but small containers (50 kg capacity) can promote aerobic fermentation and increase methanol formation (Cortés et al. 2010). Also time plays an important role due to the increased activities of Acetobacter that can affect the quality of the distillate (Silva et al. 2000; Hang and Woodams 2008).

136 VALORIZATION OF WINE MAKING BY-PRODUCTS

4.7 Conventional and Nonconventional Extraction Techniques for the Extraction of Phenolics

With beneficial impacts on health, the environment, and the economy, it is only rational that researchers and the wine industry strive to implement the extraction and the recovery of valuable phenolic compounds from wine by-products. Different components of wine waste (grape seeds, skin, and stalks) possess different phenolic compounds that differ in activities and properties. Both the nature of the occurring species (structure and the degree of polymerization) and the overall quantity could vary markedly among the wine waste. Flavonols are abundant phenolic compounds in grape skin, while grape seeds are rich in flavan-3-ol. Grape-seed extract from Merlot contains 2323.5 mg tannins/kg seeds while the skin and stalks contain 1605 and 221.3 mg tannins/kg, respectively. Wine by-products from different varieties of wine also contain different compounds, even if they are subjected to the same extraction parameters. For example, anthocyanins are present in red grapes but they are not present in white grapes. Extraction efficiency is commonly a function of process conditions. The aim of an extraction process is to obtain maximum yield of the compound of interest at the highest quality/ purity (concentration of target compounds and the efficacy of the extracts). At present, numerous and frequently contradictory results can be found regarding the optimization of the extraction process for bioactive compounds from wine by-products and ambiguous data on the methods and extraction parameters are available, particularly if different raw materials are compared. A summary of previously published research on the optimization of processing parameters for conventional solvent extraction of active compounds from grapes and wine by-products (grape pomace, seeds, skin, and stalks) is presented in Tables 4.4 through 4.6. Conventional solvent extraction is the extraction method most often investigated. Several novel and more advanced extraction technologies have been proposed, such as supercritical fluid extraction, microwave-assisted extraction, ultrasonically assisted solvent extraction, and high hydrostatic pressure, as well as a pulsed electric field (Table 4.7). However, some of these technologies are still at the experimental stages and mostly require substantial investment to set up. The most common processing

Table 4.4 Summary of Conventional Solvent Extraction Methods Used to Optimize the Extraction of Polyphenols from Grape Seeds

SAMPLE	EXTRACTION PARAMETERS	OPTIMUM EXTRACTION	TOTAL POLYPHENOLS	REFERENCES
Dried whole white grape seeds after vinification (Italian Riesling)	*Solvent:* Ethyl acetate with 3.3%, 5%, 10%, 15%, 20% water content *Time:* 1, 12, 24 h	*Solvent:* Ethyl acetate with 10% water content *Time:* 24 h	Proanthocyanins: 5.18 g/kg seeds	Pekić et al. (1998)
Dried powdered de-fatted fresh grape seeds (*V. vinifera* var. Bangalore blue grapes)	*Solvent:* Acetone, ethyl acetate, methanol, ethyl acetate with 10%–15%–20% water content *Temperature:* 60°C–70°C *Time:* 5, 10 h	*Solvent:* Ethyl acetate with 15% water content *Temperature:* 60°C–70°C *Time:* 5 h	54% total flavanols (in CE/100 g extract)	Jayaprakasha et al. (2001)
Dried powdered de-fatted fresh grape seeds (*V. vinifera* var. Bangalore blue grapes)	*Solvent:* Acetone:water:acetic acid (90:9.5:0.5), Methanol:water:acetic acid (90:9.5:0.5)	*Solvent:* Acetone:water: acetic acid (90:9.5:0.5) *Time:* 8 h	46% (w/w) (in CE/crude extract)	Jayaprakasha et al. (2003)
Dried powdered grape seeds after vinification	*Solvent:* Ethanol with 5%, 20%, 30%, 40%, 50%, 60%, 70%, 80% water content *Temperature:* 25°C, 45°C, 65°C *Time:* 1, 1.5, 2 h *L/S ratio:* 5:1, 7.5: 1, 10:1 (v/w) *Stage:* Multiple-stage extractions (single, double, and triple steps)	*Solvent:* Ethanol with 50% water content *Temperature:* 65°C *Time:* 1.5 h *L/S ratio:* 7.5:1 (v/w) *Stage:* Double-stage extraction	39.246 mg GAE/g	Shi et al. (2003a)
Dried powdered de-fatted grape seeds from fresh grapes	*Solvent:* Acetone:water:acetic acid (90:9.5:0.5), Ethyl acetate:methanol:water (60:30:10), Ethanol:water (95:5) *Temperature:* 60°C–70°C *Time:* 8 h	*Solvent:* Ethyl acetate:methanol:water (60:30:10) *Temperature:* 60°C–70°C *Time:* 8 h	667.87 mg GAE/g	Baydar et al. (2004)

(Continued)

Table 4.4 (*Continued*)   Summary of Conventional Solvent Extraction Methods Used to Optimize the Extraction of Polyphenols from Grape Seeds

SAMPLE	EXTRACTION PARAMETERS	OPTIMUM EXTRACTION	TOTAL POLYPHENOLS	REFERENCES
Dried powdered de-fatted grape seeds from fresh grapes	*Solvent:* Acetone:water:acetic acid (90:9.5:0.5), Ethyl acetate:methanol:water (60:30:10), Ethanol:water (95:5)	*Solvent:* Ethyl acetate:methanol:water (60:30:10)	667.37 mg GAE/g	Baydar et al. (2004)
	Temperature: 60°C–70°C	*Temperature:* 60°C–70°C		
	Time: 8 h	*Time:* 8 h		
Powdered seeds after vinification	*Solvent:* Ethanol with 50% water content	*Solvent:* Ethanol with 50% water content	11.43% of total seed weight	Nawaz et al. (2006)
	Temperature: Room temperature	*Temperature:* Room temperature		
	Time: 1 h	*Time:* 1 h		
	L/S ratio: 0.1:1, 0.15:1, 0.2:1, 0.25:1 (v/w)	*L/S ratio:* 0.2:1 (v/w)		
	Stage: Multiple-stage extractions (single, double, and triple steps)	*Stage:* Double-stage extraction		
	Membrane pore size: 0.22, 0.45 μm	*Membrane pore size:* 0.22 μm		
Dried powdered grape seeds after vinification (Merlot and Chardonnay) and muscadine seeds	*Solvent:* Ethanol, methanol, acetone (all with 0%, 10%, 20%, 30%, 40%, 50%, 60%, 70%, 80%, 90%, 100% water content)	*Solvent:* Ethanol, methanol, and acetone (all with 50%–70% water content)	28–53 mg GAE/g_{db}	Yilmaz and Toledo (2006)
	Temperature: Room temperature	*Temperature:* Room temperature		
	Time: 30 min	*Time:* 30 min		
	L/S ratio: 10:1 (v/w)	*L/S ratio:* 10:1 (v/w)		

(*Continued*)

Table 4.4 (Continued)   Summary of Conventional Solvent Extraction Methods Used to Optimize the Extraction of Polyphenols from Grape Seeds

SAMPLE	EXTRACTION PARAMETERS	OPTIMUM EXTRACTION	TOTAL POLYPHENOLS	REFERENCES
Dried powdered de-fatted red grape seeds (*V. vinifera* var. Syrah)	*Solvent:* Ethanol, methanol, ethyl acetate, acetone (all with 30%–50% water content) *Temperature:* 45°C, 50°C, 55°C, 60°C (ending with boiling point of each solvent) *Time:* NA *L/S ratio:* NA	*Solvent:* Methanol with 40%–50% water content *Temperature:* 65°C *Time:* 3 h *L/S ratio:* 7–8:1 (v/w)	57.16%	Youssef and El-Adawi (2006)
Dried powdered de-fatted grape seeds after vinification (from Narince grape cultivar)	*Solvent:* Acetone:water:acetic acid (90:9.5:0.5), Ethyl acetate:methanol:water (60:30:10) *Temperature:* 60°C *Time:* 8 h	*Solvent:* Acetone:water:acetic acid (90:9.5:0.5) *Temperature:* 60°C *Time:* 8 h	704 mg CE/g extract	Baydar et al. (2007)
Dried powdered white grape seeds	*Solvent:* Ethanol with 50% water content *Temperature:* 25°C, 40°C, 50°C, 60°C, 70°C, 80°C *Time:* 0–200 min *L/S ratio:* 2.5:1, 5:1, 7.5:1, 10:1 (v/w) *Particle size:* 0.125, 0.16, 0.4, 0.63, >0.63 mm	*Solvent:* Ethanol with 50% water content *Temperature:* 80°C *Time:* 40 min *L/S ratio:* 10:1 (v/w) *Particle size:* 0.16–0.125 mm	66.81 mg GAE/g_{db} of seeds	Bucić-Kojić et al. (2007)

140 VALORIZATION OF WINE MAKING BY-PRODUCTS

Table 4.5 Summary of Conventional Solvent Extraction Methods Used to Optimize the Extraction of Polyphenols from Grape Pomace/Marc

SAMPLE	EXTRACTION PARAMETERS	OPTIMUM EXTRACTION	TOTAL POLYPHENOLS	REFERENCES
Red grape marc after vinification (Cabernet Sauvignon: Tempranillo; 1:1) (w/w)	*Feed pretreatment:* Crushed, uncrushed marc *Solvent:* Ethyl acetate with 50% water content *Temperature:* 20°C *Time:* 5, 10, 20, 30 min *L/S ratio:* 10:1 (v/w)	*Feed pre-treatment:* Crushed marc *Solvent:* Ethyl acetate phase *Temperature:* 20°C *Time:* 10 min *L/S ratio:* 10:1 (v/w)	112 mg/L	Bonilla et al. (1999)
Dried powdered de-fatted marc after juice processing (*V. vinifera* var. Bangalore blue grapes)	*Solvent:* Ethyl acetate, methanol, water	*Solvent:* Methanol	35.7% (w/w) CE	Murthy et al. (2002)
Red grape marc after vinification (*V. vinifera* var. Agiorgitiko)	*Feed pretreatment:* Crushing and removal of stems *Solvent:* Methanol, ethyl acetate, KOH (3%) *Temperature:* 18°C, 20°C, 30°C, 60°C, 75°C *Time:* 8, 24 h *L/S ratio:* 3:1, 10:3 (v/w)	*Feed pretreatment:* Crushing and removal of stems *Solvent:* Ethyl acetate *Temperature:* 75°C *Time:* 8 h *L/S ratio:* 10:3 (v/w)	13.8% (w/w) GAE	Louli et al. (2004)
Pressed red grape marc after vinification *V. vinifera* var. Pinot Noir)	*Solvent:* 70% Ethanol, 70% methanol, water *Temperature:* Room temperature *Time:* 1, 12, 24 h	*Solvent:* 70% Methanol *Temperature:* Room temperature *Time:* 24 h	6717.9–6782.8 mL/g extracts	Lapornik et al. (2005)

(*Continued*)

ASPECTS OF BY-PRODUCT UTILIZATION 141

Table 4.5 (*Continued*) Summary of Conventional Solvent Extraction Methods Used to Optimize the Extraction of Polyphenols from Grape Pomace/Marc

SAMPLE	EXTRACTION PARAMETERS	OPTIMUM EXTRACTION	TOTAL POLYPHENOLS	REFERENCES
Dried red and white grape marc with two different pretreatments: pressed and distillated (Garnacha grapes)	*Solvent:* Methanol, 96% ethanol, water	*Solvent:* Methanol	493.4 mg GAE/L	Pinelo et al. (2005a)
	Temperature: 25°C, 37.5°C, 50°C	*Temperature:* 50°C		
	Time: 30, 60, 90 min	*Time:* 90 min		
	L/S ratio: 1:1, 3:1, 5:1 (v/w)	*L/S ratio:* 1:1 (v/w)		
Powdered white grape marc after distillation (Garnacha grapes)	*Solvent:* Water	*Solvent:* Water	52.7 mg GAE/L	Pinelo et al. (2005b)
	Temperature: 50°C	*Temperature:* 50°C		
	Solvent flow rate: 2, 2.5, 3 mL/min	*Solvent flow rate:* 2 mL/min		
	Sample amount: 2.5, 5, 7.5 g	*Sample amount:* 2.5 g		
	Particle size: 0.5, 3, 5.5 mm	*Particle size:* 0.5 mm continuous extraction		
Powdered white grape marc after distillation (Garnacha grapes)	*Solvent:* Ethanol, methanol, water	*Solvent:* Ethanol	Catechin: 89.31/ OD^3/min/(mg_{dw})	Pinelo et al. (2005c)
	Temperature: 50°C	*Temperature:* 50°C	Resveratrol: 91.12/OD^3/min/ (mg_{dw})	
	Time: 30 min	*Time:* 30 min		
	Storage temperature: 22°C–37°C–60°C	*Storage temperature:* 60°C		
	Storage time: 0, 5, 10, 15, 20, 25, 30 days	*Storage time:* 1–2 days		

(*Continued*)

142 VALORIZATION OF WINE MAKING BY-PRODUCTS

Table 4.5 (*Continued*) Summary of Conventional Solvent Extraction Methods Used to Optimize the Extraction of Polyphenols from Grape Pomace/Marc

SAMPLE	EXTRACTION PARAMETERS	OPTIMUM EXTRACTION	TOTAL POLYPHENOLS	REFERENCES
Dried powdered de-fatted red grape marc after vinification (*V. vinifera* var. Agiorgitiko)	*Solvent:* Methanol, ethanol, ethanol:water (1:1), isopropanol, ethyl acetate	*Solvent:* Ethanol:water (1:1) (v/v)	2.89% (w/w) (GAE)	Lafka et al. (2007)
	Temperature. Ambient temperature	*Temperature.* Ambient temperature		
	Time: 30 min, 5, 10, 17, 24 h	*Time:* 3 h		
	pH: 1.5, 2, 2.4, 3.3, 3.6, 3.7	*pH:* 1.5		
	L/S ratio: 3:1, 4:1, 5:1, 6:1, 7:1, 8:1, 9:1, 10:1. 11:1, 12:1 (v/w)	*L/S ratio:* 9:1 (v/w)		
	Stage: Continuous extraction and multiple-stage extractions (double and triple steps)	*Stage:* Triple-stage extraction		
Dried red grape marc after vinification (Barbera red grape)	Degreasing pretreatment with hexane: 24 h for marc	Degreasing is not a useful pretreatment	0.4%–0.6% (w/w) (GAE)	Spigno and De Faveri (2007)
	Solvent: Ethanol, ethyl acetate with 10% water	*Solvent:* Ethanol		
	Temperature: 28°C, 60°C	*Temperature:* 60°C		
	Time: 5, 24 h	*Time:* 5 h		
Red grape pressed marc after vinification (Barbera red grape)	First phase		2.03–4.05 g GAE/100 g_{dw}	Spigno et al. (2007)
	Solvent: Ethanol			*Solvent:* Ethanol with 0%–30% water content
	Temperature: 45°C, 60°C			*Temperature:* 6060°C

(Continued)

ASPECTS OF BY-PRODUCT UTILIZATION   143

Table 4.5 (*Continued*)   Summary of Conventional Solvent Extraction Methods Used to Optimize the Extraction of Polyphenols from Grape Pomace/Marc

SAMPLE	EXTRACTION PARAMETERS	OPTIMUM EXTRACTION	TOTAL POLYPHENOLS	REFERENCES
	Time: 1, 3, 5, 7, 9, 15, 20, 24 h	*Time:* 5–8 h		
	Second phase			
	Solvent: Ethanol with 10%–60% water content			
	Temperature: 60°C			
	Time: 5 h			
Dried red grape marc after vinification (Refošk, Merlot and Cabernet)	*Solvent:* Acetone with 30%–50% water content, ethyl acetate with 30%–50% water content, ethanol with 30%–50% water content	*Solvent:* Ethanol with 50% water content, acetone with 50% water content	17.3–20.2 mg GAE/g_{dw}	Vatai et al. (2009)
	Temperature: 20°C, 40°C, 60°C	*Temperature:* 60°C		
	Time: 2 h	Time: 2 h		
	L/S ratio: 20:1 (v/w)	*L/S ratio:* 20:1 (v/w)		

parameters investigated for extraction were as follows: pretreatment of the sample (de-oiling and size reduction), solvent-to-sample ratio, type of solvent, contact time, and the temperature at which extraction is carried out.

4.7.1 Extraction Parameters

4.7.1.1 Sample Pretreatment   The de-oiling/de-fatting of grape seeds slightly decreases the yield of total phenols of extracts as parts of the phenolic are normally removed with the oil. Numerous researchers have reported on the use of de-oiling prior to the extraction process since the presence of oil might affect the measurement of activities (Murthy et al. 2002; Baydar et al. 2007; Lafka et al. 2007). However, Spigno and De Faveri (2007) claimed

144 VALORIZATION OF WINE MAKING BY-PRODUCTS

Table 4.6 Summary of Conventional Solvent Extraction Methods Used to Optimize the Extraction of Polyphenols from Grape Skins and Stalks

SAMPLE	EXTRACTION PARAMETERS	OPTIMUM EXTRACTION	TOTAL POLYPHENOLS	REFERENCES
Skin from fresh red grapes (Cabernet Sauvignon)	*Solvent:* Methanol, 80% methanol, water, 75% acetone, methanol/12N HCl (99:1), methanol 12N HCl (98:2)	*Solvent:* Methanol/12N HCl (98:2)	267 mg/100 g of grapes	Revilla et al. (1998)
	Temperature: −25°C, 4°C, room, 25°C	*Temperature:* 4°C		
	Time: 1, 4, 12, 16, 24 h	*Time:* 24 h		
	Stage: Multiple-stage extractions (double, triple, and more)	*Stage:* 5 stage extraction		
Dried powdered grape skin from fresh red winegrapes	PLE (pressurized liquid extraction) at 10.1 MPa		179.4–189.5 mg GAE/g_{dw}	Ju and Howard (2003)
	Solvent: 0.1% HCl in deionized water, 0.1% HCl in 60% ethanol, 0.1% HCl in 60% methanol, 0.1% HCl in 40:40:20 (methanol, acetone, water, solvent mixture), 7% acetic acid in 70% methanol, 0.1% trifluoroacetic acid in 70% methanol	*Solvent:* 0.1% HCl in 40:40:20 (Methanol, acetone, water, solvent mixture)		
	Temperature: 20°C, 40°C, 60°C, 80°C, 100°C, 120°C, 140°C	*Temperature:* 120°C		
	Time & stage: 5 min × 3 extraction stages	*Time & stage:* 5 min × 3 extraction stages		
Dried red grape stalks after vinification (Barbera red grape)	Degreasing pretreatment with hexane: 6 h for stalks	Degreasing is not a useful pretreatment	≈0.2% (w/w) GAE	Spigno and De Faveri (2007)
	Solvent: Ethanol, ethyl acetate with 10% water	*Solvent:* Ethanol		
	Temperature: 28°C, 60°C	*Temperature:* 60°C		
	Time: 5, 24 h	*Time:* 5 h		

Table 4.7 Examples of Emerging Technologies for the Extraction of Polyphenols from Wine By-Products

TECHNIQUE	MATERIALS	COMPOUNDS	EXTRACTION CONDITIONS	YIELD	REFERENCES
Supercritical fluid extraction (SCE)	Grape-seed cake	Polyphenol	30 MPa, 40°C, 15% methanol, 180 min	>90% low molecular weight polyphenols	Murga et al. (2000)
		Proanthocyanidins	30 MPa, 50°C, 20% of ethanol, 60 min	Gallic acid, epigallocatechin and epigallocatechingallate were extracted at their maximum level	Yilmaz et al. (2011)
Ultrasonic (US)	Grape seed	Oil and polyphenol	25 g sample in 200 mL hexane, 20 kHz, 150 W for 30 min for oil extraction.	14% oil yield by ultrasonication was similar to 16 h maceration	Da Porto et al. (2013)
			10 g of de-fatted grape-seed flour in 100 mL of methanol, 20 kHz, 150 W for 15 min, <30°C for polyphenols	105.20 mg GAE/g flour and antioxidant activity (109 Eq αToc/g flour)	
		Polyphenol, anthocyanins	2 g of grape-seed flour in 100 mL of 52% ethanol, 55°C–60°C, 40 kHz, 250 W, 30 min in US water bath	Total phenolics (5.41 mg GAE/100 mL), antioxidant activity (12.28 mg/mL), and total anthocyanins (2.29 mg/mL)	Ghafoor et al. (2009)
Electricity-assisted extraction methods	Grape seed	Polyphenol	PEF (8–20 kV/cm, 0–20 ms); 2000 pulses with a frequency of 0.33 Hz. Time was 15 min. HVED (10 kA/40 kV, 1 ms)	Both treatments produced increase in polyphenols; Better yield at 50°C and with 50% ethanol; Maximum yield 9 g GAE/100 g seeds	Boussetta et al. (2012)

(Continued)

Table 4.7 (*Continued*)   Examples of Emerging Technologies for the Extraction of Polyphenols from Wine By-Products

TECHNIQUE	MATERIALS	COMPOUNDS	EXTRACTION CONDITIONS	YIELD	REFERENCES
		Polyphenol	Pulsed arc treatment (3–10 J, specific pulse energy was in the range 41.8–139.4 J/kg). Electric arc was 1800 pulses at 2 Hz. The total treatment time was 15 min. Streamers treatment (1–1.5 J, specific pulse energy was about 14–20.9 J/kg). Number of pulses was 60,000 at freq. of 20 Hz PEF (20 and 40 kV/cm, pulse number = 60,000, 5.2–20.9 J/kg)	Effectiveness Arc ≫ streamer ≫ PEF	Boussetta et al. (2013a)
Microwave-assisted extraction	Grape seed	Polyphenols	Ethanol (10%–90%), solid:liquid ratio 1:10–1:40 (w/v), 2–32 min, 40°C–60°C	0.072–0.169 (g extract/g FW)	Li et al. (2011)
	Wine lees	Phenolic compounds	200 W, 75% ethanol, 17 min	Total phenolic content, 36.8% (mg GAE/g FW)	Pérez-Serradilla and de Castro (2011)
	Grape skin	Anthocyanins	40% methanol, 100°C, 5 min, 500 W	Total anthocyanins (118%)	Liazid et al. (2011)

FW, fresh weight; DW, dry weight; GAE, gallic acid equivalent; W, watt.

ASPECTS OF BY-PRODUCT UTILIZATION 147

that de-oiling did not seem to be a useful pretreatment to enhance extract purity. Size reduction is another pretreatment used prior to the extraction of phenolic compounds from wine by-products. A smaller particle size increases the surface area available for mass transfer, and subsequently increases the extraction yield, especially in continuous extraction systems (Pinelo et al. 2006b). Bucić-Kojić et al. (2007) reported a higher concentration of extracted polyphenols (66.81 mg GAE/g dry basis) for a smaller particle size of grape seeds (0.16–0.125 mm) compared to the concentration of polyphenols (23.907 mg GAE/g dry basis) obtained from seeds with a larger particle size (>0.63 mm). Crushing of grape pomace also resulted in a higher extraction of phenolic compounds (Bonilla et al. 1999). Moure et al. (2001) reported that particle size reduction increased the antioxidant activity as a result of both increased extractability and enhanced enzymatic degradation of polysaccharides. However, the increment is not very large, thus crushing cannot be considered as a decisive parameter for process efficiency (Louli et al. 2004).

4.7.1.2 Solvent-to-Sample Ratio The effect of solvent-to-sample or liquid-to-solid (L/S) ratio on the extraction of phenolic compounds has been investigated by Shi et al. (2003a), Pinelo et al. (2005a), Nawaz et al. (2006), Bucić-Kojić et al. (2007), and Lafka et al. (2007). The influence of L/S ratio was more significant at higher temperatures than at room temperature (Shi et al. 2003a). According to the mass transfer principles, the higher the L/S ratio, the higher was the amount of extracts attained, regardless of the type of solvent used. However, an equilibrium between the use of high and low L/S ratios, involving a balance between high costs, solvent waste, and avoidance of saturation effects has to be reached (Pinelo et al. 2005b). A L/S ratio of 40 mL/g gave the best extraction yield from grape seeds at all tested temperatures (Bucić-Kojić et al. 2007) but other studies reported different preferable L/S ratios; 0.2/1 (v/w) (Nawaz et al. 2006), 1/1 (v/w) (Pinelo et al. 2005a), 4/1 (v/w) (Spigno et al. 2007), 8–10/1 (v/w) (Bonilla et al. 1999; Shi et al. 2003a; Yilmaz and Toledo 2006; Youssef and El-Adawi 2006; Lafka et al. 2007), and 20/1 (v/w) (Vatai et al. 2009) (Tables 4.4 through 4.6).

148 VALORIZATION OF WINE MAKING BY-PRODUCTS

4.7.1.3 Types of Solvents The type of solvent used for extraction is one of the most important variables in the extraction process. The extraction of polyphenols is dependent upon two mechanisms: (1) the dissolution of each phenolic compound in the plant material matrix and (2) their diffusion in the external solvent medium (Kallithraka et al. 1995). Previously, all extractions were performed with organic solvents such as ethanol–benzene combinations (Kofujita et al. 1999), hexane and methanol combinations (Santos-Buelga et al. 1995), and sulfur dioxide (Cacace and Mazza 2002). These solvent extraction procedures have proved to be efficient; however, the extracts were not safe for human consumption due to the potential toxic effects from the residual solvents. These health concerns have sparked research into methods that reduce or eliminate organic solvents in the extraction procedures. A U.S. patent investigated the extraction of procyanidins from grape seeds using hot water extraction at high temperature and high pressure (Nasis-Moragher et al. 1999). However, water alone can cause the undesired dissolution of proteins and polysaccharides, especially under conditions of high pressure and temperature. In recent years, alcoholic solvents have been employed to provide efficient phenolic extractions from natural sources. This provides high yields of total extract but is not highly selective for phenols (Spigno et al. 2007). Generally, mixtures of alcohols and water have been found to be more efficient in the extraction of phenolic constituents compared to a corresponding mono component solvent system (Pinelo et al. 2005a; Yilmaz and Toledo 2006). The use of a single solvent, such as acetone or methanol, gives a high yield of extract with low antioxidant activity and low reducing power, while mixtures of solvent and water give the opposite results (Jayaprakasha et al. 2001). Other than this, high concentrations of a single solvent also increase the cost of extraction and limit the operating temperature due to their volatilities (Shi et al. 2003a). Several hydro-alcoholic solvents have been claimed to be the best solvent for the maximum qualitative and quantitative extraction of polyphenols from wine by-products. These include ethanol (Shi et al. 2003a; Pinelo et al. 2005c; Spigno and De Faveri 2007), methanol (Murthy et al. 2002; Lapornik et al. 2005; Pinelo et al. 2005a), and acetone (Jayaprakasha et al. 2003; Baydar et al. 2007). Many publications have reported that 50% aqueous ethanol acts as the best and the safest extraction solvent (Shi et al. 2003a; Nawaz et al. 2006; Youssef and El-Adawi 2006; Bucić-Kojić et al. 2007; Lafka et al. 2007; Vatai et al. 2009) and is the most widely employed solvent because it has hygiene benefits and is compatible with food materials. Ethanol effectively extracts flavonoids and their glycosides, catechols, and tannins from raw materials (Bazykina et al. 2002). However, the addition of water over a limited range up to 50% can enhance the solubility of these compounds (Cacace and Mazza 2003; Shi et al. 2003a; Spigno et al. 2007). Ethyl acetate is capable of selectively extracting proanthocyanidins (5.18 g/kg seeds) and addition of water up to a certain level (10%) increased the yield of proanthocyanidins due to increased permeability of the seeds, thus enabling a better mass transfer by molecular diffusion (Pekić et al. 1998). Ethyl acetate preferentially extracts phenols that are readily dissolved in the lipid fraction due to its low boiling point that facilitates its removal (Pekić et al. 1998). On the other hand, Kallithraka et al. (1995) reported that methanol is the best solvent for qualitative extraction of (+)-catechins, (–)-epicatechin, and epigallocatechin from grape seeds. This indicates that the selective extraction of polyphenols from wine by-products by an appropriate solvent mixture is very important in obtaining a fraction with maximum polyphenols yield and selective function.

4.7.1.4 Time and Temperature Extraction time and temperature are among the most important parameters to be optimized, in order to minimize the energy cost of the process and increase process efficiency. Many authors have highlighted the fact that an increase in temperature, in conjunction with extraction time, favors increased extraction by enhancing both the diffusion coefficient of the solute and its solubility. Bucić-Kojić et al. (2007) concluded that extraction at 80°C for 40 min resulted in the maximum extractable total polyphenols (66.81 mg GAE/g dry basis). A higher temperature can enhance the mass transfer into the internal liquid phase which raises the pressure, causing centrifugal circulation of the solutes through membranes of the plants. Additionally, heat treatment can assist in breaking the phenolic-matrix bonds and modify the membrane structure of plant cells by coagulation of lipoproteins

150 VALORIZATION OF WINE MAKING BY-PRODUCTS

making them less selective (Shi et al. 2003a). However, the temperature should be maintained at a level that maintains the stability of phenolic compounds and does not promote denaturation of membranes (Cacace and Mazza 2003; Pinelo et al. 2005b; Yilmaz and Toledo 2006; Spigno and De Faveri 2007). Extraction at room temperature or moderately high temperatures (approximately 60°C–70°C) is most commonly used for the extraction of phenolics from wine by-products (Jayaprakasha et al. 2001; Shi et al. 2003a; Baydar et al. 2004, 2007; Youssef and El-Adawi 2006; Spigno and De Faveri 2007; Spigno et al. 2007; Vatai et al. 2009). There is some controversy over the optimum time for the extraction process. Some authors have found that shorter extraction times (<1 h) (Bonilla et al. 1999; Pinelo et al. 2005c; Yilmaz and Toledo 2006; Bucić-Kojić et al. 2007), while others have found that a longer extraction time (5–8 h) (Jayaprakasha et al. 2001, 2003; Baydar et al. 2004, 2007; Louli et al. 2004; Spigno et al. 2007), or even a considerably longer extraction time ($\approx$24 h) (Revilla et al. 1998; Lapornik et al. 2005) was required for maximum yields. The yield of polyphenols in grape alcohol extracts increased with the extraction time (approximately 6–7-fold increment from 1 to 24 h) (Lapornik et al. 2005). On the other hand, Spigno and De Faveri (2007) found that there was no significant difference between extraction times of 5 and 24 h. It is worth noting that while the yield may increase with the time of extraction, the antioxidant activity per mass unit may decrease as a result of oxidation. Additionally, it has been reported that a longer extraction time decreases the total phenolics extracted, possibly due to some loss of phenolic compounds via oxidation or polymerization into insoluble compounds (Shi et al. 2003a; Youssef and El-Adawi 2006). Therefore, time–temperature combinations are important and should be optimized, possibly by considering each material (e.g., a separate type of waste or a mixture of different types and sources) individually to optimize the extraction of polyphenols. From a recovery of compounds point of view, extraction at a lower temperature longer time would be the best solution. However, considering the low extraction rates and bearing in mind an industrial application of the process, it would be more rational and cost-effective to work at a higher temperature for a shorter time (possibly less than 8 h) (Spigno et al. 2007). Drying the red GP extract at 60°C did not change the

ASPECTS OF BY-PRODUCT UTILIZATION 151

total phenolics or condensed tannin contents, but drying at 100°C and 140°C resulted in 18.6% and 32.6% reduction in total phenolics and 11.1% and 16.6% reduction in condensed tannins compared to freeze-dried samples (Larrauri et al. 1997).

4.7.2 Novel Conventional Technologies for the Extraction of Bioactives from Wine By-Products

Several technologies have been investigated to improve the yield of extracted bioactive compounds from wine by-products (Table 4.7). A high recovery of proanthocyanidins can be obtained with CO_2 supercritical fluid extraction (SCFE) when 20% of ethanol was used as a modifier (Yilmaz et al. 2011). This is because individual polyphenol compounds possess varying polarities for maximum recovery from grape seeds. Maximum yields of gallic acid, epigallocatechin, and epigallocatechingallate were extracted when 30 MPa, 50°C, and 20% of ethanol was used, while the maximum recovery of catechin and epicatechin was obtained when 30 MPa, 30°C, and 20% of ethanol was used for extraction (Yilmaz et al. 2011). Maximum yield of epicatechingallate was obtained at 25 MPa, 30°C, and 15% of ethanol. According to Murga et al. (2000), an ethanol concentration of <5% was only able to extract low-molecular-weight compounds such as gallic acid, protocatechuic acid, and protocatechuic aldehyde. However, higher-molecular-weight compounds such as catechin, epigallocatechin, and epicatechin require higher concentrations ($\geq$15%) of ethanol and pressures for maximum SCFE. There are several factors that need to be considered to insure the stability of extracted compounds by SCFE. Generally, polyphenols are easily degraded due to the natural characteristics of their molecular structure and matrix composition. This instability of phenolics should be taken into account while designing the SCFE conditions such as temperature, pressure, and extraction time. The optimum SCFE conditions vary for each plant material due to different classes of polyphenols and type of matrices. In addition, the stability of extracted polyphenols is affected by oxygen, light, solvents, and the existence of enzymes. Overall, SCFE resulted in higher recovery of phenolic compounds and antiradical activity of extracts from grape pomace compared to solid–liquid extraction (Pinelo et al. 2007).

152 VALORIZATION OF WINE MAKING BY-PRODUCTS

Longer extraction times and higher temperatures in the conventional solid–liquid extraction method led to degradation of polyphenols. Exposure to light and air accelerates the deteriorative process of polyphenols (Tura and Robards 2002). SCFE operates under an inert atmosphere at a low temperature and less processing time in the absence of oxidizing agents such as oxygen and light and thus it results in highly active polyphenols. It is generally believed that temperatures above 70°C can negatively affect the antioxidant activities of polyphenol extracts. Increasing the temperature increases the number of molecular collisions and formation of polymerization and can consequently decrease the antioxidant capacity of the material (Durling et al. 2007). Ultrasound-assisted extraction (UAE) is another useful technology to improve the yield and quality of extracted wine by-products depending on the temperature, time, frequency, and solvent concentration. Ghafoor et al. (2009) used response surface methodology that involved ethanol concentration (40%, 50%, and 60%), temperature (40°C, 50°C, and 60°C), and time (20, 25, and 30 min) to maximize the extraction of polyphenols from grape seeds using ultrasound. Increasing the ethanol concentration at a fixed extraction time and temperature increased the total phenolic content. Also, increasing the extraction temperature at a constant ethanol concentration and extraction time resulted in an increased total phenolic content. Pulsed electric field (PEF) and high voltage electric discharge (HVED) are commonly investigated for the extraction of bioactive compounds from plants. PEF is a milder treatment than HVED since only electroporation (rupture) of cell membranes occurs whereas HVED is a more destructive treatment that causes physical disintegration of cell walls and membranes. The main effect of PEF on biological materials is believed to be due to electropermeabilization and electroporation of cell membranes. The application of PEF at 10 kJ/kg energy input increased the yield of anthocyanins from GP by 30%–35% compared to conventional liquid extraction (Töpfl 2006). HVED is more powerful compared with PEF and results in physical disintegration of the cell wall and cell membrane, thus it is regarded as a destructive method. The use of HVED in solutions can create a high energy arc and cause the formation of bubble cavitation (either from dissolved air in solution or vapor generated from rapid localized heating of water), high pressure (up to 10 kbar) shock-waves and liquid turbulence (Boussetta et al. 2012). Collectively, these effects can lead to destructive effects on the structure of the material and promote the extraction of compounds from the cytoplasm of the cells. Both the techniques demonstrated various levels of success in maximizing the extraction of polyphenols from plants, including grapes and their by-products (Corrales et al. 2008; Boussetta et al. 2009a,b, 2011, 2012, 2013a). PEF technology (3 kV/cm) increased the extraction of polyphenols from grape by-products fourfold compared to conventional extraction (Corrales et al. 2008). PEF was more effective in the recovery of the polyphenols than ultrasound (35 kHz) or high hydrostatic pressure (600 MPa) and demonstrated a selective extraction toward anthocyanins. The use of PEF (8–20 kV/cm, 0–20 ms) and HVED (10 kA/40 kV, 1 ms) improved the extraction of polyphenols from grape seeds (Boussetta et al. 2012). Synergistic effects with the treatment were found with temperature and use of ethanol with maximum yield of 9 g gallic acid equivalent/100 g seed at 50°C and 50% ethanol. PEF had the advantage of producing cleaner extracts that do not require subsequent extensive separation processes. In terms of energy use per unit mass of grape seeds, PEF was found to be 47 times higher than electric arc treatment whereas HVED treatment was 27 times higher than arc (Boussetta et al. 2013a). It was noted that HVED treatment increased the temperature of the extraction medium and that the increase can partially contribute to the observed faster rate of extraction. Microwave-assisted extraction (MAE) technology uses very short wave electromagnetic radiation to achieve extraction of the phenolic compounds from materials. The oscillating electromagnetic waves penetrate into the material, interact with the polar components of the material (e.g., water), and generate heat due to molecular friction. Subsequently, the generated heat acts directly on the molecules in the sample via ionic conduction (migration of ions) and dipole rotation (realignment of dipoles) leading to the release of compounds of interest. The effect of microwaves can be selective and target specific materials depending on their dielectric constant. The MAE method is very attractive for the extraction of phenolic compounds from materials due to the fact that it requires shorter extraction time, uses less solvent,

154 VALORIZATION OF WINE MAKING BY-PRODUCTS

has only a moderate capital cost, performs well under atmospheric conditions, and is not polluting (Chan et al. 2011). Generally, it has been reported that the extraction yields of the phenolic compounds obtained with MAE were higher or at least equivalent to the yields achieved with conventional extraction techniques (Chan et al. 2011). There are many parameters that strongly affect the performance and efficiency of MAE which are important for extraction of polyphenols. These factors include microwave power, temperature, extraction time, sample characteristics, solvent nature, and solvent:solid ratio. The presence of water can enhance the heating efficiency due to molecular friction and improve the penetration of solvent into the sample matrix. The organic solvent can also play an important role in dictating the yield extracted. For example, Casazza et al. (2010) reported that methanol resulted in a higher MAE polyphenols yield from grape skin and seed (30.9% and 21.3% higher yield for skin and seed, respectively) after 19 min of extraction. Polysaccharides in the cell wall of wine by-products contain hydrogen groups, aromatic and glycosidic oxygen atoms, which form hydrogen bonds and hydrophobic interactions with polyphenols making them less readily available for extraction. Enzymatic treatment of grape pomace which enhanced recovery of phenolics was reported by Meyer et al. (1998). Treatment of PG with tannase and pectinase or a combination of both for 24 h released catechin and increased gallic acid, procyanidin B2, and the antioxidant activity of the phenolic extract (Chamorro et al. 2012). The use of cellulose was not successful in improving the yield of phenolics. However, other research found moderate temperature (e.g., 40°C) at low pH required for optimum enzymatic activity may degrade polyphenols and brings about a loss of antioxidant activity (Maillard and Berset 1995). Aqueous solutions of cyclodextrins (α, β, and γ forms) were used to extract phenolics compounds from GP (Ratnasooriya and Rupasinghe 2012). The use of 2.5% (w/v) β cyclodextrin for 24 h at 60°C resulted in the extraction of 35.8 mg/100 g fresh GP. β-Cyclodextrin recovered more flavan-3-ols than flavonols. *Trans*-resveratrol, rutin, epigallocatechin, and chlorogenic acid form inclusion complexes with β-CD at a 1:1 molecular ratio. Gamma irradiation can be used to extend the shelf-life of GP as well as improving the yield of anthocyanin (Ayed et al. 1999).

4.8 Food and Non-Food Applications

Grape pomace is a rich source of high-value products including ethanol, tartrates, malates, citric acid, grape-seed oil, hydrocolloids, and dietary fiber (Valiente et al. 1995; Igartuburu et al. 1997; Bravo and Saura-Calixto 1998; Nurgel and Canbas 1998; Girdhar and Satyanarayana 2000). Pomace has been utilized in Italy to produce a spirit called *grappa*, which contains between 38% and 80% of alcohol by volume. The high phenolic and flavonol contents in grape pomace make it a good source of natural antioxidants (Revilla and Ryan 2000; Jayaprakasha et al. 2001, 2003; Murthy et al. 2002), dietary supplements, and phytochemical products (Sanchez et al. 2002; González-Paramás et al. 2004). Walter and Sherman (1974) reported grape pomace to be an excellent source of low-ash carbon whose extensive porosity would be an advantage in adsorption operations. All the aforementioned properties of grape pomace make their utilization worthwhile and support sustainable agricultural production (Kammerer et al. 2004). The recovery of phytochemical compounds from wine by-products has attracted great interest in food and nutraceutical industries (González-Paramás et al. 2004) due to the numerous beneficial health effects and the antioxidant properties. Phenolic phytochemicals are the largest category of phytochemicals in grapes, wines, and wine by-products.

4.8.1 Phenolics

The phenolic content in grape seeds ranges from 5000 to 8000 mg/kg seeds, compared to 285 to 550 mg phenols/kg skin, depending on the grape variety and the type of pretreatment (Pinelo et al. 2005a). During wine making, parts of the phenolics from the seeds and skin are passed onto the wine. The amounts of phenolics that are present in the wine are more or less dependent on the processing conditions of the wine making process. Given the fact that the major parts of phenolic compounds are found in the grape solids and that the ethanol content is not strong enough to enable efficient extraction, a high proportion of phenolics remains in the vinification residues. Thus, the use of wine by-products as a good source for the recovery of phenolics is justifiable (Kammerer et al. 2005). The general distribution and contents of phenolic compounds found in the different fractions of wine

156 VALORIZATION OF WINE MAKING BY-PRODUCTS

by-products (seeds, skin and stalks) and grape pomace are depicted in Table 4.8. Phenolics in wine by-products can be classified into two main groups: (1) Non-flavonoids phenolic acids: mainly hydroxy-cinnamic acids (found esterified with tartaric, caftaric, and coutaric acids); hydroxybenzoic acids (gallic acid); stilbenes (resveratrol and piceid). (2) Flavonoids flavanols, also known as flavan-3-ols, found in grape seeds, skin, and stalks (exist in monomeric, oligomeric, or poly-meric forms; the latter two forms are known as proanthocyanidins [PACs]) (Figure 4.6); flavonols (kaempferol, quercetin, and myricetin are present in the vacuoles of the epidermal tissues as glycosides [sugar attached]); anthocyanins, a color pigment mainly located in the grape skin (3-O-monoglucisodes and 3-O-acylated monogluco-sides of cyanidin, peonidin, delphinidin, petunidin, and malvidin) (Figure 4.6). While grape seeds are regarded as the main source of phenolics, grape skins are increasingly getting more attention and interest from researchers and are gaining ground as a valuable source of biological compounds (Yilmaz and Toledo 2006). Phenolic extracts from wine by-products could potentially play a significant role in the medical and pharmaceutical fields.

4.8.1.1 Health Supplements and Extracts Grape-seed extracts (GSEs) are a group of natural antioxidants, known to possess a broad spec-trum of pharmacological activity. GSE was found to increase the antioxidant activity in rat plasma after oral administration (Koga et al. 1999) and was reported to be useful in protecting against the oxidation of the low-density lipoproteins (LDL) by reducing lipid oxidation (Bouhamidi et al. 1998) and/or blocking the production of free radicals (Bagchi et al. 1998). GSE can also be used to improve the oxidative stability of meat products such as cooked beef (Ahn et al. 2002b) and turkey patties (Lau and King 2003) due to the fact that these extracts contain a heterogeneous mixture of mono-mers, oligomers, and polymers formed by subunits of flavan-3-ol (Revilla and Ryan 2000; Waterhouse et al. 2000) as well as other antioxidants (vitamin C and β-carotene). Grape-seed extracts are known to exhibit various properties including antiulcer (Saito et al. 1998), anticarcinogenic, antimutagenic (Liviero et al. 1994), antivi-ral (Takechi et al. 1985), hypocholesterolemic as well as antiathero-sclerotic activities (Tebib et al. 1994, 1997; Yamakoshi et al. 1999).

ASPECTS OF BY-PRODUCT UTILIZATION 157

Table 4.8 Main Phenolic Compounds in Different Fractions of Grapes

GROUPS	COMPOUNDS	POMACE (mg/g)	SEEDS (mg/g)	SKIN (mg/g)	STALKS (mg/g)
Non-flavonoids	Gallic acid	0.03–0.11[a,b]	0.10–0.11[a]	0.03[b]	—
	Coutaric acid	0–1.23[c]	—	0.03–1.23[d]	Traces[e]
	Caftaric acid	0–6.97[c]	—	0.11–6.97[c,d]	0.04[e]
	Phenolic acids	0.03–8.31	0.10–0.11	0.17–8.23	0–0.04
	Resveratrol	0–0.015[f]	Traces[h]	0.05–0.1[g]	—
Flavonoids	(1) Flavan-3-ols Catechin	0–0.18[i]	2.14–2.15[k]	0–0.16[b,i]	0.06[e]
	Epicatechin	0–0.16[i]	0.88–0.91[k]	0–0.13[b,i]	0.28[e]
	Epigallocatechin	0–0.05[e]	0.05[e]	Traces[b,i]	0.01[e]
	Epigallocatechin 3-gallate	0–0.07[k]	0.06–0.07[k]	—	—
	Epicatechin 3-gallate	0–0.03[l]	0.25–0.31[k]	0.04[i]	0.07[e]
	Procyanidin dimers B1	0.11–0.6[i,m]	0.14–0.16[k]	0.11–0.6[i,m]	—
	Procyanidin dimers B2	0.01–0.84[i,m]	0.04–0.18[k]	0.01–0.84[i,m]	—
	Tannins	0.22–2.32[e]	2.32[e]	1.61[e]	0.22–0.39[e]
	Total flavan-3-ols content	0.34–4.25	3.56–6.15	0.12–3.38	0.22–0.89
	(2) Anthocyanins Delphinidin 3-*O*-glc	0.44–1.11[d]	—	0.44–1.11[d]	—
	Cyanidin 3-*O*-glc	1.51–3.81[d]	—	1.51–3.81[d]	—
	Petunidin 3-*O*-glc	0.53–1.34[d]	—	0.53–1.34[d]	—
	Peonidin 3-*O*-glc	0.99–2.49[d]	—	0.99–2.49[d]	—
	Malvidin 3-*O*-glc	4.12–10.19[d]	—	4.12–10.19[d]	—
	Delphinidin 3-*O*-acglc	0.08–0.19[d]	—	0.08–0.19[d]	—
	Petunidin 3-*O*-acglc	0.11–0.28[d]	—	0.11–0.28[d]	—
	Peonidin 3-*O*-acglc	0.27–0.30[d]	—	0.27–0.30[d]	—
	Malvidin 3-*O*-acglc	0.62–1.74[d]	—	0.62–1.74[d]	—
	Cyanidin 3-*O*-pcmglc	0.07–0.22[d]	—	0.07–0.22[d]	—
	Petunidin 3-*O*-pcmglc	0.19–0.49[d]	—	0.19–0.49[d]	—
	Peonidin 3-*O*-pcmglc	0.43–1.37[d]	—	0.43–1.37[d]	—

(Continued)

158 VALORIZATION OF WINE MAKING BY-PRODUCTS

Table 4.8 (*Continued*) Main Phenolic Compounds in Different Fractions of Grapes

GROUPS	COMPOUNDS	POMACE (mg/g)	SEEDS (mg/g)	SKIN (mg/g)	STALKS (mg/g)
	Malvidin 3-*O*-pcmglc	2.11–6.29[d]	—	2.11–6.29[d]	—
	Total anthocyanin content[o]	11.47–29.82[a]	—	11.47–29.82[d]	—
(3) Flavonols	Quercetin 3-glucoside	0.01–0.2[d]	0.01–0.02[d]	0.15–0.2[d]	0.02[e]
	Myricetin 3-glucoside	Traces[e]	—	—	Traces[e]
	Quercetin 3-glucuronide	0.01–0.29[d]	0.01–0.02[d]	0.22–0.29[d]	0.2[e]
	Kaempferol 3-glucoside	0.01–0.14[d]	0.01[d]	0.11–0.14[d]	Traces[e]
	Myricetin 3-glucuronide	Traces[e]	—	—	Traces[e]
	Total flavonols	0.03–0.63[n]	0.02–0.05[n]	0.48–0.63[n]	0–0.22[n]

[a] Oszmianski and Sapis (1989) (unspecified variety).
[b] Yilmaz and Toledo (2004) (var. Merlot).
[c] Borbalán et al. (2003) (var. Cabernet Sauvignon).
[d] Kammerer et al. (2004) (var. Cabernet Mitos).
[e] Souquet et al. (2000) (var. Merlot).
[f] Vatai et al. (2009) (var. Refošk).
[g] Joe et al. (2002) (unspecified variety).
[h] Zhu et al. (2000).
[i] Arts et al. (2000) (unspecified variety).
[j] Freitas et al. (2000) (var. Cabernet Sauvignon).
[k] Guendez et al. (2005) (var. Cabernet Sauvignon).
[l] Mendes et al. (2013).
[m] Mateus et al. (2001) (var. Touriga Nacional).
[n] Kammerer et al. (2004) (var. Weisser Riesling).
[o] Anthocyanin content is exclusive for red grapes.

The production of laccase from grape seeds is a viable alternative for the utilization of this raw material due to its high lignocellulose content (Moldes et al. 2003). Grape-seed oil is another by-product that has commercial potential since it is rich in vitamin C, β-carotene, tocopherols (0.8%–1.5%), steroids, omega-6 fatty acids (69%–78%), omega-9 fatty acids (15%–20%), omega-3 fatty acids (0.3%–1%), and several other fatty acids (Arvanitoyannis et al. 2006). Grape skins are good sources of phytochemicals such as gallic acid, catechin and epicatechin, and polyphenolic tannins that provide the astringent taste to wine. The main constitutive units of skin tannins are (+)-catechin, (–)-epicatechin

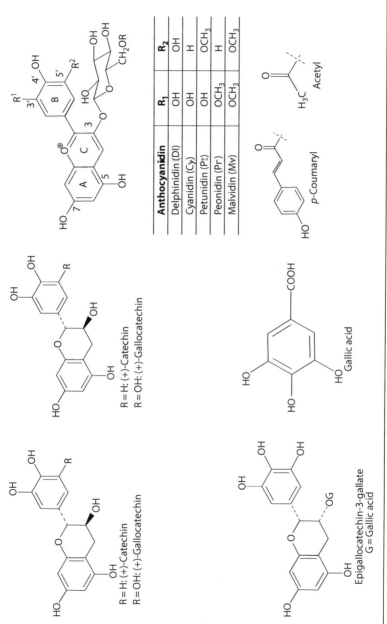

Figure 4.6 Structures of the main constitutive units of flavan-3-ols identified in grape seeds and skin and the general structure of anthocyanin compounds. (From Corrales, M. et al., *Innov. Food Sci. Emerg. Technol.*, 9, 85, 2008.)

160 VALORIZATION OF WINE MAKING BY-PRODUCTS

and (–)-epicatechin gallate, although (+)-gallocatechin and (–)-epigallocatechin are present in minor quantities (Shi et al. 2003a; Yilmaz and Toledo 2006). Structures of the main constitutive units of skin tannins are shown in Figure 4.6 (Shi et al. 2003a). Grape skin extracts are considered as a valuable source of anthocyanins and procyanidins (Shrikhande 2000). Anthocyanins (Figure 4.6) are a group of phenolic compounds that belong to the flavonoid family and are considered the most valuable phenolic compounds which can be used as a natural food colorant (Shrikhande 2000; Arvanitoyannis et al. 2006). Anthocyanins in grape skins are the pigments responsible for the red grape and wine colors and these include delphinidin, cyanidin, petunidin, peonidin and malvidin 3-glucosides, 3-(6-acetyl)-glucosides and 3-(6-*p*-coumaroyl)-glucosides, and some pyruvates (Monagas et al. 2006). The anthocyanins from winery waste, known as enocyanin, used as a natural food colorant, have been commercialized since 1879 (Alonso et al. 2002; Arvanitoyannis et al. 2006). Procyanidins (Figure 4.6) are polymers from catechin and flavan-3,4-diols and can be found in dimer, trimer, tetramer, or other polymer forms. Procyanidins could inhibit the oxidation of LDL and display an antiatherosclerotic activity by reacting with reactive species in plasma and interstitial fluid of the arterial wall (Yamakoshi et al. 1999).

4.8.1.2 Bioactivity of Phenolic Compounds   In recent years, growing interest in phenolics from grapes and grape products has focused on their bioactivities associated with human health benefits such as anticancer, anti-inflammatory, anticarcinogenic, antiallergic, and cardioprotective properties as well as antioxidant, antibacterial, antifungal, and antiviral activities. A summary of the biological activities of phenolic compounds commonly isolated from grapes, wines, and wine by-products is shown in Table 4.9.

4.8.1.3 Antioxidant Activity   The antioxidative properties of grapes and its products, being the most notable bioactivity, have been widely studied in different biological or food systems. Phenolic compounds extracted from different fractions of grapes, such as catechins, flavonols, anthocyanins, tannins, resveratrol, and proanthocyanidins, have been considered to be powerful antioxidants *in vitro* (Fernandez-Panchon et al. 2008) and have been proved to be more potent antioxidants than vitamin C,

ASPECTS OF BY-PRODUCT UTILIZATION 161

Table 4.9 Biological Properties of Some Phenolic Compounds from Grapes, Wines, and Wine By-Products

PHENOLIC COMPOUNDS	BIOLOGICAL ACTIVITY	REFERENCES
Anthocyanins	Vasorelaxing and antioxidant activities	Mullen et al. (2002)
	Antiangiogenic activity	Bagchi et al. (1998)
	Antitumor activity	Kamei et al. (1998)
	Antibacterial activity	Radovanović et al. (2009)
	Anti-inflammatory and antiedema activities	Wagner (1985)
	Apoptosis-inducing activity	Lazze et al. (2004)
	Inhibitory activity on lipid peroxidation and platelet aggregation	Ghiselli et al. (1998)
Flavan-3-ols (Catechin, epicatechin, PACs)	Free radical scavenging activity	Castillo et al. (2000) and Cos et al. (2004)
	Anticancer activity	Isemura et al. (2000) and Faria et al. (2006)
	Antiviral activity	Song et al. (2005)
	Antibacterial activity	Mabe et al. (1999)
	Anti-inflammatory activity	Cos et al. (2004) and Li et al. (2011)
	Apoptosis-inducing activity	Mantena et al. (2006)
	Regulating activity on lipid metabolism	Auger et al. (2005)
	Antiulcer activity	Saito et al. (1998)
Flavone	Antiproliferative activity	Wenzel et al. (2000)
Flavonol	Free radical scavenging activity	Makris et al. (2006)
Gallic acid	Free radical scavenging activity	Yilmaz and Toledo (2006)
Resveratrol	Anticancer activity	Kuwajerwala et al. (2002) and Qian et al. (2009)
	Inhibitory activity on cardiovascular disease, lipid metabolism regulating activity, inhibitory activity on lipid peroxidation and platelet aggregation, copper chelating activity	Frémont (2000) and Auger et al. (2005)
	Apoptosis-inducing activity	Surh et al. (1999)
	Antioxidant and antimicrobial activities	Filip et al. (2003)
	Free radical scavenging activity	Leonard et al. (2003)
	Antiviral activity	Palamara et al. (2005)
	Antiaging and anti-inflammatory activities	de la Lastra and Villegas (2005)
Quercetin	Antibacterial activity	Rodríguez Vaquero et al. (2007)
	Antimutagenic and antioxidant activities	Geetha et al. (2005)

162 VALORIZATION OF WINE MAKING BY-PRODUCTS

vitamin E, or carotenoids (Rice-Evans et al. 1997). Several mechanisms exerted by these phenolic antioxidants include scavenging of free radicals by donating hydrogen or an electron, protecting biological molecules from oxidation, inhibition of lipid oxidation, reduction of hydroperoxide formation, and by metal-ion chelation (Robak and Gryglewski 1988; Hamilton et al. 1997).

4.8.1.4 Antibacterial and Antifungal Activities Phenolic compounds extracted from grapes and their products have long been demonstrated to exhibit antimicrobial activity against several microorganisms (Masquelier 1959). Table 4.10 shows the investigation of microorganisms inhibited by isolated phenolic compounds or extracts obtained from grapes, wines, and wine by-products. Various microbial species have different sensitivities toward phenolic compounds, for example, *Staphylococcus aureus* was found to be the most sensitive to wine extracts, followed by *Escherichia coli* and the least effect of inhibition was detected in *Candida albicans* (Papadopoulou et al. 2005). In another study by Radovanović et al. (2009), the diameter of the growth inhibition zone of various wine samples at the concentration of 50 µL/disc for *S. aureus* and for *E. coli* were 16–22 and 12–20 mm, respectively. The diameter of the inhibition zone for *S. aureus* was larger than that for *E. coli* strain, indicating that the gram-(+) strain was more sensitive than the gram-(–) strain. Grape-seed extract was more potent than cranberry juice in suppressing the growth of gram-positive and gram-negative bacteria, but only grape-seed extract inhibited *Candida albicans*. Thus grape-seed extract, alone or in combination with cranberry juice, can be used for prolonging the shelf-life of aqueous hypromellose gels. Grape-seed extract (5%) alone with the activity independent of pH, or 0.7% grape-seed extract combined with 10% cranberry juice active only in pH 2.5–5, has been suggested. A combination of a grape-seed extract with an amine fluoride (Fluorinol®) displayed an inhibitory action on dental plaque and biofilm formation and oxidative damage caused by oral bacteria (Furiga et al. 2009). Phenolic compounds from different fractions of grapes also displayed different antimicrobial activities. Grape pomace exhibited either similar or better antimicrobial activity against *Streptococcus mutans* compared with extracts from whole grapes (Thimothe et al. 2007). Seeds were found to display better inhibitory effects than

ASPECTS OF BY-PRODUCT UTILIZATION 163

Table 4.10 Summary of Inhibition Effects of Phenolic Compounds or Extracts Obtained from Grapes, Wines, and Wine By-Products against Microorganisms

EXTRACTS/PHENOLIC COMPOUNDS	MICROORGANISMS INHIBITED	REFERENCES
Grape seeds	*Bacillus cereus, Bacillus coagulans, Bacillus subtilis, Staphylococcus aureus, Escherichia coli, Pseudomonas aeruginosa*	Jayaprakasha et al. (2003)
Grape pomace	*Aeromonas hydrophila, Bacillus cereus, Enterobacter aerogenes, Enterococcus feacalis, Escherichia coli, Escherichia coli 0157-H7, Mycobacterium smegmatis, Proteus vulgaris, Pseudomonas aeruginosa, Pseudomonas fluorescens, Salmonella enteriditis, Salmonella typhimurium, Staphylococcus aureus, Yersinia enterocolitica*	Özkan et al. (2003)
Grape seeds and bagasse	*Aeromonas hydrophila, Bacillus brevis, Bacillus cereus, Bacillus megaterium, Bacillus subtilis, Enterobacter aerogenes, Enterococcus feacalis, Escherichia coli, Klebsiella pheumoniae, Listeria monocytogenes, Mycobacterium smegmatis, Proteus vulgaris, Pseudomonas aeruginosa, Staphylococcus aureus*	Baydar et al. (2004)
Red and white wine	*Escherichia coli, Staphylococcus aureus, Candida albicans*	Papadopoulou et al. (2005)
Grape extracts (red, white, black and muscadine grape skin, muscadine grape seeds, and muscadine pomace)	*Helicobacter pylori*	Brown et al. (2009)
Red, white, and rosé wine, ferulic acid, methyl gallate, epicatechin, synaptic acid, *p*-hydroxybenzoic, gallic acid, vanillic acid, tryptophol, cumaric acid, caffeic acid	*Campylobacter jejuni*	Gañan et al. (2009)
Red wine	*Staphylococcus aureus, Escherichia coli*	Radovanović et al. (2009)
Grape skin	*Staphylococcus aureus, Bacillus cereus, Escherichia coli 0157:H7, Salmonella infantis, Campylobacter coli*	Katalinić et al. (2010)

164 VALORIZATION OF WINE MAKING BY-PRODUCTS

other parts of the grapes (Anastasiadi et al. 2009; Brown et al. 2009). Therefore, the potent function of phenolic compounds extracted from grapes and grape products as natural preservatives and antimicrobial agents for food is promising. The *in vitro* antifungal activity of Cabernet Sauvignon, Carmènere, and Syrah extracts against *Botrytis cinerea* was investigated by Mendoza et al. (2013). The extraction was carried out using hexane, chloroform, and ethyl acetate. The highest inhibitory activity (IC_{50} = 40 ppm) was found in hexane or chloroform fractions. Liquid–liquid extraction generated better antifungal activity than soxhlet extraction, but the activity was affected by the form (whole versus ground) and the solvent used for extraction. The antifungal activity of various phenolics against *Botrytis cinerea* are in the following order: kaempferol = *p*-coumaric acid > quercetin > syringic acid = catechin > epicatechin = vanillic acid > gallic acid > ellagic = protocatechuic acid. *In vitro* and *in situ* (in apple and orange juices) growth inhibition activity against *Zygosaccharomyces rouxii* and *Z. bailii* was reported for Gamay and Kalecik varieties GP (Sagdic et al. 2011). The activity was lower in extracts from other grape cultivars (Emir, Narince, and Okuzgozu) due to variations in polyphenol composition.

4.8.1.5 Antiviral Activity Studies have been carried out to investigate the antiviral activity of phenolic compounds extracted from grapes and grape products (Konowalchuk and Speirs 1976, 1978; Nair et al. 2002) as well as from wine by-products (Matias et al. 2010). Grape extracts (skin and whole black grapes), grape juice, and wine were reported to inactivate various enteric viruses and herpes simplex virus (HSV) type 1 (Konowalchuk and Speirs 1976). Grape juice had a higher antiviral activity than wine, and red wine was more active than white wine (Konowalchuk and Speirs 1976). It has also been demonstrated that black table grapes had higher antiviral activity than green table grapes and the antiviral activity of red grapes resides in the skin of the grapes. Wine residues exhibit antiadenoviral activity (Matias et al. 2010). Isolated pure compounds from plant materials such as resveratrol (Palamara et al. 2005), catechins (Song et al. 2005), tannins (Nakashima et al. 1992), and proanthocyanidins (Iwasawa et al. 2009) were shown to possess antiviral activity. These pure compounds can also be extracted from grapes and grape products.

ASPECTS OF BY-PRODUCT UTILIZATION   165

Since the early study by Green (1949) which first reported the inhibition of influenza virus replication in egg embryos by tea extracts, numerous studies have reported anti-influenza virus activity for plant phenolics (Cos et al. 2004; Droebner et al. 2007; Liu et al. 2008; Mori et al. 2008; Saladino et al. 2008). Polyphenol compounds which are found in tea, grape products, and other plant sources such as epigallocatechin gallate (ECGC), epicatechin gallate (ECG), epigallocatechin (EGC), and theaflavin digallate exhibit several mechanisms which promote the prevention of the virus infectivity, such as by binding to the haemagglutinin (HA) of influenza virus (Nakayama et al. 1993) or by altering the physical properties of the viral membrane (Song et al. 2005). Hudson (1990) reported that viral inactivation *in vitro* is attributed to preferential binding of phenolics to the protein coat of the virus, thus arresting virus absorption. Sakagami et al. (1995) suggested that the antiviral activity in phenolics could be derived from their direct inactivation of the virus and/or from inhibition of the virus binding to the cells. Several investigations have drawn attention to possible antiviral activity attributable to other phenolic compounds such as proanthocyanidins, which are the oligomer or polymer form of flavan-3-ol units and resveratrol. Proanthocyanins (PACs) have been shown to exhibit antiviral activity against poliomyelitis virus (Konowalchuk and Speirs 1976). Three PAC compounds existing in dimer, trimer, and tetramer form showed pronounced antiviral properties against herpes simplex and coxsackie viruses (Fukuchi et al. 1989; Balde et al. 1990). Several potential mechanisms have been reported for the antiviral activity of PACs. For instance, PACs have been shown to inhibit enzymes involved in the replication of rhino virus (common cold) and HIV virus (Hocman 1989; Amouroux et al. 1998). Furthermore, Cheng et al. (2005) found that PCAs A-1 purified from Vaccinium vitis-idaea had the ability to suppress HSV-2 infection through the inhibition of viral attachment and penetration. Resveratrol has been found to affect influenza virus replication both *in vitro* and *in vivo* by several modes of action as follows: (1) by blockade of the nuclear-cytoplasmic translocation of the viral ribonucleoprotein complex, (2) by reducing the expression of late viral proteins, and (3) by inhibition of protein kinase C (PKC) activity and PKC-dependent pathways (Palamara et al. 2005). Resveratrol is able to inhibit the replication of HSV types 1 and 2 in a dose-dependent and

166 VALORIZATION OF WINE MAKING BY-PRODUCTS

reversible manner (Docherty et al. 1999). Resveratrol also synergistically enhances the anti-HIV activity of a number of nucleoside analogs in combating infection in peripheral white blood cells (Heredia et al. 2000). Pflieger et al. (2013) reported that grape stilbenoid compounds encompassing E-resveratrol, E-ε-viniferin glucoside, E-pterostilbene, and E-piceid inhibited HIV-1 integrase. Bekhit and Bekhit (2012) demonstrated the anti-H1N1 influenza activity of GP extracts.

4.8.2 Food Ingredients

Tartrates can be recovered from wine waste (Braga et al. 2002). An estimated 100–150 and 50–75 kg of calcium tartrate can be recovered from a ton of wine lees and GP, respectively.

GP is a good source of soluble dietary fiber. Llobera and Canéllas (2007) found Manto Negro GP and stems had 74.5% and 77.2% (dry matter) total dietary fibers of which 10.8% and 3.8% total soluble fibers were in GP and stem, respectively. White and red GP from varieties grown in the United States had 17.3%–28.0% DM and 51.1%–56.3% DM, respectively (Deng et al. 2011). The content of water-soluble pectin was in the following order: Merlot > Pinot Noir > Morio Muscat > Cabernet Sauvignon = Muller Thurgau, but total extractable pectins and dietary fibers were the highest in Merlot and Cabernet Sauvignon and the lowest in Morio Muscat. Grape and wine by-products offer the potential of having fibers with high antioxidant activity due to the natural presence of polyphenols and other bioactive compounds (Saura-Calixto 1998; Llobera and Canéllas 2007; Deng et al. 2011). These high antioxidant fibers can be used in novel healthy products such as yoghurt and salad dressing (Tseng and Zhao 2013). Wine by-products have also been used for the production of organic acids (e.g., lactic and citric acids), and improving the quality of wine (e.g., color and phenolics profile) before bottling. The use of red and white grape skins can significantly increase anthocyanins and low-molecular-weight phenolic compound content. Important compounds that are located in the skin (e.g., gallic acid, anthocyanins, resveratrol, and catechins) can be translocated to wine by further treatment (Pedroza et al. 2013). Improved volatiles from the skins can also be released leading to the ability to tailor the final product sensory profile. Grape skins can also be used for the production of sensory acceptable tea (Cheng et al. 2010).

4.8.3 Health Applications

These include anti-obesity, antihyperlipidemic, and anti-inflammatory activities, inhibitory activity on heterocyclic amine production and metabolic effects. Grape-seed extract and resveratrol are potent inhibitors of fatty acid synthase and they may be useful for treating obesity. In diet-induced obese mice, grape seed proanthocyanidin extract prevented weight gain, mitigated hepatic lipid infiltration, and lowered serum lipid levels. In obese mice with collagen-induced arthritis, grape seed proanthocyanidin extract inhibited the development of autoimmune arthritis. In both types of mice, grape seed proanthocyanidin extract suppressed Th17 cells and stimulated regulatory T cells. Thus grape seed proanthocyanidin extract shows promise for use in treating immunologic diseases associated with enhanced STAT3 activity such as autoimmune diseases and metabolic disorders (Jhun et al. 2013). Wild grape seed procyanidins counteracted inflammation by suppressing COX-2 and iNOS through the regulation of NF κB and p38 MAPK pathway. Grape seed proanthocyanidin extract suppressed plasma levels of triglycerides, free fatty acids, glycerol, and urea in rats receiving oral administration of lard oil. The extract enhanced oxygen consumption using pyruvate as substrate in skeletal muscle mitochondria. It upregulated the expression of genes involved in energy metabolism like peroxisome proliferator-activated receptor gamma, coactivator 1 alpha, and modified the enzyme activity of proteins in the electron transport chain and tricarboxylic acid cycle in brown adipose tissue (Pajuelo et al. 2011). In rats, addition of oil rich in docosahexaenoic acid to lard oil in the diet augmented insulin sensitivity and redirected fatty acids toward skeletal muscle, stimulated fatty acid oxidation, and improved adipose mitochondrial function and uncoupling. Grape seed proanthocyanidin extract reduced lipidemia, suppressed generation of muscle reactive oxygen species and the consequent damage, and stimulated mitochondrial biogenesis and lipogenesis in adipose tissue. The addition of both grape seed proanthocyanidin extract and oil rich in docosahexaenoic acid to lard produced a less dramatic profile. Polyphenols that were present in grape seeds, green tea, oranges, and grapefruits inhibited adipogenesis and induced lipolysis. Rats that were fed a diet containing fructose and an excess of fat, manifested inflammation, hepatic steatosis, elevated lipid storage, lipogenesis, and suppressed lipolysis. Grape seed proanthocyanidins intake in these rats produced a decline in hepatic triglyceride level, mRNA expression of sterol regulatory element binding protein 1c and hepatic lipid droplet proteins that was more effective than metformin, a standard anti-diabetic drug (Yogalakshmi et al. 2013). The flavonoids in the Concord grape *Vitis labrusca* exerted a vasodilatory action probably via their antioxidant effect. The consumption of grape seed proanthocyanidin extract (100 mg/kg/day) might help reduce cisplatin-induced nephrotoxicity which was a result of oxidative damage. Grape seed proanthocyanidin extract prevented cisplatin A-induced nephropathy and that this effect was achieved by antiapoptotic and antioxidant activity. In overweight or obese female subjects given grape-seed oil (consuming 15% of energy from grape-seed oil) within a weight loss program for 2 months, a homeostatic model assessment of insulin resistance scores, high sensitive C-reactive protein, and tumor necrosis factor-α decreased, indicating an improvement in the inflammatory condition and insulin resistance (Irandoost et al. 2013). In genetically obese Zucker rats, grape seed procyanidin extract inhibited the accumulation of oxidized glutathione, elevated the total reduced glutathione/ oxidized glutathione ratio in liver and reduced the activation of antioxidant enzymes, including glutathione S-transferase, glutathione reductase, and glutathione peroxidase, and enhanced the total cellular antioxidant capacity (Fernández-Iglesias et al. 2014). De-fatted milled grape-seed extract, a wine by-product from the extraction of grape-seed oil, protected cell membranes from oxidative damage and proteins and lipids from oxidation. The extract ameliorated changes in the levels of adenosine triphosphate and glutathione in liver cells brought about by adriamycin (Valls-Belles et al. 2006). Daily treatment with grape-seed extract concurrent with oral ethanol administration for 10 weeks reduced the deleterious changes in the following parameters brought about by the oxidative damage of ethanol: serum testosterone level; epididymal spermatozoal analysis; and weight, histopathology, glutathione level and malondialdehyde level in the brain, liver, and testis. Grape polyphenols maintained endothelial function and protected against the oxidation of low-density lipoprotein. Grape seed and skin extracts exerted a protective action against high-fat diet-induced obesity, relative heart

ASPECTS OF BY-PRODUCT UTILIZATION 169

mass, cardiac oxidative stress and accumulation of triglycerides and total cholesterol, and hepatic oxidative stress of male rats (Charradi et al. 2013). Polyphenols from Concord grape pomace, stabilized by complexation to soy protein isolate, facilitated the capture of grape pomace polyphenols in a protein-rich food matrix. It manifested hyperglycemic activity in obese and hyperglycemic C57BL/6 mice (Roopchand et al. 2013). Grape-seed extract improved renal function and ameliorated the alterations in the circulatory levels of inflammatory cytokines and reactive oxygen species production and Nox activity, and the protein expression levels of the NADPH subunits (Nox$_2$, p47phox, and Nox$_4$) brought about by arsenic administration. TGF-β/Smad signaling was attenuated, as seen in the decreased protein levels of pSmad2/3 and TGF-β1 in kidney tissue (Zhang et al. 2014). In streptozotocin-induced diabetic rats, grape seed proanthocyanidins improved the abnormal peripheral nerve function and impaired nervous tissues (Ding et al. 2014). A combination therapy utilizing grape seed proanthocyanidin extract and a niacin-bound chromium lowered the levels of total cholesterol, low-density lipoprotein (LDL) cholesterol, and oxidized LDL in hypercholesterolemic human patients. The extract, used alone and in conjunction with niacin-bound chromium, reduced oxidative lipid damage as seen in the inhibition of production of thiobarbituric acid reactive substances (Vinson et al. 2002). Grape seed proanthocyanidins mitigated carrageenan-induced edema in rat paws and croton oil-induced swelling in murine ears, lowered malondialdehyde content in inflamed paws, suppressed N-acetyl-beta-D-glucosaminidase and nitric oxide synthase activity, and lowered nitric oxide, interleukin-1beta, tumor necrosis factor-alpha, and prostaglandin E2 levels in edema exudates. The mechanisms of anti-inflammatory action involved scavenging of oxygen free radicals, attenuation of lipid peroxidation, and suppression of the generation of inflammatory cytokines (Li et al. 2011).

4.8.4 Applications in Meat Products

Replacing pork fat with a mixture of grape-seed oil (0%, 5%, 10%, and 15%) and 2% rice bran fiber can improve the healthiness of pork sausage. The ash content, moisture content, uncooked pH value, cooked

170 VALORIZATION OF WINE MAKING BY-PRODUCTS

pH value, yellowness, cohesiveness, chewiness, gumminess, and sarcoplasmic protein solubility were increased in the grape-seed oil and rice bran fiber samples compared to control samples. The low-fat meat sausage mix with escalating grape-seed oil concentrations were characterized by lower cooking loss, emulsion stability, and apparent viscosity. The changes were all desirable. Similarly, decreasing pork fat levels in frankfurters from 30% to 10% and partly replacing the pork fat with grape-seed oil ensued in higher moisture and ash content; lowered caloric values, cholesterol level, trans-fat level and fat content; increased pH and cooking yield; and comparable sensory properties compared with control frankfurters containing pork fat (Choi et al. 2010). The use of grape-seed extracts can improve the nutritional and the quality of meat products. Lau and King (2003) reported that the addition of grape-seed extract to dark poultry meat patties at 1.0% and 2.0% effectively inhibited the development of TBARS by about 10-fold compared to untreated control. Similar results were reported by Pazos et al. (2005) in fish oil-in-water emulsions and GP extracts were as effective an antioxidant as propyl gallate. Natale et al. (2013) examined the suppressive action of liposome-encapsulated grape-seed extract on the generation of heterocyclic aromatic amines during frying of beef patties. PhIP (2-amino-1-methyl-6-phenylimidazo[4,5b] pyridine), MeIQx (2-amino-3,8-dimethylimidazo[4,5-f]quinoxaline), Norharman, and Harman were detected following marinade application and frying. There was a decline in PhIP level after marination with grape-seed extract (0.1%) and grape-seed extract–containing liposomes (1% and 5%) MeIQx level declined but no alterations in β-carboline levels were discernible. Liposomal encapsulation of grape-seed extract failed compared with grape-seed extract, to produce further inhibition of the generation of MeIQx and PhIP. The data appeared to be contradictory to earlier findings that liposomal encapsulation, and also a water-in-oil marinade with grape-seed extract being encapsulated in water droplets suspended in an oil phase, augmented the efficacy of polyphenols to suppress radical reactions and lipid oxidation in model systems. Natale et al. (2014) attributed the discrepancy in the observations to mechanisms affected by the structural and physical features of the encapsulation system employed and its interaction with the application matrix involving diffusion and partitioning behavior and interaction behavior (Natale et al. 2014). Grape-seed extract suppressed the production of

ASPECTS OF BY-PRODUCT UTILIZATION 171

heterocyclic amines including 2-amino-3,8-dimethylimidazo[4,5-f] quinoxaline and 2-amino-1-methyl-6-phenylimidazo[4,5b]pyridine in fried beef patties (Gibis and Weiss 2012). The production of these compounds was decreased upon marination with grape-seed extract (0.1%) and grape-seed extract–containing liposomes (1% and 5%). A correlation between inhibition of the formation of heterocyclic amines 2-amino-3,8-dimethylimidazo[4,5-f]quinoxaline, 2-amino-1-methyl-6-phenylimidazo[4,5b]pyridine, and the antioxidant activity of grape-seed extract (expressed as Trolox-equivalents) was found. Sensory tests showed a high acceptance of flavor and color for controls and samples (Gibis and Weiss 2012). Inclusion of grape-seed extract in the diet of Merino Branco lambs protected the meat against lipid oxidation but did not influence the color or sensory characteristics of the meat. Although addition of grape-seed extract, grape seed flour, and grape-seed oil had some undesirable consequences on the sensory characteristics of frankfurters, all three of them demonstrated different positive effects in frankfurter production. Grape-seed extract was the best with regard to β-oxidation and overall acceptability (Ozvural and Vural 2014).

4.8.5 Grape-Seed Oil

Grape-seed oil is rich in unsaturated fatty acids with linoleic acid contributing to 72%–76%, w/w (Martinello et al. 2007; Da Porto et al. 2013; Fiori et al. 2014) and may contain significant amounts (63–1208 ppm) of tocopherols and tocotrienols (Crews et al. 2006; Fernández et al. 2010). The oil can be extracted by pressing, solvent extraction, or combination of both techniques. Mechanical pressing generate a low yield and therefore it is used only for economic reasons (small production or boutique products). Different solvents can be used for the extraction of oil from grape seed (Figure 4.7). Diethyl ether and hexane produced the highest oil yields (20.8% and 18.4%, respectively), whereas ethanol and methanol produced the lowest yields (11.2% and 12.9%, respectively) (Fernández et al. 2010). The oxidation stability of oil obtained from these solvents was 8.4, 3.8, 15.8, and 25 h for diethyl ether, hexane, ethanol, and methanol, respectively. Hexane is used for conventional oil extraction, but the use of CO_2 supercritical extraction (SC–CO_2) system (Fiori 2007) can be extremely useful to generate oil

Figure 4.7 Grape-seed oil extracted using hexane.

with high purity or rich in antioxidants. Supercritical extraction technology offers many advantages (see above) compared to hexane. The cost associated with the step-up of the technology, however, is prohibitive for small-scale operations. Fiori et al. (2014) investigated the lipid profile and bioactive concentration in oil extracted from six grape varieties using hexane extraction to SC–CO_2. The grape-seed oil yield was most comparable from both extraction methods (11.0–15.0 and 11.1–16.6 for SC–CO_2 and hexane, respectively), but significant seasonal effects were observed. The concentration ranges (mg/kg oil) of α-tocopherol, α-tocotrienol, γ-tocopherol, and γ-tocotrienol of the six grape seed varieties were 51–196, 81–170, 18–55, and 110–253 for SC–CO_2 and 27–114, 68–124, 11–62, and 52–224 for hexane-extracted oils, suggesting higher levels of tocols in SC–CO_2 oils compared with conventional hexane-extracted oil. The use of the CO_2 supercritical extraction system (Beveridge et al. 2005; Fiori 2007) results in higher tocol contents from GP and grape seed. It is worth noting that the concentrations of tocols in mechanically extracted oils were not different from those found in SC–CO_2 (Fiori et al. 2014). The concentrations of tocols vary widely depending on the extraction system and location. Crews et al. (2006) reported a wide range of tocol contents in hexane-extracted grape-seed oil from various European countries (63–1208 mg/kg). Fernández et al. (2010) found slight changes in the tocol content due to refining (total tocol was 529 and 555 mg/kg oil in refined and raw oils, respectively). The tocol contents in that study were 345–368, 145–147, and 39.7–40.1 mg/kg oil for α-tocotrienol, β-tocotrienol, and α-tocopherols, respectively. This profile for Spanish grape-seed oil is different from that reported by Fiori et al. (2014) for Italian grape-seed oil, confirming the impact of environmental effects and inter-variety variations. The use of ultrasound-assisted extraction in the extraction of oil from grape seeds can improve the efficiency of extraction greatly. Ultrasound treatment (20 kHz, 150 W for 30 min) of grape seed in hexane produced the same yield as that generated by soxhlet extraction for 6 h (Da Porto et al. 2013). Another method used to enhance the yield of oil is treatment with enzymes (cellulose, xylanase, and proteases) to release the oil from the seed structure. A long treatment time was very effective in enhancing the oil yield (Passos et al. 2009), whereas a short treatment time had limited success (Rosenthal et al. 1996). The use of methanol/ethanol in addition to hexane to extract oil produces oil that has high oxidation stability due to the extraction of phenolics or due to synergetic effects of the extracted polar and nonpolar compounds. Grape-seed oil is employed for culinary, pharmaceutical, and cosmetic applications as well as its potential use as a biofuel. The presence of squalene, an organic compound that consists of 30 carbon atoms, in red and white wine lees was confirmed (Naziri et al. 2014). Squalene is an all-trans linear triterpenoid hydrocarbon, which has important biological activities (Kamimura et al. 1992; Kohno et al. 1995; Newmark 1997; Smith 2000). The yeast strain used in wine making has the greatest impact on squalene formation. Squalene contents ranged from 2.43 to 5.90 g/kg dry lees in white wine lees and 0.54 to 1.54 g/kg dry lees in red wine lees.

4.8.6 Non-Food Applications

Grape seed alcoholic extracts have been shown to have molluscidal and insecticidal activities. They brought about mortality in the snail (*Biomphalaria alexandrina, Lymnea cailliaudi*) but did not exert ovicidal actions in snails or larvicidal effects on *Culex pipiens* (Taher et al. 2012).

Grape pomace (pulp and skin) could speedily and concurrently remove *in vitro* a number of mycotoxins from liquid media, with the biosorbent efficacy in the order aflatoxin B1 > zearalenone, ochratoxin A and fumonisin B1 >> deoxynivalenol (Avantaggiato et al. 2014).

174 VALORIZATION OF WINE MAKING BY-PRODUCTS

Several heavy metals can bind lignin and the fibers of wine by-products such as copper and nickel ions (Villaescusa et al. 2004). In cosmetics, a mixture of grape-seed oil, linseed oil, retinyl palmitate, tocopheryl acetate, and coenzyme Q10 encapsulated in nanoparticles produced a significant wrinkle-reducing effect after topical application for 3 weeks (Felippi et al. 2012). Grape-seed extract is promising for dermatological applications. Grape-seed oil has an abundance of the essential fatty acid linoleic acid, vitamins and minerals, which are beneficial to the skin. Grape oil has a wound-healing potential as revealed by a study using an excision wound rat model (Shivananda Nayak et al. 2011). Grape skins were used to make low-density (0.38 g/cm^3) boards (Mendes et al. 2013). The boards were fabricated by hot-pressing grape skins into a mold and using urea–formaldehyde (UF) as a binder with a loading level of 8% of dry weight of the board. The produced boards had a moderate bending strength compared to low-density particle boards from agricultural residues, but had good dimensional stability under humid conditions. The thermal conductivity of the boards (0.09 W/m/K) at 40°C was similar to commercial insulating materials, such as cork and polymeric foams, suggesting a potential use as insulating material. The authors estimated a potential of 1.25 m^3 of thermal insulation boards to be produced from 1 ton of absolutely dry grape skins (Mendes et al. 2013). GP extracts can be used in adhesive applications due to their reactivity toward formaldehyde, as demonstrated by Rondeau et al. (2013). The use of GP as an additive in brick production was investigated by Muñoz et al. (2014). The maximum amount of GP that can be used was 5%, after which the water absorption failed the industry standards. The addition at a 5% level reduced the thermal conductivity by 10% (better insulation). Grape stalks were used to generate activated carbon using phosphoric acid as activating agent (Deiana et al. 2009).

4.9 Fermentation

Grape and wine by-products represent a good source of carbon and have been used to generate a large number of high-value products such as citric acid, lactic acid, gluconic acid, xanthan, ethanol, methanol, and carotenoids through submerged and solid-state fermentation techniques. Karpe et al. (2014) used *Trichoderma harzianum*, *Aspergillus niger*, *Penicillium chrysogenum*, and *P. citrinum* to degrade

ASPECTS OF BY-PRODUCT UTILIZATION **175**

winery biomass. The use of *A. niger*, *P. chrysogenum*, *T. harzianum*, and *P. citrinum* at the ratio 60:14:4:2 was the optimum condition for the degradation of the biomass. Several valuable compounds were generated such as stigmasterol, glycerol, citric acid, maleic acid, and xylitol. Zepf and Jin (2013) reported an efficient process to increase the protein content of grape marc from 7% to up to 27% in 5 days using a solid-state fermentation process. Three fungal strains of *A. oryzae* DAR 3699, *A. oryzae* RIB 40, and *T. reesei* RUT C30 were used under optimum experimental conditions including inoculum size 1×10^5 CFU/g substrate, 5 days of fermentation time, 28°C–30°C fermentation temperature, moisture content in the range of 60%–66.7%, steam treatment for 60 min, $(NH_4)_2SO_4$ supplement at a concentration of 0.60%, and wine lees supplement at a concentration of 25%. The protein-rich product generated can be used as feedstock for animals.

Submerged fermentation of grape waste using *Monascus purpureus* produced a red pigment suitable for the food applications (Silveira et al. 2008). The use of *Lactococcus lactis* and *Lactobacillus pentosus* is useful to produce lactic acid and *Trametes pubescens* has been used to produce laccase.

4.10 Green Material

4.10.1 Grape Stalks

Prozil et al. (2012) investigated the composition of grape stalks from red grape varieties and found that the major components were cellulose (30.3%), hemicellulose (21.0%), lignin (17.4%), tannins (15.9%), and protein (6.1%). Xylose and glucose were the main monosaccharides in grape stalks (62.7% and 20.4%, respectively). Grape stalks are characterized by a low proportion of 4-O-methyl-α-D-glucuronosyl residues attached to xylan leading to strong binding of the latter to cellulose and resistance to degradation by rumen microorganisms and chemical treatments (Prozil et al. 2012). The stalk contains considerable amounts of condensed and hydrolysable tannins (Makris et al. 2006). The major tannins in grape stalks are procyanidins and prodelphinidins (Prozil et al. 2012). Stem extracts are threefold more powerful antioxidants compared with GP extracts. Vine shoots consist of lignocellulose material. Chemical and/or enzymatic hydrolysis can degrade the polymers to monomers that can be used for the production of lactic acid, which is an

176 VALORIZATION OF WINE MAKING BY-PRODUCTS

important compound in food and pharmaceutical applications. Lactic acid is used as a buffering agent, flavoring agent, or preserver to inhibit spoilage. The production of lactic acid can be achieved by fermentation using *Lactobacillus pentosus*. The same technology can be used to produce biosurfactants that have wide applications as emulsifiers.

4.10.2 Grape Leaves

Little attention has been paid to the utilization of grape leaves. Several medicinal and pharmacological properties of vine leaves were cited by Fernandes et al. (2013) including antinociceptive, antiviral, and antibacterial, antifungal, anti-inflammatory, spasmolytic, hypoglycemic, vasorelaxant, and hepatoprotective activities. Several bioactive compounds have been found in vine leaves such as phenolics, vitamins, carotenoids, and terpenes that may contribute to some of the observed activities. Fernandes et al. (2013) investigated the chemical composition and the antioxidant activity of grape leaves from 20 red and white Portuguese cultivars. Quercetin-3-O-galactoside and kaempferol-3-O-glucoside were the most predominant found in the aqueous extracts in addition to the availability of trans-caffeoyltartaric and trans-coumaroyltartaric acids, myricetin-3-O-glucoside, and quercetin-3-O-glucoside. The yield of the aqueous extracts varied from 8% to 18% depending on the grape cultivars. The total reducing capacity was between 174 and 573 mg GAE/g extract. These results demonstrated the potential use of the leaves to obtain useful bioactive compounds. Vine leaves are used as culinary ingredients for several meals. One of the most famous dishes in the Mediterranean, many East European, and Asian countries is stuffed grape leaves (known as Yaprak Dolma in Turkey, dolmeh barg mo in Iran, and warraq enab in Egypt and Lebanon). The leaves used for food have to be young and tender since mature leaves can have an extensive fibrous structure and a high level of bitterness and a sour taste due to phenolics and oxalates. Oleuopein and secoiridoids are found in the leaves and contribute significantly (about 51%) to the antioxidant activity of the leaf extracts (Naziri et al. 2014). About 5.1% w/w oleuropein can be extracted from the leaves and the efficiency of extraction can be enhanced by using $SC–CO_2$, microwave-assisted or ultrasound-assisted extraction techniques.

ASPECTS OF BY-PRODUCT UTILIZATION 177

4.11 Challenges and Opportunities

Despite the great prospective for the utilization of grape and wine by-products, there are several issues that might influence the viability of commercial use and full utilization. For example, the production of bioactives/secondary metabolites in grapes, and consequently the wine by-products, is dictated by environmental factors and agricultural practices. Therefore, by-products from certain sites can have higher levels of bioactives and can be advantageous for commercial utilization compared to others such as the case of tocols in grape-seed oil or the expression of phytoalxins for example. Processing and extraction conditions can play an important role in determining the contents of bioactives and their stability during storage, therefore optimized conditions need to be established for individual production regions. It is worth mentioning that the level of bioactivity and the purity of a bioactive will carry different processing costs, thus the recovery of bioactives from wine by-products should target defined industries (e.g., food versus pharmaceutical). The utilization of an integrated processing strategy should be considered for maximum benefit from by-products and to reduce production costs. Utilization of combination of treatments (e.g., mechanical, biochemical, and chemical) could enhance yields and maximize profitability, but this will vary depending on the by-product fraction (seed, stalks, GP, and so on).

The seasonality of wine processing and production of large amounts of by-products in a short time is a challenge for the wine industry. However, the diversity of products that can be generated from the available by-products can contribute to greater level of utilization. Biotransformation of wine by-products to high-value and high-nutrition products by anaerobic digestion and fermentation is very promising but require the establishment of cooperation-type businesses and radical integration of the traditional winery model into a multi agroeconomic model system.

Acknowledgments

The authors acknowledge funding received from the New Zealand Ministry for Environment (Community Environment Fund and Waste Minimisation Fund, Deed Number 20398), and the Sustainable Farm Fund (Project Number 09/099). This work is part of the

178 VALORIZATION OF WINE MAKING BY-PRODUCTS

New Zealand Grape and Wine Research programme, a joint investment by the Plant and Food Research and NZ Winegrowers.

References

Ahn, J.H., Grun, I.U., Fernando, L.N. (2002b). Antioxidant properties of natural plant extract containing polyphenolic compounds in cooked ground beef. *Journal of Food Science*, 67, 1364–1369.

Alipour, D., Rouzbehan, Y. (2007). Effects of ensiling grape pomace and addition of polyethylene glycol on *in vitro* gas production and microbial biomass yield. *Animal Feed Science and Technology*, 137, 138–149.

Alonso, A., Guillean, D., Barroso, C., Puertas, B., Garcia, A. (2002). Determination of antioxidant activity of wine byproducts and its correlation with polyphenolic content. *Journal of Agricultural Food and Chemistry*, 50, 5832–5836.

Amouroux, P., Jean, D., Lamaison, J.-L. (1998). Antiviral activity *in vitro* of *Cupressus sempervirens* on two human retroviruses HIV and HTLV. *Phytotherapy Research*, 12, 367–368.

Anastasiadi, M., Chorianopoulos, N.G., Nychas, G.-J.E., Haroutounian, S.A. (2009). Antilisterial activities of polyphenol-rich extracts of grapes and vinification byproducts. *Journal of Agricultural and Food Chemistry*, 57, 457–463.

Andrés, L., Riera, F., Alvarez, R. (1997). Recovery and concentration by electrodialysis of tartaric acid from fruit juice industries waste waters. *Journal of Chemical Technology and Biotechnology*, 70, 247–252.

Arts, I.C.W., Putte, B., Hollman, P.C.H. (2000). Catechin contents of foods commonly consumed in the Netherlands. 1. Fruits, vegetables, staple foods and processed foods. *Journal of Agricultural and Food Chemistry*, 48, 1746–1751.

Arvanitoyannis, I.S., Ladas, D., Mavromatis, A. (2006). Potential uses and applications of treated wine waste: A review. *International Journal of Food Science and Technology*, 41, 475–487.

Auger, C., Teissedre, P.L., Gerain, P., Lequeux, N., Bornet, A., Serisier, S., Besançon, P., Caporiccio, B., Cristol, J.P., Rouanet, J.M. (2005). Dietary wine phenolics catechin, quercetin and resveratrol efficiently protect hypercholesterolemic hamsters against aortic fatty streak accumulation. *Journal of Agricultural and Food Chemistry*, 53, 2015–2021.

Avantaggiato, G., Greco, D., Damascelli, A., Solfrizzo, M., Visconti, A. (2014). Assessment of multi-mycotoxin adsorption efficacy of grape pomace. *Journal of Agriculture and Food Chemistry*, 62(2), 497–507.

Ayed, N., Lu, H.L., Lacroix, M. (1999). Improvement of anthocyanin yield and shelf-life extension of grape pomace by gamma irradiation. *Food Research International*, 32, 539–543.

Bagchi, D., Garg, A., Krohn, R.L., Bagchi, M., Bagchi, D.J., Balmoori, J., Stohs, S.J. (1998). Protective effects of grape seed proanthocyanidins and selected antioxidants against TPA-induced hepatic and brain lipid peroxidation and DNA fragmentation and peritoneal macrophage activation in mice. *General Pharmacology*, 30, 771–776.

ASPECTS OF BY-PRODUCT UTILIZATION 179

Bahrami, Y., Foroozandeh, A.-D., Zamani, F., Modarresi, M., Eghbal-Saeid, S., Chekani-Azar, S. (2010). Effect of diet with varying levels of dried grape pomace on dry matter digestibility and growth performance of male lambs. *Journal of Animal & Plant Sciences*, 6, 605–610.

Balde, A.M., van Hoof, L., Pieters, L.A., Vanden Berghe, D.A., Vlietinck, A.J. (1990). Plant antiviral agents. VII. Antiviral and antibacterial proanthocyanidins from the bark of *Pavetta owariensis*. *Phytotherapy Research*, 4, 182–188.

Bartolomé, B., Nuñéz, V., Monagas, M., Gómez-Cordovés, C. (2004). *In vitro* antioxidant activity of red grape skins. *European Food Research Technology*, 218, 173–177.

Basalan, M., Gungor, T., Owens, F.N., Yalcinkaya, I. (2011). Nutrient content and *in vitro* digestibility of Turkish grape pomaces. *Animal Feed Science and Technology*, 169, 194–198.

Baumgartel, T., Kluth, H., Epperlein, K., Rodehutscord, M. (2007). A note on digestibility and energy value for sheep of different grape pomace. *Small Ruminants Research*, 67, 302–306.

Baydar, N.G., Özkan, G., Sağdiç, O. (2004). Total phenolic contents and antibacterial activities of grape (*Vitis vinifera* L.) extracts. *Food Control*, 15, 335–339.

Baydar, N.G., Özkan, G., Samim Yasar, S. (2007). Evaluation of the antiradical and antioxidant potential of grape extracts. *Food Control*, 18, 1131–1136.

Bazykina, N.I., Nikolaevskii, A.N., Filippenko, T.A., Kaloerova, V.G. (2002). Optimization of conditions for the extraction of natural antioxidants from raw plant materials. *Pharmaceutical Chemistry Journal*, 36, 100–103.

Beef Central. (2011). Wine waste-product cuts cattle emissions. http://www.beefcentral.com/lotfeeding/wine-waste-product-cuts-cattle-emissions/. Accessed on 11/11/2014.

Bekhit, A.E.D., Bekhit, A.A. (July 8–12, 2012). The potential use of wine lees in health applications. *Proceedings of the Third CIGR International Conference of Agricultural Engineering* (*CIGR-AgEng2012*), Valencia, Spain, Electronic edition.

Bekhit, A.E.D., Cheng, V.J., McConnell, M., Zhao, J.H., Sedcole, R., Harrison, R. (2011a). Antioxidant activities, sensory and anti-influenza activity of grape skin tea infusion. *Food Chemistry*, 129, 837–845.

Bekhit, A.E.D., Richardson, A., Grant, C. (2011b). Effect of wine lees on faecal egg count and lamb performance. *Proceedings of the New Zealand Society of Animal Production*, 71, 309–313.

Bertran, E., Sort, X., Soliva, M., Trillas, I. (2004). Composting winery waste: Sludges and grape stalks. *Bioresource Technology*, 95, 203–208.

Beveridge, T.H.J., Girard, B., Kopp, T., Drover, J.C.G. (2005). Yield and composition of grape-seed oils extracted by supercritical carbon dioxide and petroleum ether: Varietal effects. *Journal of Agricultural and Food Chemistry*, 53, 1799–1804.

Bhale, P.V., Nishikant, V., Deshpande, N.V., Thombre, S.B. (2009). Improving the low temperature properties of biodiesel fuel. *Renewable Energy*, 34, 794–800.

Bonilla, F., Mayen, M., Merida, J., Medina, M. (1999). Extraction of phenolic compounds from red grape marc for use as food lipid antioxidants. *Food Chemistry*, 66, 209–215.

180 VALORIZATION OF WINE MAKING BY-PRODUCTS

Borbalán, A.M.A., Zorro, L., Guillén, D.A., Barroso, C.G. (2003). Study of the polyphenol content of red and white grape varieties by liquid chromatography-mass spectrometry and its relationship to antioxidant power. *Journal of Chromatography A*, 1012, 31–38.

Bouhamidi, R., Prevost, V., Nouvelot, A. (1998). High protection by grape seed proanthocyanidins (GSPC) of polyunsaturated fatty acids against UV-C induced peroxidation. *Life Sciences*, 321, 31–38.

Boussetta, N., Lanoisellé, J.-L., Bedel-Cloutour, C., Vorobiev, E. (2009a). Extraction of soluble matter from grape pomace by high voltage electrical discharges for polyphenol recovery: Effect of sulphur dioxide and thermal treatments. *Journal of Food Engineering*, 95, 192–198.

Boussetta, N., Lebovka, N., Vorobiev, E., Adenier, H., Bedel-Cloutour, C., Lanoisellé, J.L. (2009b). Electrically assisted extraction of soluble matter from Chardonnay grape skins for polyphenol recovery. *Journal of Agricultural and Food Chemistry*, 57, 1491–1497.

Boussetta, N., Lesaint, O., Vorobiev, E. (2013a). A study of mechanisms involved during the extraction of polyphenols from grape seeds by pulsed electrical discharges. *Innovative Food Science and Emerging Technologies*, 19, 124–132.

Boussetta, N., Turka, M., De Taeye, C., Larondelle, Y., Lanoisellé, J.L., Vorobieva, E. (2013b). Effect of high voltage electrical discharges, heating and ethanol concentration on the extraction of total polyphenols and lignans from flaxseed cake. *Industrial Crops and Products*, 49, 690–696.

Boussetta, N., Vorobiev, E., Deloison, V., Pochez, F., Falcimaigne-Cordin, A., Lanoisellé, J.L. (2011). Valorization of grape pomace by the extraction of phenolic antioxidants: Application of high voltage electrical discharges. *Food Chemistry*, 128, 364–370.

Boussetta, N., Vorobiev, E., Le, L.H., Cordin-Falcimaigne, A., Lanoisellé, J.-L. (2012). Application of electrical treatments in alcoholic solvent for polyphenols extraction from grape seeds. *LWT—Food Science and Technology*, 46, 127–134.

Braga, F.G., Fernando, A., Silva, L., Alves, A. (2002). Recovery of winery by-products in the Douro Demarcated Region: Production of calcium tartrate and grape pigments. *American Journal of Enology and Viticulture*, 53(1), 41–45.

Bravo, L., Saura-Calixto, F. (1998). Characterization of dietary fiber and the *in vitro* indigestible fraction of grape pomace. *American Journal of Enology and Viticulture*, 49, 135–141.

Brown, J.C., Huang, G., Haley-Zitlin, V., Jiang, X. (2009). Antibacterial effects of grape extracts on *Helicobacter pylori*. *Applied and Environmental Microbiology*, 75, 848–852.

Bucić-Kojić, A., Planinić, M., Tomas, S., Bilić, M., Velić, D. (2007). Study of solid-liquid extraction kinetics of total polyphenols from grape seeds. *Journal of Food Engineering*, 81, 236–242.

Bustamante, M.A., Paredes, C., Morales, J., Mayoral, A.M., Moral, R. (2009). Study of the composting process of winery and distillery wastes using multivariate techniques. *Bioresource Technology*, 100, 4766–4772.

Bustamante, M.A., Pérez-Murcia, M.D., Paredes, C., Moral, R., Pérez-Espinosa, A., Moreno-Caselles, J. (2007). Short-term carbon and nitrogen mineralisation in soil amended with winery and distillery organic wastes. *Bioresource Technology*, 98, 3269–3277.

Bustos, G., de la Torre, N., Moldes, A.B., Cruz, J.M., Domínguez, J.M. (2007). Revalorization of hemicellulosic trimming vine shoots hydrolyzates trough continuous production of lactic acid and biosurfactants by *L. pentosus*. *Journal of Food Engineering*, 78, 405–412.

Cacace, J.E., Mazza, G. (2002). Extraction of anthocyanins and other phenolics from black currants with sulphured water. *Journal of Agricultural and Food Chemistry*, 50, 5939.

Cacace, J.E., Mazza, G. (2003). Optimization of extraction of anthocyanins from black currants with aqueous ethanol. *Journal of Food Science*, 68, 240–248.

Carmona, E., Moreno, M.T., Avilés, M., Ordovás, J. (2012). Use of grape marc compost as substrate for vegetable seedlings. *Scientia Horticulturae*, 137, 69–74.

Carson, R.G., Patel, K., Carlomusto, M., Bosko, C.A., Pillai, S., Santhanam, U., Weinkauf, R.L., Iwata, K., Palanker, L.R. (2001). Cosmetic compositions containing resveratrol. U.S. Patent 6,270,780.

Casazza, A.A., Aliakbarian, B., Mantegna, S., Cravotto, G., Perego, P. (2010). Extraction of phenolics from *Vitis vinifera* wastes using nonconventional techniques. *Journal of Food Engineering*, 100(1), 50–55.

Castillo, J., Benavente-García, O., Lorente, J., Alcaraz, M., Redondo, A., Ortuño, A., Del Rio, J.A. (2000). Antioxidant activity and radioprotective effects against chromosomal damage induced in vivo by X-rays of flavan-3-ols (procyanidins) from grape seeds (*Vitis vinifera*): Comparative study versus other phenolic and organic compounds. *Journal of Agricultural and Food Chemistry*, 48(5), 1738–1745.

Cavani, C., Maiani, A., Manfredini, M., Zarri, M.C. (1988). The use of dehulled grape seed meal in the fattening of rabbits. *Annales De Zootechnie*, 37, 1–12.

Chamorro, S., Viveros, A., Alvarez, I., Vega, E., Brenes, A. (2012). Changes in polyphenol and polysaccharide content of grape seed extract and grape pomace after enzymatic treatment. *Food Chemistry*, 133, 308–314.

Chan, C.-H., Yusoff, R., Ngoh, G.C., Kung, F.W.L. (2011). Microwave-assisted extractions of active ingredients from plants. *Journal of Chromatography A*, 1218(37), 6213–6225.

Charradi, K., Mahmoudi, M., Elkahoui, S., Limam, F., Aouani, E. (2013). Grape seed and skin extract mitigates heart and liver oxidative damage induced by a high-fat diet in the rat: Gender dependency. *Canadian Journal of Physiology and Pharmacology*, 91, 1076–1085.

Cheng, H.Y., Lin, T.C., Yang, C.M., Shieh, D.E., Lin, C.C. (2005). *In vitro* anti-HSV-2 activity and mechanism action of proanthocyanidins A-1 from Vaccinium vitis-idaea. *Journal of the Science of Food and Agriculture*, 85, 10–15.

Cheng, V.J., Bekhit, A.E.D., McConnell. M., Mros, S., Zhao, J.H. (2012). Effect of extraction solvent, waste fraction and grape variety on the antimicrobial and antioxidant activities of extracts from wine residue from cool climate. *Food Chemistry*, 134, 474–482.

182 VALORIZATION OF WINE MAKING BY-PRODUCTS

Cheng, V.J., Bekhit, A.E.D., Sedcole, R., Hamid, N. (2010). The impact of information of grape skin bioactive functionality information on the acceptability of tea infusions from wine by-products. *Journal of Food Science*, 75, S167–S172.

Choi, Y.S., Choi, J.H., Han, D.J., Kim, H.Y., Lee, M.A., Jeong, J.Y., Chung, H.J., Kim, C.J. (2010). Effects of replacing pork back fat with vegetable oils and rice bran fiber on the quality of reduced-fat frankfurters. *Meat Science*, 84, 557–563.

Corrales, M., Toepfl, S., Butz, P., Knorr, D., Tauscher, B. (2008). Extraction of anthocyanins from grape by-products assisted by ultrasonics, high hydrostatic pressure or pulsed electric fields: A comparison. *Innovative Food Science and Emerging Technologies*, 9, 85–91.

Cortés, S., Salgado, J.M., Rodríguez, N., Domínguez, J.M. (2010). The storage of grape marc: Limiting factor in the quality of the distillate. *Food Control*, 21, 1545–1549.

Cos, P., De Bruyne, T., Hermans, N., Apers, S., Vanden Berghe, D., Vlietinck, A.J. (2004). Proanthocyanidins in health care: Current and new trends. *Current Medicinal Chemistry*, 10, 1345–1359.

Crews, C., Hough, P., Godward, J., Brereton, P., Lees, M., Guiet, S., Winkelmann, W. (2006). Quantitation of the main constituents of some authentic grape seed oils of different origin. *Journal of Agricultural and Food Chemistry*, 54, 6261–6265.

Da Porto, C. (1998). Grappa and grape-spirit production. *Critical Reviews in Biotechnology*, 18, 13–24.

Da Porto, C., Freschet, G. (2003). Study of some volatile compounds of raw grappa obtained using marc treated in different ways and stored in plastic tunnels. *Journal of Wine Research*, 14, 139–146.

Da Porto, C., Porretto, E., Deborha Decorti, D. (2013). Comparison of ultrasound-assisted extraction with conventional extraction methods of oil and polyphenols from grape (*Vitis vinifera* L.) seeds. *Ultrasonics Sonochemistry*, 20, 1076–1080.

Dairy, N.Z. (2014). Feed values. http://www.dairynz.co.nz/feed/supplements/feed-values/. Accessed on 28/11/2014.

Deiana, A.C., Sardella, M.F., Silva, H., Amaya, A., Tancredi, N. (2009). Use of grape stalk, a waste of the viticulture industry, to obtain activated carbon. *Journal of Hazardous Materials*, 172, 13–19.

de la Lastra, C.A., Villegas, I. (2005). Review: Resveratrol as an anti-inflammatory and anti-aging agent: Mechanisms and clinical implications. *Molecular Nutrition and Food Research*, 49, 405–430.

Del Castilho, P., Chardon, W., Salomons, W. (1993). Influence of cattle-manure slurry application on the solubility of Cd, Cu, and Zn in a manured acidic, loamy sand soil. *Journal of Environmental Quality*, 22, 689–697.

Deng, Q., Penner, M.H., Zhao, Y. (2011). Chemical composition of dietary fiber and polyphenols of five different varieties of wine grape pomace skins. *Food Research International*, 44, 2712–2720.

Devesa-Rey, R., Vecino, X., Varela-Alende, J.L., Barral, M.T., Cruz, J.M., Moldes, A.B. (2011). Valorization of winery waste vs. the costs of not recycling. *Waste Management*, 31, 2327–2335.

ASPECTS OF BY-PRODUCT UTILIZATION 183

Di Blasi, C.D., Signorelli, G., Russo, C.D., Rea, G. (1999). Product distribution from pyrolysis of wood and agricultural residues. *Industrial and Engineering Chemical Research*, 38, 2216–2224.

Diéguez, S.C., De La Peña, M.L.G., Gómez, E.F. (2001). Concentration of volatiles in marc distillates from Galicia according to storage conditions of the grape pomace. *Chromatographia*, 53, S-406–S-411.

Ding, Y., Dai, X., Jiang, Y., Zhang, Z., Li, Y. (2014). Functional and morphological effects of grape seed proanthocyanidins on peripheral neuropathy in rats with type 2 diabetes mellitus. *Phytotheraphy Research*, 28(7), 1082–1087.

Docherty, J.J., Fu, M.M.H., Stiffler, B.S., Limperos, R.J., Pokabla, C.M., DeLucia, A.L. (1999). Resveratrol inhibition of herpes simplex virus replication. *Antiviral Research*, 43, 243–244.

Droebner, K., Ehrhardt, C., Poetter, A., Ludwig, S., Planz, O. (2007). CYCTUS052, a polyphenol-rich plant extract, exerts anti-influenza virus activity in mice. *Antiviral Research*, 76, 1–10.

Durling, N.E., Catchpole, O.J., Grey, J.B., Webby, R.F., Mitchell, K.A., Foo, L.Y. (2007). Extraction of phenolics and essential oil from dried sage (*Salvia officinalis*) using ethanol–water mixtures. *Food Chemistry*, 101(4), 1417–1424.

EPA (2014). Completed prosecution and civil penalties. available in http://www.epa.sa.gov.au/what_we_do/public_register_directory/completed_prosecutions_and_civil_penalties. Accessed on December 10, 2014.

Eleonora, N., Dobrei, A., Alina, D., Bampidis, V., Valeria, C. (2014). Grape pomace in sheep and dairy cows feeding. *Journal of Horticulture, Forestry and Biotechnology*, 18(2), 146–150.

Famuyiwa, O., Ough, C.S. (1982). Grape pomace: Possibilities as animal feed. *American Journal of Enology and Viticulture*, 33, 44–46.

Faria, A., Calhau, C., De Freitas, V., Mateus, N. (2006). Procyanidins as antioxidants and tumor cell growth modulators. *Journal of Agricultural and Food Chemistry*, 54, 2392–2397.

Farinella, N.V., Matos, G.D., Arruda, M.A.Z. (2007). Grape bagasse as a potential biosorbent of metals in effluent treatments. *Bioresource Technology*, 98, 1940–1946.

Felippi, C.C., Oliveira, D., Ströher, A., Carvalho, A.R., Van Etten, E.A., Bruschi, M., Raffin, R.P. (2012). Safety and efficacy of antioxidants-loaded nanoparticles for an anti-aging application. *Journal of Biomedecial and Nanotechnology*, 8, 316–321.

Fernandes, F., Ramalhosa, E., Pires, P., Verdial, J., Valentao, P., Andrade, P., Bento, A., Pereira, J.A. (2013). *Vitis vinifera* leaves towards bioactivity. *Industrial Crops and Products*, 43, 434–440.

Fernández, C.M., Ramos, M.J., Pérez, Á., Rodríguez, J.F. (2010). Production of biodiesel from winery waste: Extraction, refining and transesterification of grape seed oil. *Bioresource Technology*, 101, 7019–7024.

Fernández-Iglesias, A., Pajuelo, D., Quesada, H., Díaz, S., Bladé, C., Arola, L., Salvadó, M.J., Mulero, M. (2014). Grape seed proanthocyanidin extract improves the hepatic glutathione metabolism in obese Zucker rats. *Molecular Nutrition and Food Research*, 58(4), 727–737.

184 VALORIZATION OF WINE MAKING BY-PRODUCTS

Fernandez-Panchon, M.S., Villano, D., Troncoso, A.M., Garcia-Parrilla, M.C. (2008). Antioxidant activity of phenolic compounds: From *in vitro* results to *in vivo* evidence. *Critical Reviews of Food Science and Nutrition*, 48, 649–671.

Ferrer, J., Paez, G., Marmol, Z., Ramones, E., Chandler, C., Marn, M., Ferrer, A. (2001). Agronomic use of biotechnologically processed grape wastes. *Bioresource Technology*, 76, 39–44.

Filip, V., Plocková, M., Šmidrkal, J., Špičková, Z., Melzoch, K., Schmidt, S. (2003). Resveratrol and its antioxidant and antimicrobial effectiveness. *Food Chemistry*, 83, 585–593.

Fiori, L. (2007). Grape seed oil supercritical extraction kinetic and solubility data: Critical approach and modelling. *Journal of Supercritical Fluid*, 43, 43–54.

Fiori, L., Florio, L. (2010). Gasification and combustion of grape marc: Comparison among different scenarios. *Waste Biomass Valorization*, 1, 191 200.

Fiori, L., Lavelli, V., Duba, K.S., Sri Harsha, P.S.C., Mohamed, H.B., Guellad, G. (2014). Supercritical CO_2 extraction of oil from seeds of six grape cultivars: Modelling of mass transfer kinetics and evaluation of lipid profiles and tocol contents. *Journal of Supercritical Fluids*, 94, 71–80.

Flavel, T.C., Murphy, D.V., Lalor, B.M., Fillery, I.R.P. (2005). Gross N mineralization rates after application of composted grape marc to soil. *Soil Biology and Biochemistry*, 37, 1397–1400.

Forbes, S.L., Cohen, D.A., Cullen, R., Wratten, S.D., Fountain, J. (2009). Consumer attitudes regarding environmentally sustainable wine: An exploratory study of the New Zealand market place. *Journal of Cleaner Production*, 17, 1195–1199.

Freitas, V.A.P., Glories, Y., Monique, A. (2000). Developmental changes of procyanidins in grapes of red *Vitis vinifera* varieties and their composition in respective wines. *American Journal of Enology and Viticulture*, 51, 397–403.

Frémont, L. (2000). Biological effects of resveratrol. *Life Sciences*, 66, 663–673.

Furiga, A., Lonvaud-Funel, A., Badet, C. (2009). *In vitro* study of antioxidant capacity and antibacterial activity on oral anaerobes of a grape seed extract. *Food Chemistry*, 113, 1037–1040.

Fukuchi, K., Sagagami, H., Okuda, T., Hatano, T., Tanuma, S., Kitajima, K., Inoue, Y., Ichikawa, S., Nonoyama, M., Kono, K. (1989). Inhibition of herpes simplex infection by tannins and related compounds. *Antiviral Research*, 11, 285–298.

Galletto, L., Rossetto, L. (2005). The Market of grappa in LSR: An analysis of scanner data. In: *Food, Agriculture and the Environment: Economic Issues*, Defrancesco, E., Galletto, L., Thiene, M. (Eds.), FrancoAngli s.r.l., Milano, Italy, p. 150.

Gañan, M., Martínez-Rogríguez, A.J., Carrascosa, A.V. (2009). Antimicrobial activity of phenolic compounds of wine against *Campylobacter jejuni*. *Food Control*, 20, 739–742.

Geetha, T., Malhotra, V., Chopra, K., Kaur, I.P. (2005). Antimutagenic and antioxidant/prooxidant activity of quercetin. *Indian Journal of Experimental Biology*, 43, 61–67.

ASPECTS OF BY-PRODUCT UTILIZATION 185

Gerogiannaki-Chrisopoulou, M., Kyriakidis, N.V., Panagiotis, E., Athanasopoulos, M. (2004). Effect of grape variety (*Vitis vinifera* L.) and grape pomace fermentation conditions on some volatile compounds of the produced grape pomace distillate. *Journal Internationale des Sciences de la Vigne et du Vin*, 38, 225–230.

Ghafoor, K., Choi, Y.H., Jeon, J.Y., Jo, I.H. (2009). Optimization of ultrasound-assisted extraction of phenolic compounds, antioxidants, and anthocyanins from grape (*Vitis vinifera*) seeds. *Journal of Agricultural and Food Chemistry*, 57(11), 4988–4994.

Ghiselli, A., Nardini, M., Baldi, A., Scaccini, C. (1998). Antioxidant activity of different phenolic fractions from an Italian red wine. *Journal of Agriculture and Food Chemistry*, 46, 361–367.

Gibis, M., Weiss, J. (2012). Antioxidant capacity and inhibitory effect of grape seed and rosemary extract in marinades on the formation of heterocyclic amines in fried beef patties. *Food Chemistry*, 134, 766–774.

Gil, M.V., Oulego, P., Casal, M.D., Pevida, C., Pis, J.J., Rubiera, F. (2010). Mechanical durability and combustion characteristics of pellets from biomass blends. *Bioresource Technology*, 101, 8859–8867.

Girdhar, N., Satyanarayana, A. (2000). Grape waste as a source of tartrates. *Indian Food Packer*, 54, 59–61.

González-Paramás, A., Esteban-Ruano, S., Santos-Buelga, C., Pascual-Teresa, S., Rivas-Gonzalo, J. (2004). Flavanol content and antioxidant activity in winery by-products. *Journal of Agricultural and Food Chemistry*, 52, 234–238.

Green, R.H. (1949). Inhibition of multiplication of influenza virus by extracts of tea. *Proceedings of the Society for Experimental Biology and Medicine*, 72, 84–85.

Guendez, R., Kallithraka, S., Makris, D.P., Kefalas, P. (2005). Determination of low molecular weight polyphenolic constituents in grape (*Vitis vinifera* sp.) seed extracts: Correlation with antiradical activity. *Food Chemistry*, 89, 1–9.

Hamilton, R.J., Kalu, C., Prisk, E., Padley, F.B., Pierce, H. (1997). Chemistry of free radicals in lipids. *Food Chemistry*, 60, 193–199.

Hang, Y.D., Edward, E., Woodams, E.E. (2008). Methanol content of grappa made from New York grape pomace. *Bioresource Technology*, 99, 3923–3925.

Hang, Y.D., Lee, C.Y., Woodams, E.E. (1986). Solid-state fermentation of grape pomace for ethanol production. *Biotechnology Letters*, 8, 53–56.

Henry, F., Pauly, G., Moser, P. (2001). Extracts from residues left in the production of wine and usage in cosmetic and pharmaceutical compositions. International Patent Application WO 200158412 A2, p. 28.

Heredia, A., Davis, C., Redfield, R. (2000). Synergistic inhibition of HIV-1 in activated and resting peripheral blood mononuclear cells, monocyte-derived macrophages, and selected drug-resistant isolates with nucleoside analogues combined with a natural product, resveratrol. *Journal of Acquired Immune Deficiency Syndromes*, 25, 246–255.

Hocman, G. (1989). Prevention of cancer: Vegetables and plants. *Comparative Biochemistry and Physiology*, 93B, 201–212.

Hudson, J.B. (1990). *Antiviral Compounds from Plants*. CRC Press, Boca Raton, FL.

Hwang, J.-Y., Shyu, Y.-S., Hsu, C.-K. (2009). Grape wine lees improves the rheological and adds antioxidant properties to ice cream. *LWT—Food Science and Technology*, 42, 312–318.

Igartuburu, J.M., Pando, E., Rodriguez Luis, F., Gil-Serrano, A. (1997). An acidic xyloglucan from grape skins. *Phytochemistry*, 46, 1307–1312.

Irandoost, P., Ebrahimi-Mameghani, M., Pirouzpanah, S. (2013). Does grape seed oil improve inflammation and insulin resistance in overweight or obese women? *International Journal of Food Science and Nutrition*, 64, 706–710.

Isemura, M., Saeki, K., Kimura, T., Hayakawa, S., Minami, T., Sazuka, M. (2000). Tea catechins and related polyphenols as anti-cancer agents. *Biofactors*, 13, 81–85.

Israilides, C., Smith, A., Harthill, J., Barnett, C., Bambalov, G., Scanlon, B. (1998). Pullulan content of the ethanol precipitate from fermented agro-industrial wastes. *Applied Microbiology and Biotechnology*, 49, 613–617.

Iwasawa, A., Niwano, Y., Mokudai, T., Kohno, M. (2009). Antiviral activity of proanthocyanidin against feline calicivirus used as a surrogate for noroviruses, and coxsackievirus used as a representative enteric virus. *Biocontrol Science*, 14, 107–111.

Jayaprakasha, G.K., Selvi, T., Sakariah, K.K. (2003). Antibacterial and antioxidant activities of grape (*Vitis vinifera*) extracts. *Food Research International*, 36, 117–122.

Jayaprakasha, G.K., Singh, R.P., Sakariah, K.K. (2001). Antioxidant activities of grape seed (*Vitis vinifera*) extracts on peroxidation models *in vitro*. *Food Chemistry*, 73, 285–290.

Jhun, J.Y., Moon, S.J., Yoon, B.Y., Byun, J.K., Kim, E.K., Yang, E.J., Ju, J.H. et al. (2013). Grape seed proanthocyanidin extract-mediated regulation of STAT3 proteins contributes to Treg differentiation and attenuates inflammation in a murine model of obesity-associated arthritis. *PLoS One*, 8(11), e78843.

Joe, A.K., Liu, H., Suzui, M., Vural, M.E., Xiao, D., Weinstein, I.B. (2002). Resveratrol induces growth inhibition, S-phase arrest, apoptosis, and changes in biomarker expression in several human cancer cell lines. *Clinical Cancer Research*, 8, 893–903.

Ju, Z.Y., Howard, L.R. (2003). Effects of solvent and temperature on pressurized liquid extraction of anthocyanins and total phenolics from dried red grape skin. *Journal of Agricultural and Food Chemistry*, 51, 5207–5213.

Kallithraka, S., Garcia-Viguera, C., Bridle, P., Bakker, J. (1995). Survey of solvents for the extraction of grape seed phenolics. *Phytochemical Analysis*, 6, 265–267.

Kamei, H., Hashimoto, Y., Koide, T., Kojima, T., Hasegawa, M. (1998). Anti-tumor effect of methanol extracts from red and white wines. *Cancer Biotherapy and Radiopharmacology*, 13, 447–452.

Kamimura, H., Koga, N., Oguri, K., Yoshimura, H. (1992). Enhanced elimination of theophylline, phenobarbital and strychnine from the bodies of rats and mice by squalane treatment. *Journal of Pharmacobio-Dynamics*, 15, 215–221.

Kammerer, D., Claus, A., Carle, R., Schieber, A. (2004). Polyphenol screening of pomace from red and white grape varieties (*Vitis vinifera* L.) by HPLC-DAD-MS/MS. *Journal of Agricultural and Food Chemistry*, 52, 4360–4367.

Kammerer, D., Claus, A., Schieber, A., Carle, R. (2005). A novel process for the recovery of polyphenols from grape (*Vitis vinifera* L.) pomace. *Journal of Food Science*, 70, 157–163.

Karaka, A. (2004). Effect of organic wastes of the extractability of cadmium, copper, nickel and zinc in soil. *Geoderma*, 122, 297–303.

Karpe, A.V., Beale, D.J., Harding, I.H., Enzo, A., Palombo, E.A. (2014). Optimization of degradation of winery-derived biomass waste by Ascomycetes. *Journal of Chemical Technology and Biotechnology*, http://onlinelibrary.wiley.com/doi/10.1002/jctb.4486/epdf. DOI: 10.1002/jctb.4486.

Katalinić, V., Možina, S.S., Skroza, D., Generalić, I., Abramović, H., Miloš, M., Ljubenkov, I. et al. (2010). Polyphenolic profile, antioxidant properties and antimicrobial activity of grape skin extracts of 14 *Vitis vinifera* varieties grown in Dalmatia (Croatia). *Food Chemistry*, 119, 715–723.

Khanna, S., Venojarvi, M., Roy, S., Sharma, N., Trikha, P., Bagchi, D. (2002). Dermal wound healing properties of redox-active grape seed proanthocyanidins. *Free Radical Biology and Medicine*, 33, 1089–1096.

Kofujita, H., Ettyu, K., Ota, M. (1999). Characterization of the major components in bark from five Japanese tree species for chemical utilization. *Wood Science and Technology*, 33, 223–228.

Koga, T., Moro, K., Nakamori, K., Yamakoshi, J., Hosoyama, H., Kataoka, S., Ariga, T. (1999). Increase of antioxidative potential of rat plasma by oral administration of proanthocyanidin-rich extract from grape seeds. *Journal of Agricultural and Food Chemistry*, 47, 1892–1897.

Kohno, Y., Egawa, Y., Itoh, S., Nagaoka, S., Takahashi, M., Mukai, K. (1995). Kinetic study of quenching reaction of singlet oxygen and scavenging reaction of free radical by squalene in n-butanol. *Biochimica et Biophysica Acta*, 1256, 52–56.

Konowalchuk, J., Speirs, J.I. (1976). Virus inactivation by grapes and wines. *Applied and Environmental Microbiology*, 36, 798–801.

Konowalchuk, J., Speirs, J.I. (1978). Antiviral effect of commercial juices and beverages. *Applied and Environmental Microbiology*, 35, 1219–1220.

Kuwajerwala, N., Cifuentes, E., Gautam, S., Menon, M., Barrack, E.R., Reddy, G.P.V. (2002). Resveratrol induces prostate cancer cell entry into S phase and inhibits DNA synthesis. *Cancer Research*, 62, 2488–2492.

Lafka, T.-I., Sinanoglou, V., Lazos, E.S. (2007). On the extraction and antioxidant activity of phenolic compounds from winery wastes. *Food Chemistry*, 104, 1206–1214.

Lapornik, B., Prošek, M., Wondra, A.G. (2005). Comparison of extracts prepared from plant by-products using different solvents and extraction time. *Journal of Food Engineering*, 71, 214–222.

Larrauri, J.A., Ruperez, P., Saura-Calixto, F. (1996). Antioxidant activity of wine pomace. *American Journal of Enology and Viticulture*, 47, 369–372.

Larrauri, J.A., Ruperez, P., Saura-Calixto, F. (1997). Effect of drying temperature on the antioxidant activity of red grape pomace peel. *Journal of Agriculture and Food Chemistry*, 45, 1390–1393.

Lau, D.W., King, A.J. (2003). Pre- and post-mortem use of grape seed extract in dark poultry meat to inhibit development of thiobarbituric acid reactive substances. *Journal of Agricultural and Food Chemistry*, 51, 1602–1607.

Laufenberg, G., Kunz, B., Nystroem, M. (2003). Transformation of vegetable waste into value added products: (A) the upgrading concept; (B) practical implementations. *Bioresource Technology*, 87, 167–198.

Law, I. (2010). Production of biobutanol from white grape pomace by *Clostridium saccharobutylicum* using submerged fermentation. MSc thesis, Auckland University of Technology, Auckland, New Zealand.

Lazze, M.C., Savio, M., Pizzala, R., Cazzalini, O., Perucca, P., Scovassi, A.I., Stivala, L.A., Bianchi, L. (2004). Anthocyanins induce cell cycle perturbations and apoptosis in different human cell lines. *Carcinogenesis*, 25, 1427–1433.

LeDuy, A., Boa, J.M. (1983). Pullulan production from peat hydrolyzate. *Canadian Journal of Microbiology*, 29, 143–146.

Lekha, P.K., Lonsane, B.K. (1997). Production and application of tannin acyl hydrolase: State of the art. *Advances in Applied Microbiology*, 44, 215–260.

Lempereur, V., Penavayre, S. (2014). Grape marc, wine lees and deposit of the must: How to manage oenological by-products? *BIO Web of Conferences*, 3, 01011.

Leonard, S.S., Xia, C., Jiang, B.-H., Stinefelt, B., Klandorf, H., Harris, G.K., Shi, X. (2003). Resveratrol scavenges reactive oxygen species and effects radical-induced cellular responses. *Biochemical and Biophysical Research Communications*, 309, 1017–1026.

Li, Y., Skouroumounis, G.K., Elsey, G.M., Taylor, D.K. (2011). Microwave-assistance provides very rapid and efficient extraction of grape seed polyphenols. *Food Chemistry*, 129(2), 570–576.

Liu, A.L., Liu, B., Qin, H.L., Lee, S.M., Wang, Y.T., Du, G.H. (2008). Anti-influenza virus activities of flavonoids from the medicinal plant *Elsholtzia rugulosa*. *Planta Medica*, 74, 847–851.

Liviero, L., Puglisi, P.P., Morazzoni, P., Bombardelli, E. (1994). Antimutagenic activity of procyanidins from *Vitis vinifera*. *Fitoterapia*, 65, 203–209.

Llobera, A., Canéllas, J. (2007). Dietary fibre content and antioxidant activity of Manto Negro red grape (*Vitis vinifera*): Pomace and stem. *Food Chemistry*, 101, 659–666.

Lo, K.V., Liao, E.H. (1986). Methane production from fermentation of winery waste. *Biomass*, 9, 19–27.

Lo Curto, R.B., Tripodo, M.M. (2001). Yeast production from virgin grape marc. *Bioresource Technology*, 78, 5–9.

Louli, V., Ragoussis, N., Magoulas, K. (2004). Recovery of phenolic antioxidants from wine industry by-products. *Bioresource Technology*, 2, 201–208.

Mabe, K., Yamada, M., Oguni, I., Takahashi, T. (1999). *In vitro* and in vivo activities of tea catechins against *Helicobacter pylori*. *Antimicrobial Agents and Chemotherapy*, 43, 1788–1791.

ASPECTS OF BY-PRODUCT UTILIZATION 189

Maillard M.-N., Berset C. (1995). Evolution of antioxidant activity during kilning: Role of insoluble bound phenolic acids of barley and malt. *Journal of Agricultural and Food Chemistry*, 43, 1789–1793.

Makris, D.P., Kallithraka, S., Kefalas, P. (2006). Flavonols in grapes, grape products and wines: Burden, profile and influential parameters. *Journal of Food Composition and Analysis*, 19, 396–404.

Manios, T. (2004). The composting potential of different organic solid wastes: Experience from the island of Crete. *Environment International*, 29, 1079–1089.

Mantena, S.K., Baliga, M.S., Katiyar, S.K. (2006). Grape seed proanthocyanidins induce apoptosis and inhibit metastasis of highly metastatic breast carcinoma cells. *Carcinogenesis*, 27, 1682–1691.

Martin-Carron, N., Garcia-Alonso, A., Goni, I., Saura-Calixto, F. (1997). Nutritional and physiological properties of grape pomace as a potential food ingredient. *American Journal of Enology and Viticulture*, 48, 328–332.

Martinello, M., Hecker, G., Pramparo, M.C. (2007). Grape seed oil deacidification by molecular distillation: Analysis of operative variables influence using the response surface methodology. *Journal of Food Engineering*, 81, 60–64.

Masquelier, J. (1959). The bactericidal action of certain phenolics of grapes and wine. In: *The Pharmacology of Plant Phenolics*, Fairbairn, J.W. (Ed.), Academic Press, New York, pp. 123–131.

Mateus, N., Marques, S., Goncalvez, A.C., Machado, J.M., Freitas, V. (2001). Proanthocyanidin composition of red *Vitis vinifera* varieties from the douro valley during ripening: Influence of cultivation altitude. *American Journal of Enology and Vitculture*, 52, 115–121.

Matias, A.A., Serra, A.T., Silva, A.C., Perdigão, R., Ferrerira, T.B., Marcelino, I., Silva, S., Coelho, A.V., Alves, P.M., Duarte, C.M.M. (2010). Portuguese winemaking residues as a potential source of natural anti-adenoviral agents. *International Journal of Food Sciences and Nutrition*, 61, 357–368.

Mendes, J.A.S., Prozil, S.O., Evtuguin, D.V., Lopes, L.P.C. (2013). Towards comprehensive utilization of winemaking residues: Characterization of grape skins from red grape pomaces of variety Touriga Nacional. *Industrial Crops and Products*, 43, 25–32.

Mendes, J.A.S., Xavier, A.M.R.B., Evtuguin, D.V., Lopes, L.P.C. (2013). Integrated utilization of grape skins from white grape pomaces. *Industrial Crops and Products*, 49, 286–291.

Mendoza, L., Yañez, K., Vivanco, M., Melo, R., Cotoras, M. (2013). Characterization of extracts from winery by-products with antifungal activity against *Botrytis cinerea*. *Industrial Crops and Products*, 43, 360–364.

Meyer, A.S., Jepsen, S.M., Sorencsen, N.S. (1998). Enzymatic release of antioxidants for human low-density lipoprotein from grape pomace. *Journal of Agricultural and Food Chemistry*, 46, 2439–2446.

Mirzaei-Aghsaghali, A., Maheri-Sis, N. (2008). Nutritive value of some agro-industrial by-products for euminants: A review. *World Journal of Zoology*, 3, 40–46.

Moldes, D., Gallego, P., Rodriguez-Couto, S., Sanroman, A. (2003). Grape seeds: The best lignocellulosic waste to produce laccase by solid state cultures of *Trametes hirsuta*. *Biotechnology Letters*, 25, 491–495.

Molero, A., Pereyra, C., Martínez, E. (1995a). Caracterización del aceite de semilla de uva extraído con dióxido de carbono supercrítico (Characterisation of grape seed oil extracted with supercritic carbon dioxide) *Grasas y Aceites*, 46, 29–34.

Molero, A., Pereyra, C., Martínez, E. (1995b). Optimización del proceso de extracción del aceite de semilla de uva con dióxido de carbono líquido y supercrítico (Optimisation of the extraction process of grape seed oil with liquid and supercritical carbon dioxide). *Alimentación, Equipos y Tecnología*, 3, 35–40.

Molina-Alcaide, E., Moumen, A., Martín-García, A.I. (2008). By-products from viticulture and the wine industry: Potential as sources of nutrients for ruminants. *Journal of the Science of Food and Agriculture*, 8, 597–604.

Monagas, M., Garrido, I., Bartolome, B., Gomez-Cordoves, C. (2006). Chemical characterization of commercial dietary ingredients from *Vitis vinifera* L. *Analytica Chimica Acta*, 463, 401–410.

Moote, P.E., Church, J.S., Schwartzkopf-Genswein, K.S., Van Hamme, J.D. (2014). Effect of fermented winery by-product supplemented rations on the temperament and meat quality of Angus-Hereford X steers during feeding in a British Columbia feedlot. *Journal of Food Research*, 3, 124–135.

Mori, A., Nishino, C., Enoki, N., Tawata, S. (1987). Antibacterial activity and mode of action of plant flavonoids against *Proteus vulgaris* and *Staphylococus aureus*. *Phytochemistry*, 26, 2231–2234.

Mori, S., Miyake, S., Kobe, T., Nakaya, T., Fuller, S.D., Kato, N., Kaihatsu, K. (2008). Enhanced anti-influenza A virus activity of (-)-epigallotcatechin-3-O-gallate fatty acid monoester derivatives: Effect of alkyl chain length. *Bioorganic and Medicinal Chemistry Letters*, 18, 4249–4252.

Moure, A., Cruz, J.M., Franco, D., Dominguez, J.M., Sineiro, J., Dominguez, H., Nunez, M.J., Parajo, J.C. (2001). Natural antioxidants from residual sources. *Food Chemistry*, 72, 145–171.

Muñoz, P., Morales, M.P., Mendívil, M.A., Juárez, M.C., Muñoz, L. (2014). Using of waste pomace from winery industry to improve thermal insulation of fired clay bricks. Eco-friendly way of building construction. *Construction and Building Materials*, 71, 181–187.

Mullen, W., McGinn, J., Lean, M.E.J., MacLean, M.R., Gardner, P., Duthie, G.G., Yokota, T., Crozier, A. (2002). Ellagitannins, flavonoids, and other phenolics in red raspberries and their contribution to antioxidant capacity and vasorelaxation properties. *Journal of Agricultural and Food Chemistry*, 50, 5191–5196.

Murga, R., Ruiz, R., Beltran, S., Cabezas, J.L. (2000). Extraction of natural complex phenols and tannins from grape seeds by using supercritical mixtures of carbon dioxide and alcohol. *Journal of Agricultural and Food Chemistry*, 48(8), 3408–3412.

Murthy, K.N.C., Singh, R.P., Jayaprakasha, G.K. (2002). Antioxidant activity of grape (*Vitis vinifera*) pomace extracts. *Journal of Agricultural and Food Chemistry*, 50, 5909–5914.

Nakashima, H., Murakami, T., Yamamoto, N., Sakagami, H., Tanuma, S., Hatano, T., Yoshida, T., Okuda, T. (1992). Inhibition of human immunodeficiency viral replication by tannins and related compounds. *Antiviral Research*, 18, 91–103.

Nakayama, M., Suzuki, K., Toda, M., Okubo, S., Hara, Y., Shimamura, T. (1993). Inhibition of the infectivity of influenza virus by tea polyphenols. *Antiviral Research*, 21, 289–299.

Nakayama, M., Toda, M., Okubo, S., Shimamura, T. (1990). Inhibition of influenza virus infection by tea. *Letters in Applied Microbiology*, 11, 38–40.

Nasis-Moragher, K., Svanoe, T.T., Seroy, W.A. (1999). Method for extraction of proanthocyanidins from plant materials. U.S. Patent 5,912,363.

Natale, D., Gibis, M., Rodriguez-Estrada, M.T., Weiss, J. (2014). Inhibitory effect of liposomal solutions of grape seed extract on the formation of heterocyclic aromatic amines. *Journal of Agriculture and Food Chemistry*, 62(1), 279–287.

Nair, N., Mahajan, S., Chawda, R., Kandaswami, C., Shanahan, T.C., Schwartz, S.A. (2002). Grape seed extract activates Th1 cells in vitro. *Clinical and Diagnostic Laboratory Immunology*, 9, 470–476.

Nawaz, H., Shi, J., Mittal, G.S., Kakuda, Y. (2006). Extraction of polyphenols from grape seeds and concentration by ultrafiltration. *Separation and Purification Technology*, 48, 176–181.

Naziri, E., Nenadis, N., Mantzouridou, F.T., Tsimidou, M.Z. (2014). Valorization of the major agrifood industrial by-products and waste from Central Macedonia (Greece) for the recovery of compounds for food applications. *Food Research International*, 65, 350–358.

Negro, C., Tommasi, L., Miceli, A. (2003). Phenolic compounds and antioxidant activity from red grape marc extracts. *Bioresource Technology*, 87, 41–44.

Newmark, H.L. (1997). Squalene, olive oil, and cancer risk: A review and hypothesis. *Cancer Epidemiology, Biomarkers & Prevention*, 6, 1101–1103.

News. (2014). Farmers battle over winery waste water. http://www.stuff. co.nz/marlborough-express/news/7735746/Farmers-battle-over-winery-waste-water. Accessed on October 21, 2014.

Nistor, E., Dobrei, A., Dobrei, A., Bampidis, V., Ciolac, V. (2014). Grape pomace in sheep and dairy cows feeding. *Journal of Horticulture, Forestry and Biotechnology*, 18, 146–150.

Nicolini, L., Volpe, C., Pezzotti, A., Carilli, A. (1993). Changes in in-vitro digestibility of orange peels and distillery grape stalks after solid state fermentation by higher fungi. *Bioresource Technology*, 45, 17–20.

192 VALORIZATION OF WINE MAKING BY-PRODUCTS

Nurgel, C., Canbas, A. (1998). Production of tartaric acid from pomace of some Anatolian grape cultivars. *American Journal of Enology and Viticulture*, 49, 95–99.

Oszmianski, J., Sapis, J.C. (1989). Fractionation and identification of some low molecular weight grape seed phenolics. *Journal of Agricultural and Food Chemistry*, 37, 1293–1297.

Özkan, G., Sagdiç, O., Baydar, N.G. (2003). Antibacterial effect of Narince grape (*Vitis vinifera* L.) pomace extract. S.Ü. *Ziraat Fakültesi Dergisi*, 17(32), 53–56.

Ozvural, E.B., Vural, H. (2014). Which is the best grape seed additive for frankfurters: Extract, oil or flour? *Journal of the Science of Food and Agriculture*, 94(4), 792–797.

Pajuelo, D., Díaz, S., Quesada, H., Fernández-Iglesias, A., Mulero, M., Arola-Arnal, A., Salvadó, M.J., Bladé, C., Arola, L. (2011). Acute administration of grape seed proanthocyanidin extract modulates energetic metabolism in skeletal muscle and BAT mitochondria. *Journal of Agriculture and Food Chemistry*, 59, 4279–4287.

Palamara, A.T., Nencioni, L., Aquilano, K., De Chiara, G., Hernandez, L., Cozzolino, F., Ciriolo, M.R., Garaci, E. (2005). Inhibition of influenza A virus replication by resveratrol. *The Journal of Infectious Diseases*, 191, 1719–1729.

Papadopoulou, C., Soulti, K., Roussis, I.G. (2005). Potential antimicrobial activity of red and white wine phenolic extracts against strains of *Staphylococcus aureus*, *Escherichia coli* and *Candida albicans*. *Food Technology and Biotechnology*, 43, 41–46.

Passos, C.P., Yilmaz, S., Silva, C.M., Coimbra, M.A. (2009). Enhancement of grape seed oil extraction using a cell wall degrading enzyme cocktail. *Food Chemistry*, 115, 48–53.

Pazos, M., Gallardo, J.M., Torres, J.L., Medina, I. (2005). Activity of grape polyphenols as inhibitors of the oxidation of fish lipids and frozen fish muscle. *Food Chemistry*, 92, 547–557.

Pedroza, M.A., Carmona, M., Alonso, G.L., Salinas, M.R. (2013). Prebottling use of dehydrated waste grape skins to improve colour, phenolic and aroma composition of red wines. *Food Chemistry*, 136, 224–236.

Pekić, B., Kovač, V., Alonso, E., Revilla, E. (1998). Study of the extraction of proanthocyanidinds from grape seeds. *Food Chemistry*, 61, 201–206.

Pérez-Serradilla, J., Luque de Castro, M. (2011). Microwave-assisted extraction of phenolic compounds from wine lees and spray-drying of the extract. *Food Chemistry*, 124(4), 1652–1659.

Pflieger, A., Waffo Teguo, P., Papastamoulis, Y., Chaignepain, S., Subra, F., Munir, S., Delelis, O. et al. (2013). Natural stilbenoids isolated from grapevine exhibiting inhibitory effects against HIV-1 integrase and eukaryote MOS1 transposase *in vitro* activities. *PLoS One*, 8(11), e81184.

Pinamonti, F., Stringari, G., Gasperi, F., Zorzi, G. (1997). The use of compost: Its effect on heavy metal levels in soil and plants. *Resources, Conversation and Recycling*, 21, 129–143.

ASPECTS OF BY-PRODUCT UTILIZATION 193

Pinelo, M., Arnous, A., Meyer, A.S. (2006a). Upgrading of grape skins: Significance of plant cell-wall structural components and extraction techniques for phenol release. *Trends in Food Science and Technology*, 17, 579–590.

Pinelo, M., Fabbro, P.D., Manzocco, L., Nunez, M.J., Nicoli, M.C. (2005b). Optimization of continuous phenol extraction from *Vitis vinifera* byproducts. *Food Chemistry*, 92, 109–117.

Pinelo, M., Rodriguez, A.R., Sineiro, J., Senorans, F.J., Reglero, G., Nunez, M.J. (2007). Supercritical fluid and solid–liquid extraction of phenolic antioxidants from grape pomace: a comparative study. *European Food Resource Technology*, 226, 199–205.

Pinelo, M., Rubilar, M., Jerez, M., Sineiro, J., Nunez, M.J. (2005a). Effects of solvent, temperature, and solvent-to-solid ratio on the total phenolic content and antiradical activity of extracts from different compounds of grape pomace. *Journal of Agricultural and Food Chemistry*, 53, 2111–2117.

Pinelo, M., Rubilar, M., Sineiro, J., Nunez, M.J. (2005c). A thermal treatment to increase the antioxidant capacity of natural phenols: Catechin, resveratrol and grape extract cases. *European Food Research and Technology*, 221, 284–290.

Pinelo, M., Sineiro, J., Nunez, M.J. (2006b). Mass transfer during continuous solid-liquid extraction of antioxidants from grape by-products. *Journal of Food Engineering*, 77, 57–63.

Prokop, M. (1979). Dried winery pomace as energy source in cattle finishing rations. California Feeders Day (El Centro), Department of Animal Science, University of California, Berkeley, CA.

Prozil, S.O., Evtuguina, D.V., Lopes, L.P.C. (2012). Chemical composition of grape stalks of *Vitis vinifera* L. from red grape pomaces. *Industrial Crops and Products*, 35, 178–184.

Pykett, M.A., Craig, A.H., Galley, E., Smith, C. (2001). Skin care composition against free radicals. International Patent Application WO 2001017495 A1, p. 59.

Qian, Y.P., Cai, Y.J., Fan, G.J., Wei, Q.Y., Yang, J., Zheng, L.F., Li, X.Z., Fang, J.G., Zhou, B. (2009). Antioxidant-based lead discovery for cancer chemoprevention: The case of resveratrol. *Journal of Medicinal Chemistry*, 52, 1963–1974.

Radovanović, A., Radovanović, B., Jovančićević, B. (2009). Free radical scavenging and antibacterial activities of southern Serbian red wines. *Food Chemistry*, 117, 326–331.

Ramos, M.J., Fernández, C.M., Casas, A., Rodríguez, L., Pérez, Á. (2009). Influence of fatty acid composition of raw materials on biodiesel properties. *Bioresource Technology*, 100, 261–268.

Ratnasooriya, C.C., Rupasinghe, H.P.V. (2012). Extraction of phenolic compounds from grapes and their pomace using β-cyclodextrin. *Food Chemistry*, 134, 625–631.

Ray, S.D., Bagchi, D. (2001). Prevention and treatment of acetaminophen toxicity with grape seed proanthocyanidin extract. U.S. Patent 6,245,336.

194 VALORIZATION OF WINE MAKING BY-PRODUCTS

Revilla, E., Ryan, J.M. (2000). Analysis of several phenolic compounds with potential antioxidant properties in grape extracts and wines by high performance liquid chromatography-photodiode array detection without sample preparation. *Journal of Chromatography A*, 881, 461–469.

Revilla, E., Ryan, J.M., Martin-Ortega, G. (1998). Comparison of several procedures used for the extraction of anthocyanins from red grapes. *Journal of Agricultural and Food Chemistry*, 46, 4592–4597.

Rice-Evans, C.A., Miller, N.J., Papaganga, G. (1997). Antioxidant properties of phenolic compounds. *Trends in Plant Science*, 4, 152–159.

Robak, J., Gryglewski, R.J. (1988). Flavonoids are scavengers of superoxide anions. *Biochemical Pharmacology*, 37, 837–841.

Rodrigo Senér, A., Pascual Vidal, A. (2001). Oportunidades de valorizaciön de los residuos de la industria vinícola (Valorisation opportunities of the residues from the winery industry). I Encuentro Internacional de Gestion de residuos organicos en el ambito rural mediterraneo. Catedra Zurich de Medio Ambiente de la Universidad de Navarra, Pamplona, Spain.

Rodríguez Vaquero, M.J., Alberto, M.R., Manca de Nadra, M.C. (2007). Antibacterial effect of phenolic compounds from different wines. *Food Control*, 18, 93–101.

Romero, E., Plaza, C., Senesi, N., Nogales, R., Polo, A. (2007). Humic acid-like fractions in raw and vermicomposted winery and distillery wastes. *Geoderma*, 139, 397–406.

Rondeau, P., Gambier, F., Jolibert, F., Nicolas Brosse, N. (2013). Compositions and chemical variability of grape pomaces from French vineyard. *Industrial Crops and Products*, 43, 251–254.

Roopchand, D.E., Kuhn, P., Krueger, C.G., Moskal, K., Lila, M.A., Raskin, I. (2013). Concord grape pomace polyphenols complexed to soy protein isolate are stable and hypoglycemic in diabetic mice. *Journal of Agriculture and Food Chemistry*, 61, 11428–11433.

Rosenthal, A., Pyle, D.L., Niranjan, N. (1996). Aqueous and enzymatic processes for edible oil extraction. *Enzyme and Microbial Technology*, 19, 402–420.

Sagdic, O., Ozturk, I., Ozkan, G., Yetim, H., Ekici, L., Yilmaz, M.T. (2011). RP-HPLC-DAD analysis of phenolic compounds in pomace extracts from five grape cultivars: Evaluation of their antioxidant, antiradical and antifungal activities in orange and apple juices. *Food Chemistry*, 126, 1749–1758.

Saito, M., Hosoyama, H., Ariga, T., Kataoka, S., Yamaji, N. (1998). Antiulcer activity of grape seed extract and procyanidins. *Journal of Agricultural and Food Chemistry*, 46, 1460–1464.

Sakagami, H., Sakagami, T., Takeda, M. (1995). Antiviral properties of polyphenols. *Polyphenol Actualites*, 12, 30–32.

Saladino, R., Gualandi, G., Farina, A., Crestini, C., Nencioni, L., Palamara, A.T. (2008). Advances and challenges in the synthesis of highly oxidised natural phenols with antiviral, antioxidant and cytotoxic activities. *Current Medicinal Chemistry*, 15, 1500–1519.

Salgado, J.M., Rodríguez, N., Cortés, S., Domínguez, J.M. (2012). Effect of nutrient supplementation of crude or detoxified concentrated distilled grape marc hemicellulosic hydrolysates on the xylitol production by *Debaryomyces hansenii*. *Preparative Biochemistry & Biotechnology*, 42, 1–14.

Sánchez, A., Ysunza, F., Beltran-Garcia, M., Esqueda, M. (2002). Biodegradation of viticulture wastes by Pleurotus: A source of microbial and human food and its potential use in animal feeding. *Journal of Agricultural and Food Chemistry*, 50, 2537–2542.

Sánchez-Alonso, I., Jiménez-Escrig, A., Saura-Calixto, F., Borderías, A.J. (2007). Effect of grape antioxidant dietary fibre on the prevention of lipid oxidation in minced fish: Evaluation by different methodologies. *Food Chemistry*, 101, 372–378.

Santos-Buelga, C., Francis Aricha, E.M., Escribano-Bailón, M.T. (1995). Comparative flavan-3-ol composition of seeds from different grape varieties. *Food Chemistry*, 53, 197–201.

Saura-Calixto, F. (1998). Antioxidant dietary fiber product: A new concept and a potential food ingredient. *Journal of Agricultural and Food Chemistry*, 46, 4303–4306.

Shi, J., Yu, J.M., Pohorly, J.E., Kakuda, Y. (2003b). Polyphenolics in grape seeds-biochemistry and functionality. *Journal of Medicinal Food*, 6, 291–299.

Shi, J., Yu, J.M., Pohorly, J.E., Young, J.C., Bryan, M., Wu, Y. (2003a). Optimization of extraction of polyphenols from grape seed meal by aqueous ethanol solution. *Food Agriculture and Environment*, 1, 42–47.

Shivananda Nayak, B., Dan Ramdath, D., Marshall, J.R., Isitor, G., Xue, S., Shi, J. (2011). Wound-healing properties of the oils of *Vitis vinifera* and *Vaccinium macrocarpon*. *Phytotheraphy Research*, 25, 1201–1208.

Shrikhande, A. (2000). Wine byproducts with health benefits. *Food Research International*, 33, 469–474.

Silva, M.L., Macedo, A.C., Malcata, F.X. (2000). Review: Steam distilled spirits from fermented grape pomace. *Food Science and Technology International*, 6, 285–300.

Silva, M.L., Malcata, F.X., De Revel, G. (1996). Volatile contents of grape marcs in Portugal. *Journal of Food Composition and Analysis*, 9, 72–80.

Silveira, S.T., Daroita, D.J., Brandelli, A. (2008). Pigment production by *Monascus purpureus* in grape waste using factorial design. *LWT—Food Science and Technology*, 41, 170–174.

Smagge, F., Mourgues, J., Escudier, J., Conte, T., Molinier, J., Malmary, C. (1992). Recovery of calcium tartrate and calcium malate in effluents from grape sugar production by electrodialysis. *Bioresource Technology*, 39, 85–189.

Smith, T.J. (2000). Squalene: Potential chemopreventive agent. *Expert Opinion on Investigational Drugs*, 9, 1841–1848.

Song, J.M., Lee, K.H., Seong, B.L. (2005). Antiviral effect of catechins in green tea on influenza virus. *Antiviral Research*, 68, 66–74.

Souquet, J-M., Labarbe, B., Le Guerneve, C., Cheynier, V., Moutounet, M. (2000). Phenolic composition of grape stems. *Journal of Agricultural and Food Chemistry*, 48, 1076–1080.

196 VALORIZATION OF WINE MAKING BY-PRODUCTS

Sousa, E.C., Uchŏa-Thomaz, A.M.A., Carioca, J.O.B., de Morais, S.M., de Lima, A., Martins, C.G., Alexandrino, C.D. et al. (2014). Chemical composition and bioactive compounds of grape pomace (*Vitis vinifera* L.), Benitaka variety, grown in the semiarid region of Northeast Brazil. *Food Science and Technology: Campinas*, 34, 135–142.

Spanghero, M., Salem, A.Z.M., Robinson, P.H. (2009). Chemical composition, including secondary metabolites, and rumen fermentability of seeds and pulp of Californian (USA) and Italian grape pomaces. *Animal Feed Science and Technology*, 152, 243–255.

Spigno, G., De Faveri, D.M. (2007). Antioxidants from grape stalks and marc: Influence of extraction procedure on yield, purity and antioxidant power of the extracts. *Journal of Food Engineering*, 78, 793–801.

Spigno, G., Pizzorno, T., De Faveri, D.M. (2008). Cellulose and hemicelluloses recovery from grape stalks. *Bioresource Technology*, 99, 1329–1337.

Spigno, G., Tramelli, L., De Faveri, D.M. (2007). Effects of extraction time, temperature and solvent on concentration and antioxidant activity of grape marc phenolics. *Journal of Food Engineering*, 81, 200–208.

Surh, Y.-J., Hurh, Y.-J., Kang, J.-Y., Lee, E., Kong, G., Lee, S.J. (1999). Resveratrol, an antioxidant present in red wine, induces apoptosis in human promyelocytic leukemia (HL-60) cells. *Cancer Letters*, 140, 1–10.

Taher, E., Mahmoud, N., Mahmoud, M. (2012). Laboratory evaluation of the effect of Egyptian native plants against some parasitic vectors. *Turkiye Parazitol Derg*, 36, 160–165.

Takechi, M., Tanaka, Y., Nonaka, G.I., Nishioka, I. (1985). Structure and antiherpatic activity among the tannins. *Phytochemistry*, 24, 2245–2250.

Tebib, K., Bitri, L., Besancon, P., Rouanet, J. (1994). Polymeric grape seed tannins prevent plasma cholesterol changes in high cholesterol fed rats. *Food Chemistry*, 49, 403–406.

Tebib, K., Rouanet, J.M., Besancon, P. (1997). Antioxidant effects of dietary polymeric grape seed tannins in tissues of rats fed a high cholesterol-vitamin E-deficient diet. *Food Chemistry*, 59, 135–141.

Thimothe, J., Bonsi, I.A., Padilla-Zakour, O.I., Koo, H. (2007). Chemical characterization of red wine grape (*Vitis vinifera* and *Vitis* interspecific hybrids) and pomace phenolic extracts and their biological activity against *Streptococcus* mutans. *Journal of Agricultural and Food Chemistry*, 55, 10200–10207.

Töpfl, S. (2006). Pulsed electric fields (PEF) for permeabilization of cell membranes in food- and bioprocessing: Applications, process and equipment design and cost analysis. PhD thesis, Universität Berlin, Berlin, Germany.

Torre, M., Rodriguez, A.R., Saura-Calixto, F. (1995). Interactions of Fe(II), Ca(II) and Fe(III) with high dietary fibre materials: A physicochemical approach. *Food Chemistry*, 54, 23–31.

Torres, J., Bobet, R. (2001). New flavanol derivatives from grape (*Vitis vinifera*) byproducts. Antioxidant aminoethylthio-flavan-3-ol conjugates from a polymeric waste fraction used as a source of flavanols. *Journal of Agricultural Food and Chemistry*, 49, 4627–4634.

ASPECTS OF BY-PRODUCT UTILIZATION 197

Tseng, A., Zhao, Y. (2013). Wine grape pomace as antioxidant dietary fibre for enhancing nutritional value and improving storability of yogurt and salad dressing. *Food Chemistry*, 138, 356–365.

Tsiplakou, E., Zervas, G. (2008). The effect of dietary inclusion of olive tree leaves and grape marc on the content of conjugated linoleic acid and vaccenic acid in the milk of dairy sheep and goats. *Journal of Dairy Research*, 75, 270–278.

Tura, D., Robards, K. (2002). Sample handling strategies for the determination of biophenols in food and plants. *Journal of Chromatography A*, 975(1), 71–93.

Valiente, C., Arrigoni, E., Esteban, R.M., Amado, R. (1995). Grape pomace as a potential food fiber. *Journal of Food Science*, 60, 818–820.

Valls-Belles, V., Torres, M.C., Muñiz, P., Beltran, S., Martinez-Alvarez, J.R., Codoñer Franch, P. (2006). De-fatted milled grape seed protects adriamycin-treated hepatocytes against oxidative damage. *European Journal of Nutrition*, 45, 251–258.

van de Wiel, A., van Golde, P.H.M., Hart, H.C. (2001). Blessings of the grape. *European Journal of Internal Medicine*, 12, 484–489.

Van Dyk, J.S., Gama, R., Morrison, D., Swart, S., Pletschke, B.I. (2013). Food processing waste: Problems, current management and prospects for utilisation of the lignocellulose component through enzyme synergistic degradation. *Renewable and Sustainable Energy Reviews*, 26, 521–531.

Vasta, V., Yáñez-Ruiz, D.R., Mele, M., Serra, A., Luciano, G., Lanza, M., Biondi, L., Priolo, A. (2010). Bacterial and protozoal communities and fatty acid profile in the rumen of sheep fed a diet containing added tannins. *Applied and Environmental Microbiology*, 76, 2549–2555.

Vatai, T., Skerget, M., Knez, Z. (2009). Extraction of phenolic compounds from elder berry and different grape marc varieties using organic solvents and/or supercritical carbon dioxide. *Journal of Food Engineering*, 90, 246–254.

Villaescusa, I., Fiol, N., Martinez, M., Mirrales, N., Poch, J., Seralocs, J. (2004). Removal of copper and nickel ions from aqueous solutions by grape stalks wastes. *Water Research*, 38, 992–1002.

Vinson, J.A., Mandarano, M.A., Shuta, D.L., Bagchi, M., Bagchi, D. (2002). Beneficial effects of a novel IH636 grape seed proanthocyanidin extract and a niacin-bound chromium in a hamster atherosclerosis model. *Molecular and Cellular Biochemistry*, 240, 99–103.

Vlissidis, A., Zouboulis A.I. (1993). Thermophilic anaerobic digestion of alcohol distillery wastewaters. *Bioresource Technology*, 43, 131–140.

Wadhwa, M., Bakshi, M.P.S., Makkar, H.P.S. (2013). Utilization of fruit and vegetable wastes as livestock feed and as substrates for generation of other value-added products. FAO publication. ISBN 978-92-5-107631-6, p. 15.

Wagner, H. (1985). New plant phenolics of pharmaceutical interest. In: *Annual Proceedings of the Phytochemical Society of Europe*, van Sumere, Lea, P.J. (Ed.), Clarendon Press, Oxford, UK, pp. 409–425.

Walter, R.H., Sherman, R.M. (1974). Fuel value of grape and apple processing wastes. *Journal of Agricultural and Food Chemistry*, 24, 1244–1245.

Waterhouse, A.L., Ignelzi, S., Shirley, J.R. (2000). A comparison of methods for quantifying oligometric proanthocyanidins from grape seed extract. *American Journal of Enology and Viticulture*, 51, 383–389.

Wenzel, U., Kuntz, S., Brendel, M.D., Daniel, H. (2000). Dietary flavone is a potent apoptosis inducer in human colon carcinoma cells. *Cancer Research*, 60, 3823–3831.

Yamakoshi, J., Kataoka, S., Koga, T., Ariga, T. (1999). Procyanidin-rich extract from grape seeds attenuates the development of aortic atherosclerosis in cholesterol fed rabbits. *Atherosclerosis*, 142, 139–149.

Ye, Z., Bekhit, A.E.D., Harrison, R. (November 30, 2011). Characterization of bioactive compounds in wine lees. *The 11th Functional Foods Symposium Foods for health & Wellness: Perspectives for industry*, Auckland, New Zealand, p. 67.

Yilmaz, E.E., Özvural, E.B., Vural, H. (2011). Extraction and identification of proanthocyanidins from grape seed (*Vitis vinifera*) using supercritical carbon dioxide. *The Journal of Supercritical Fluids*, 55(3), 924–928.

Yilmaz, Y., Toledo, R.T. (2006). Oxygen radical absorbance capacities of grape/wine industry byproducts and effect of solvent type on extraction of grape seed polyphenols. *Journal of Food Composition and Analysis*, 19, 41–48.

Yogalakshmi, B., Sreeja, S., Geetha, R., Radika, M.K., Anuradha, C.V. (2013). Grape seed proanthocyanidin rescues rats from steatosis: A comparative and combination study with metformin. *Journal of Lipids*, 2013, 1–11.

Youssef, D., El-Adawi, H. (2006). Study on grape seeds extraction and optimization: An approach. *Journal of Applied Sciences*, 6, 2944–2947.

Zeller, B.L. (1999). Development of porous carbohydrate food ingredients for the use in flavour encapsulation. *Trends in Food Science and Technology*, 9, 389–394.

Zepf, F., Jin, B. (2013). Bioconversion of grape marc into protein rich animal feed by microbial fungi. *Chemical Engineering and Process Technology*, 1, 1011.

Zhang, J., Pan, X., Li, N., Li, X., Wang, Y., Liu, X., Yin, X., Yu, Z. (2014). Grape seed extract attenuates arsenic-induced nephrotoxicity in rats. *Experimental and Therapeutic*, 7, 260–266.

Zheng, Y., Lee, C., Yu, C., Cheng, Y-S., Simmons, C.W., Zhang, R., Jenkins, B.M., VanderGheynst, J.S. (2012). Ensilage and bioconversion of grape pomace into fuel ethanol. *Journal of Agriculture and Food Chemistry*, 60, 11128–11134.

Zhu, Y.X., Coury, L.A., Long, H., Duda, C.T., Kissinger, C.B., Kissinger, P.T. (2000). Liquid chromatography with multichannel electrochemical detection for the determination of resveratrol in wine, grape juice, and grape seed capsules with automated solid phase extraction. *Journal of Liquid Chromatography & Related Technologies*, 23, 1555–1564.

Zocca, F., Lomolino, G., Spettoli, P., Anna Lante, A. (2008). A study on the relationship between the volatile composition of Moscato and Prosecco grappa and enzymatic activities involved in its production. *Journal of Institute of Brewing*, 114, 262–269.

5

REGULATORY AND LEGISLATIVE ISSUES

MATTEO BORDIGA

Contents

5.1 Food Industry Waste	199
5.2 Reduce, Reuse, and Recycle	203
5.3 California's Issue	203
5.3.1 Solid Waste Reduction and Management	205
5.4 New Zealand's Issue	206
5.5 European Union's Issue	209
5.5.1 Turning Waste into a Resource	209
5.5.2 European Commission Legislation Related to Wastewater Management	210
5.5.3 Community Legislation in the Field of Waste Management	214
5.6 South Africa's Issue	217
5.7 Australia's Issue	218
References	221

5.1 Food Industry Waste

In the United States, official surveys indicate that every year more than 160 billion kg of edible food is available for human consumption. Of that total, nearly 30% (45 billion kg)—including fresh vegetables, fruits, milk, and grain products—is lost to waste (FAO 2012). According to the South Australian Environmental Protection Agency (EPA), it costs the United States around $1 billion every year just to dispose of all its food waste (South Australian EPA 2004). Proportionally, in recent years, when it comes to food waste, the United Kingdom and Japan have been among the worst offenders, discarding between 30% and 40% of their food produce annually (e.g., about 7 million tons of household food is wasted each year in the United Kingdom). In total, 20 million tons of food waste is produced annually in the United Kingdom, with the producers, processors, and others involved outside the household accounting for 13 million tons (Eurostat 2013). The report claims that this involved about 18 million tons of CO_2 emissions, because every ton of food waste means about 4.5 tons of CO_2 emissions. In the Western world, about 30% of the time household food is wasted in the landfill where methane gas (more destructive than CO_2) is released into the environment. According to the United Kingdom's Waste & Resources Action Program (WRAP) report, fresh fruits, vegetables, and salads make up the largest category of waste, reaching about 1.5 million tons per year. Every year, the European food processing industry produces huge volumes of aqueous waste. This waste includes fruit and vegetable residues; molasses and bagasse from sugar refining; bones and flesh from meat and fish processing; stillage and other residue from wineries, distilleries, and breweries; dairy waste such as cheese whey; and wastewater from washing, blanching, and cooling operations. Many of these contain low levels of suspended solids (SSs) and low concentrations of dissolved materials. In addition to the environmental challenges, such flows represent considerable amounts of potentially reusable materials. Much of the material generated by the food processing industries throughout Europe contains components that could be utilized as substrates in a great variety of processes (e.g., microbial/enzymatic), to give rise to new value-added products. These products, which are actually produced from food industry waste include animal feed, single-cell proteins (SCPs), and other fermented edible products, yeast, organic acids, amino acids, enzymes, flavors and pigments, microbial gums, and polysaccharides (VIVA 2011). According to the European Landfill Directive (1999/31/EC), the amount of biodegradable waste sent to landfills (in member countries) by 2016 must reach 35% of the levels reached in 1995. Consequently, the food processing industry operations have to comply with increasingly more stringent European Union (EU) environmental regulations related to the disposal or utilization of by-products and waste (Resolution VITI 2/2003, 02/2006; Resolution OENO 1/2005; OIV Resolution CST 1/2004, 1/2008; OIV 2012–2014; International Organization for Standardization 2004). These regulations include growing restrictions on land, spraying with agro-industrial waste, disposal within landfill operations, and requirements to obtain products that are both stabilized and hygienic.

To increase the quantity of food, waste that is biologically treated (by either aerobic composting or anaerobic digestion) represents a good alternative to reduce the flux of waste to the landfills. While programs and facilities to manage yard waste are well established, the management of food waste in composting facilities is less developed. There is nonetheless considerable interest in food waste composting and the desire to increase food waste diversion is likely to grow. When discussing the environmental impact of food production, it is important to use a holistic approach, which can integrate the environmental aspects into the product development and food production. As the food supply chain is complex, environmental impacts can occur in both different places and times for a single food product. Life cycle assessment (LCA) provides a way of addressing this situation. LCA gives businesses the opportunity to anticipate environmental issues, thereby integrating the environmental dimension into products and processes. Crucial issues directly related to food processing result in energy and waste management. Food production in general uses significant amounts of energy and produces relatively large amounts of waste, mainly packaging waste. The waste management hierarchy (Figure 5.1) represents one of the guiding principles of the zero waste practice. Similarly, the development of green production processes can be achieved following the short-, medium-, and long-term goals. Short-term goals involve waste minimization by the reduction and recycling of valuable substances, by-products, and residues and reducing emissions and risk. Medium-term goals include the development of efficient production processes, thereby adding value to the by-products. The outcome for the companies is their higher environmental responsibility accompanied by

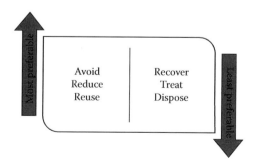

Figure 5.1 The food waste management hierarchy. (From Laufenberg, G. et al., *Bioresour. Technol.*, 87, 167, 2003.)

202 VALORIZATION OF WINE MAKING BY-PRODUCTS

competitive advantages. Long-term goals consist of systematic implementation of environmentally friendly manufacturing, thus developing innovative products. The ultimate outcome is the design of innovative food products such as functional foods, which can open new markets and meet green productivity objectives. Fruit and vegetable waste (FVW), produced in large quantities in markets, constitutes a source of disturbance in municipal landfills because of its high biodegradability. In Barcelona for instance, in the central distribution market for food (meat, fish, fruit, and vegetables), the total amount of waste that comes from the fruit and vegetables is around 90 tons per day during 250 days per year. The FVW's whole production collected from the market of Tunisia has been measured and estimated to be 180 tons per month. In India, FVWs constitute about 5.6 million tons annually and currently disposed by dumping on the outskirts of cities. The waste from fruit and vegetable processing industries generally contains large amounts of suspended solids (SSs) and high values of biological (BOD) and chemical oxygen demand (COD). Table 5.1 summarizes some indicative parameters, such as BOD, COD, SS, and pH for the processing of some fruit and vegetables. On average, the total initial solid concentration of FVW is between 8% and 18%, with a total volatile solid (VS) content of about 87%. The organic fraction includes about 75% sugars and hemicellulose, 9% cellulose, and 5% lignin. In general, this waste consists of hydrocarbons and relatively small amounts of proteins and fat with an acidic pH and a moisture content of 80%–90%. It must be highlighted that related wastewater contains dissolved compounds such as pesticides, herbicides, and cleaning chemicals.

Table 5.1 Fruit and Vegetable Waste Characteristics

FRUIT/VEGETABLE	BOD (mg/L)	COD (mg/L)	SS (mg/L)	pH
Apples	9600	18,700	450	5.9
Carrots	1350	2,300	4120	8.7
Cherries	2500	2,500	400	6.5
Corn	1500	2,500	210	6.9
Grapefruits	1000	1,900	250	7.4
Green peas	800	1,650	260	6.9
Tomatoes	1025	1,500	950	7.9
Winery wastewater	9000	14,000	2800	4.9

Source: Modified from Saez, L. et al., *Water Res.*, 26, 1261, 1992.

5.2 Reduce, Reuse, and Recycle

Reduce, reuse, and recycle represent current and mainly future guiding principles for vineyards and wineries. Along with reducing environmental impacts and disposal costs of waste, reusing and recycling conserve raw materials and energy, by diverting several by-products from the waste stream and turning them into a beneficial use. "Reducing" appears the best strategy for beginning to gain control over the amount of materials and supplies purchased for different operations. If the amount used cannot be reduced, another option might be focused on the containers and packaging associated with the materials and supplies coming in. Suppliers might be contacted to evaluate if there is some other way to deliver the materials and supplies with less packaging, and less waste. "Reusing" supplies whenever possible appears to be better than recycling. Vendors should be encouraged to begin reusing their packaging, thus allowing them to save money and develop a service along with their products. Many companies are already doing this common practice by providing winery supplies, including capsule, cork, and label manufacturers. If it is not possible to reduce or reuse, "recycling" is the next best step. Most of the materials used in the wine industry can be recycled but this approach does require labor and training to ensure that employees are properly using containers at the operations to divert solid waste out of the waste stream and into the recycling stream.

5.3 California's Issue

Wine production, like all farm production, generates waste. Nevertheless, many of the waste products from vineyard and winery operations can be assets to the viticulturist. Vine pruning, grape stalks, and marc (skins and seeds) can be composted and applied to the vineyard as mulch. This approach is able to improve the soil's moisture retention, to keep weeds down, to increase soil organic matter, and to reduce soil temperature variations. Grape seeds can be processed to create extracts potentially suitable for the health and cosmetics industries. Recycling and reusing packaging materials results in a central point, with various initiatives focused toward different matrices such as agrichemical containers, cardboard, plastic wrap, glass, batteries, solvents, and waste oils. Management of wastewater also appears to be an important issue.

204 VALORIZATION OF WINE MAKING BY-PRODUCTS

Wineries must manage their wastewater to ensure they do not cause nutrient enrichment of downstream water sources, degrade soil structure, contaminate soils, or emit odors during treatment. Reducing and recycling solid waste help to conserve natural resources, reduce greenhouse gases, and decrease costs. In this context, California proves its leadership in the nation, in large part due to AB939 enacted in 1989. While the state saw a 58% diversion rate of the solid waste stream in 2010, the majority of this was achieved through recycling programs in the residential sector. As a comparison, large office buildings only divert 7% of their waste streams to recycling. This indicates that there is still a large untapped recycling opportunity in the commercial sector, which comprises two-thirds of California's solid waste generation. In July 2012, the state passed AB341 making commercial recycling mandatory for any business that generates 4 cubic yards or more of commercial solid waste per week. One metric the state uses to measure solid waste generation is pounds of material thrown away per employee per day. In 2010, the per employee disposal rate reached a historic low of 11.7 lb per employee per day. Overall, each American generates about 4.34 lb of solid waste per day according to a 2009 survey on Municipal Solid Waste Generation, Recycling, and Disposal in the United States, conducted by the U.S. Environmental Protection Agency. The five main materials that make up most of the solid waste stream are paper, food, metal, plastic, and lumber. Organics, such as food, are the largest component of the solid waste stream. The wine industry is in a unique position because much of the solid waste generated at the winery (e.g., pomace, lees, cardboard, paper, glass) can be reused or recycled. Many wineries are composting pomace for use in vineyards, and a few are composting their paper and cardboard as well. As the largest source of organic waste at the winery, composting pomace can divert 50% or more of the waste stream. California's Waste Reduction Award's Program (WRAP) has recognized several wineries, with one winery recognized for 12 consecutive years. Many others are realizing that the most cost-effective strategy is to work with suppliers to reduce packaging that comes with the materials and supplies they purchase. This direct communication of environmental requirements can motivate suppliers to develop systems for reusable containers, recyclable packaging, or reprocessing of waste material (SIP 2008).

5.3.1 *Solid Waste Reduction and Management*

The purpose of Chapter 12 of the voluntary California Code of Sustainable Winegrowing Workbook is to help vintners understand the full cost of solid waste generation and the multiple benefits of implementing reduction measures (Table 5.2). Moreover, vintners are motivated to improve existing or develop new solid waste reduction and recycling plans to target the biggest problem areas while optimizing the overall efficiency of winery operations. Table 5.3 summarizes the 16 criteria to self-assess included in the chapter. Following these criteria, vintners are able to evaluate the state of their solid waste reduction planning, monitoring, goals, and results. They can assess both the total solid waste generated and the extent of solid waste generated per major operation. Then vintners can estimate the extent of management support needed for employee training in solid waste reduction efforts; likewise the opportunities to optimize solid waste reduction in their operations (California Sustainable Winegrowing Alliance 2004).

Table 5.2 The California Code of Sustainable Winegrowing Workbook Results Organized into the Following 15 Chapters (Beginning with Chapter 2), Including 191 Self-Assessment Questions

Chapter 2	Sustainable Business Strategy
Chapter 3	Viticulture
Chapter 4	Soil Management
Chapter 5	Vineyard Water Management
Chapter 6	Pest Management
Chapter 7	Wine Quality
Chapter 8	Ecosystem Management
Chapter 9	Energy Efficiency
Chapter 10	Winery Water Conservation and Water Quality
Chapter 11	Material Handling
Chapter 12	Solid Waste Reduction and Management
Chapter 13	Environmentally Preferable Purchasing
Chapter 14	Human Resources
Chapter 15	Neighbors and Community
Chapter 16	Air Quality

Source: Modified from CSWA, Sustainable winegrowing program, 2013, http://www.sustainablewinegrowing.org/, accessed on May 2014.

206 VALORIZATION OF WINE MAKING BY-PRODUCTS

Table 5.3 A List of Solid Waste Reduction and Management Criteria

1	Planning, Monitoring, Goals, and Results	9	Plastic
2	Pomace and Lees	10	Packaging (Incoming and Outgoing)
3	Diatomaceous Earth	11	Metals
4	Plate and Frame Filters	12	Natural Cork
5	Cooperage	13	Pallets, Wood Packaging, Bins
6	Glass	14	Capsules
7	Cardboard	15	Landscape Residuals
8	Paper	16	Food Waste

Source: Modified from CSWA, Sustainable winegrowing program, 2013, http://www.sustainablewinegrowing.org/, accessed on May 2014.

5.4 New Zealand's Issue

The New Zealand Government Waste Minimization Act (2008) provides a legislative framework for waste management. All parties such as producers, owners, and consumers take responsibility for the environmental effects of their products. Together with these government initiatives, other regional and industry actions for responsible waste management practices are present in New Zealand. Sustainable Winegrowing New Zealand (SWNZ) collaborates with district and regional councils to identify recycling options for vineyards and wineries in order to communicate potential information, provided by a single district, for example, to all members in all regions to encourage members to continue to search and implement solutions. At the same time, SWNZ provides a checklist so members can record and measure recycled materials and waste. The New Zealand Winegrowers Code of Practice for Management of Winery Waste provides guidance on practical solutions, thus aiming to develop practices to minimize waste and recycle/reuse. Moreover, the Code provides guidance for winemakers on cleaner production and sound environmental practices, including waste management and disposal and new strategies are provided to manage solid and liquid waste from wineries (New Zealand Wines 2014). The disposal of grape marcs to landfills is discouraged, conversely, disposal of sludge and solids is recommended because of high nutrient loading of these by-products. The SWNZ recommends the pretreatment of wastewater, even outlining the issues to be considered when establishing land-based disposal and treatment systems. The program is aimed at, wherever possible, reducing, reusing, and recycling the management of in-house waste and by-products while also taking into consideration that the collection and disposal of by-products should not impact the receiving environment. As a good example in this direction, glass recycling and reuse contribute significantly to reducing glass packaging's carbon footprint. The use of recycled glass in batch materials shows beneficial impact because every 1 kg of cullet used replaces 1.2 kg of virgin raw materials that would otherwise need to be extracted. Moreover, every 10% of recycled glass used in production results in an approximate 5% reduction in carbon emissions and energy savings of about 3%. Glass, mainly used by wineries, is resource efficient, and can be reused in its original form more than other packaging materials. Additionally, several initiatives currently underway in the glass industry will further increase the efficiency of glass packaging. Several studies are focused to improve recovery and recycling of glass containers leading to a decrease in energy use and global warming potential. On the other hand, other projects have the purpose to improve the light-weight glass containers, so reducing raw material usage, emissions, energy used, and overall weight. Even paper and cardboard are reused, and collected for recycling, similarly plastics, but the amounts from vineyards and wineries tend to be small. Most regions have recycling options for these products. Timber, where possible, is reused as landscaping and fencing, and if not treated it may be used for firewood. Finally, vehicle waste (e.g., tires, oil) is taken to transfer stations or collection points. Different waste management options exist, which cover many matrices. Vineyard waste primarily includes pruning materials and marc while winery waste includes marc, lees, winery sludge, and processing aids. Generally, pruning is mulched within the vineyard or composted. However, there are different initiatives, such as burning these under controlled conditions for fuel/heat in wineries. Marc (the pressed residue from grapes) is often applied directly to the vineyard, or composted, but must meet allowable nitrogen limits (in compliance with the Regional Council and SWNZ). In addition, sludge (the thick mixture removed after the fining process) is composted or sent to landfills. SWNZ requires that winery waste must not affect the environment, and in this direction monitoring is required. For example, if the waste incorrectly spreads to the land, this may impact soil or water quality and interfere with wastewater management. Staff training is implemented and records are conserved to

208 VALORIZATION OF WINE MAKING BY-PRODUCTS

ensure that the SWNZ requirements for the management of waste are met. All members have systems in place to reduce the risk of spills of by-products, chemicals, and potentially toxic petrochemicals. SWNZ provides an emergency procedures flipchart to all its members. This has been developed with the Environmental Protection Authority (EPA), including information on training, first aid guidelines, and emergency contacts. All this documentation is maintained onsite and checked during the auditing process. All members engage in training staff in storage, handling, and use of chemicals and chemical containers, as well as spill prevention. Small containers of oil and petrochemicals should be preferred; this approach reduces the potential for unmanageable spills. They should ensure that all storage containers, tanks, pipes, and valves are secure (avoiding leaks). All members are encouraged to store large amounts of fuel and oil in a manner that complies with the local regional plan requirements for vineyards and to collect waste oils for recycling programs. They have to pay attention both that any spillage will not reach waterways or drains (including storm water drains) and in ensuring management systems are in place to prevent future spills. The entire approach must be documented and all actions taken must be recorded. Considering that the volume of wastewater is directly proportional to the amount of water use, waste water management appears to be a major issue. Consequently, it is evident that minimizing water use minimizes the wastewater content. Members are encouraged to conduct wastewater system reviews frequently, and these are examined during the auditing process. Evidence of reviews includes a detailed and updated winery waste checklist, audits undertaken by different councils, or consultant reports. Members have standard operating procedures and maintain staff training records about regulating and reducing products in wastewater.

Wastewater treatment systems have to be designed to complement other disposal treatment systems, and the volume and components of winery wastewater, thus minimizing any impact on the environment. The system should be reviewed at the beginning, during, and end of the season.

In order to have a low impact on the environment, members should focus on the primary wastewater pretreatment, which complies with local authority trade waste bylaws, following all monitoring requirements (e.g., allowable levels). Reused water can be defined as water that has been used in the winery as part of the wine production process, and is then reused for an additional purpose (e.g., reusing moderately clean water for dirty washing operation). Conversely, recycled water can be defined as using treated winery wastewater for a secondary purpose, such as irrigation, vineyard, or maintaining winery grounds or wetlands. Members have strategies in place to address any problems that may occur related to liquid waste (wastewater disposal practices). Measuring water use to ensure that the wastewater system is able to manage the volume generated (storm water included) represents one of these strategies. The same considerations are concerned with the liquid waste generated by a winery. Members have to ensure that the actual volume of liquid waste falls within consented limits, and if not, proper action should be taken. Members have to prove robust design calculations showing the volume limits of their wastewater system to ensure it can cope with peak flow, and their irrigation system. Generally, winery wastewater is not considered harmful, but it can interrupt the normal function of treatment disposal systems. Engaging into the program, SWNZ members ensure that wash additives are strictly monitored and limited to minimize the effects on the environment. They prove that water use is controlled and limited in accordance with the residence time of waste within treatment systems, and the resting time between land applications of wastewater. They guarantee that their staff are properly trained in order to limit the amount of water used per wash, which also includes fixing valves, joints, and lines. Employees should also be prepared to prevent water from flowing unsupervised during cleaning operations.

5.5 European Union's Issue

5.5.1 Turning Waste into a Resource

Each year in the European Union (EU), 2.7 billion tons of waste is thrown away, 98 million tons of which is hazardous. On average, only 40% of our solid waste is reused or recycled, the rest goes to landfills or incineration (Eurostat 2013). Overall, waste generation is stable in the EU, however, the generation of some waste streams like construction and demolition waste to sewage sludge and marine litter continues to increase. Waste from electrical and electronic equipment alone is expected to increase by roughly 11% between 2008 and 2014. In some

210 VALORIZATION OF WINE MAKING BY-PRODUCTS

member states, more than 80% of waste is recycled, indicating the possibilities of using waste as one of the EU's key resources. Improving waste management makes for better use of resources and can open up new markets and jobs, as well as encourage less dependence on imports of raw materials and have a lower impact on the environment. If waste is to become a resource to be fed back into the economy as a raw material, then a much higher priority needs to be given to reusing and recycling. A combination of policies would help create a full recycling economy, such as a product design integrating a life cycle approach, better cooperation among all market actors along the value chain, better collection processes, appropriate regulatory frameworks, incentives for waste prevention and recycling, as well as public investment in modern facilities for waste treatment and high-quality recycling.

By 2020, waste is to be managed as a resource. Waste generated per capita will in an absolute decline. Recycling and reusing waste will be economically attractive options for public and private actors due to widespread separate collection and the development of functional markets for secondary raw materials. More materials, including materials having a significant impact on the environment and critical raw materials, will be recycled. Waste legislation will be fully implemented. Illegal shipments of waste will have been eradicated. Energy recovery will be limited to nonrecyclable materials, landfilling will be virtually eliminated, and high-quality recycling will be ensured.

The objectives of the European Commission are reported in Table 5.4. Member states should ensure full implementation of the EU waste acquisition including minimum targets through their national waste prevention and management strategies.

5.5.2 European Commission Legislation Related to Wastewater Management

At the European Commission level, Council Directive 76/464/EEC, on pollution caused by certain dangerous substances discharged into the aquatic environment of the community, establishes that member states shall take the appropriate steps to eliminate or reduce the pollution of water by dangerous substances in List I and List II, respectively. In both lists, there are substances, which could be present in winery wastewater, such as metals and inorganic compounds of phosphorous and elemental phosphorous. In addition, agro-food industries

REGULATORY AND LEGISLATIVE ISSUES 211

Table 5.4 European Commission's Purpose

1	To stimulate the secondary materials market and demand for recycled materials through economic incentives and developing end-of-waste criteria
2	To review existing prevention, reuse, recycling, recovery, and landfill diversion targets to move toward an economy based on reuse and recycling, with residual waste close to zero
3	To assess the introduction of minimum recycled material rates, durability and reusability criteria, and extensions of producer responsibility for key products
4	To assess areas where legislation on the various waste streams could be aligned to improve coherence
5	To continue working within the EU and with international partners to eradicate illegal waste shipments with a special focus on hazardous waste
6	To ensure that public funding from the EU budget gives priority to activities higher up the waste hierarchy as defined in the Waste Framework Directive (e.g., priority to recycling plants over waste disposal)
7	To facilitate the exchange of best practices on collection and treatment of waste among member states and develop measures to combat more effectively breaches of EU waste rules

Source: Modified from OIV, Strategic plan, 2012–2014, http://www.oiv.int/oiv/info/enplanstrategique, accessed on May 2014.

are influenced by Directive 91/271/EEC regarding the collection, treatment, and discharge of urban wastewater; and the treatment and discharge of wastewater from certain industrial sectors, including the production of alcohol and alcoholic beverages. This directive indicates a group of requirements for discharge for industries that are not connected to the urban wastewater treatment plants have to obey. Commission Directive 98/15/EC amends Council Directive 91/271/EEC with respect to certain requirements established in Annex I thereof. On the other hand, Directive 2000/60/EC establishes a framework for community action in the field of water policy. Its purpose is to establish a framework for the protection of inland waters, transitional waters, coastal waters, and groundwater. Along these lines, Directive 2006/118/EC establishes specific measures in order to prevent and control groundwater pollution (Commission Decision 2000/532/EC; Council Decision 2001/573/EC). These measures include particular criteria for the assessment of good groundwater chemical status and criteria for the identification and reversal of significant and sustained upward trends and for the definition of starting points for trend reveals (Council Directive 91/676/EEC, 1999/31/EC, 2006/12/EC). Finally, Directive 2008/1/EC, concerning integrated pollution prevention and control, lays down measures designed to prevent or, where that is not practicable, to reduce emissions in the air, water, and land from certain activities. This directive includes measures concerning waste, in order to achieve a high level of protection of the environment taken as a whole. It is important to mention also the European framework law, namely the Common Market Organization for Wine (CMO), issued in Regulation (EC) 479/2008 of the Council of 29 April 2008, which aims to give due effect to Article 33 of the Treaty establishing the European Community regarding the objectives of the Common Agricultural Policy (CAP) (Council Regulation (EC) No. 491/2009, (EEC) No. 822/87, (EEC) No. 823/87). The major objectives include increased agricultural productivity by promoting technical progress and ensuring the rational development of agricultural production, the optimum use of production factors, or the stabilization of markets, taking into account the need for the appropriate adjustments, fighting structural and natural disparities between the various agricultural regions (Figure 5.2). The CMO is integrated with the following rules: Regulation (EC) 607/2009 of the Commission of

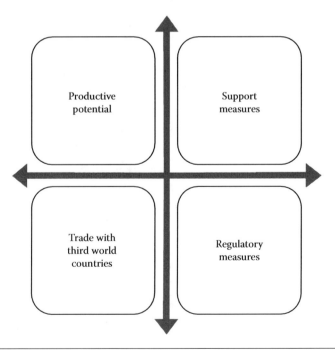

Figure 5.2 The Common Market Organization for Wine (CMO) framework for regulations toward different issues. (From OIV Resolution CST 1/2008, OIV guidelines for sustainable vitiviniculture: Production, processing and packaging of products, http://www.oiv.int/oiv/info/enresolution, accessed on January 2014.)

REGULATORY AND LEGISLATIVE ISSUES   213

14 July 2009 lays down provisions for implementing Regulation (EC) No. 479/2008 of the Council with regard to the designation of origin and protected geographical indications, in traditional terms, labeling and presentation of certain wine products. Regulation (EC) 606/2009 of the Commission of 10 July 2009 provides provisions for implementing Regulation (EC) No. 479/2008 of the Council with regard to the categories of wine products, oenological practices, and applicable restrictions. Regulation (EC) 415/2009 of the Commission of 20 May 2009 by amending Directive 2007/68/EC that modifies Annex III bis of Directive 2000/13/EC of the European Parliament and Council with regard to certain food ingredients. Regulation (EC) 981/2008 of the Commission of 7 October 2008 amending Regulation (EC) No. 423/2008 lays down detailed rules for implementing Regulation (EC) No. 1493/1999 and establishing a community code of oenological practices and processes. Regulation (EC) 3/2008 of the Council of 17 December 2007 on information provisions and promotion of agricultural products on the local market and in third world countries. Regulation (EC) 1234/2007 of the Council of 22 October 2007 establishes a common organization of agricultural markets with specific provisions for certain agricultural products (Single CMO Regulation). Regulation (EC) 1924/2006 of the European Parliament and of the Council of 20 December 2006 provides provisions for nutrition and health claims on foods. Regulation (EC) 1507/2006 of the Commission of 11 October 2006 lays down provisions for implementing Regulation (EC) No. 1493/1999 on the common organizations of the wine market with regard to the use of pieces of oak wood in wine making and the designation and presentation of wines so treated. Regulation (EC) 1782/2003 of the Council of 29 September 2003 establishes common rules for direct support schemes under the common agricultural policy, establishing certain support schemes for farmers. Regulation (EC) 753/2002 of the Commission of 29 April 2002 lays down certain rules for applying Council Regulation (EC) No. 1493/1999 with regard to the description, designation, presentation, and protection of certain wine products. Regulation (EC) 884/2001 of the Commission of 24 April 2001 lays down rules for the documents accompanying the freight transportation of wine products and the records to be kept in that sector. Regulation (EC) 2868/95 of the Commission of 13 December 1995 lays down rules for implementing Regulation

214 VALORIZATION OF WINE MAKING BY-PRODUCTS

(EC) No. 40/94 of the Council on the community brand. Directive 2007/68/EC of the Commission of 27 November 2007 amends Annex III bis to Directive 2000/13/EC of the European Parliament and the Council with regard to certain food ingredients. Directive 2007/45/EC of the European Parliament and of the Council of 5 September 2007 lays down rules on nominal quantities for pre-packed products, repealing Directives 75/106/EEC and 80/232/EEC and amending Council Directive 76/211/EEC of the Council. Directive 2005/25/EC of the Council of 14 March 2005 amends Annex VI to Directive 91/414/EEC with regard to plant protection (phytosanitary) products containing microorganisms

5.5.3 Community Legislation in the Field of Waste Management

In 1990, the Council adopted a proposal by the Commission for a Community Strategy for waste management, which rested heavily on the principle of a hierarchy of preferred options for waste management. According to this, the waste production should be prevented or reduced at source, particularly by the use of clean or low waste technologies and products, wherever possible. Waste that cannot be recycled or reused has to be disposed of in the most environmentally safe manner. Successively, the Community Strategy for waste management was revised through the Council Resolution of 24 February 1997. According to this Resolution, all economic actors (producers, importers, distributors, and consumers) bear their specific share of responsibility with regard to the prevention, recovery, and disposal of waste. In addition, it insists on the need for promoting waste recovery with a review to reducing the quantity of waste for disposal and saving natural resources. Since its adoption, considerable legislative progress has taken place in the waste area. Directive 1999/31/EC establishes the general requirements for all classes of landfills, waste acceptance criteria and procedure and, finally, control and monitoring procedures in operation and aftercare phases of the landfill. The aim of the Directive is to prevent or reduce negative effects on the environment as well as any risk to human health, from waste landfills. Within the accepted guidelines, waste in landfills includes biodegradable, nonhazardous waste (winery waste). Subsequently, the legislative framework for the handling of waste in the community

REGULATORY AND LEGISLATIVE ISSUES

has been established by Directive 2006/12/EC. The latter defines key concepts such as waste, recovery, and disposal and puts in place the essential requirements for the management of waste. Handling waste in a way that does not have a negative impact on the environment or human health is one of the major principles established by this Directive. It also encourages applying the waste hierarchy, which requires that the costs of disposing waste must be borne by the holder of the waste, by previous holder, or by the producers of the product from which the waste came. In that direction, member states must prohibit the abandonment, dumping, or uncontrolled disposal of waste, and must promote waste prevention, recycling, and processing for reuse. Finally, in 2008, the European Parliament published Directive 2008/98/EC on waste. This directive clarifies issues such as which objects are to be considered a by-product, with end-of-waste status, and the distinction between waste and nonwaste. In addition, this directive lays down measures to protect the environment and human health by preventing or reducing the adverse impact of the generation and management of waste. It also aims to both reduce the overall impact of resource use and improve the efficiency of such use. In this directive, a waste hierarchy is established with a priority order for waste prevention and management legislation (Figure 5.3).

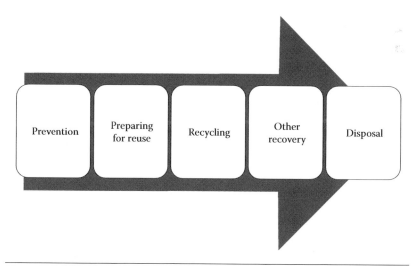

Figure 5.3 Established waste hierarchy as a priority order in waste prevention and management legislation. (From OIV Resolution CST 1/2008, OIV guidelines for sustainable vitiviniculture: Production, processing and packaging of products, http://www.oiv.int/oiv/info/enresolution, accessed on January 2014.)

Concerning winery waste, Article 22 reports that member states shall take measures to encourage the separate collection of bio-waste with a view to the composting and digestion of these. They should encourage the bio-waste treatment in a proper way, thus fulfilling a high level of environmental protection. In order to comply with the objectives of this directive and move toward a European recycling society with a high level of resource efficiency, member states shall take the necessary measures designed to achieve some relevant targets. The first example is to be prepared for the reuse and recycling of waste materials (paper, metal, plastic, and glass) from households and possibly from other origins as far as these waste streams are similar to waste from households, shall be increased to a minimum of overall 50% by weight. The second example is to be prepared for reuse, recycling, and other material recovery, including backfilling operations using waste to substitute other materials of nonhazardous construction and demolition waste excluding naturally occurring material, shall be increased to a minimum of 70% by weight. The European Waste Catalogue and Hazardous Waste List is used for the classification of all waste and hazardous waste and is designed to form a consistent waste classification system across the EU. In this catalogue, a wide variety of waste materials is classified as biodegradable. Specifically included under Category 2 is the waste produced in the alcoholic industries (winery and brewery industry). Council Regulation EC 1493/1999 on the common organization of the market for wine controls the use of the pomace and lees. This establishes that in view of the poor quality of wine obtained by overpressing, this practice should be prohibited and provisions should be made, in order to prevent it, for the compulsory distillation of marc and lees. Regulation EC 555/2008 lays down detailed rules about the common organization of the market in wine. This regulation establishes conditions under which producers shall withdraw the by-products of wine making or any other processing of grapes. The by-products must be withdrawn immediately and no later than the end of the same year of vintage in which they were obtained; withdrawal, together with an indication of the estimated quantities, shall be entered and kept in the registers. Withdrawal must respect applicable community legislation, in particular with regard to the environment. It also establishes that the withdrawal of wine lees shall be regarded as having taken place once the lees have been denatured to make their use in wine making impossible and where the delivery of the denatured lees to third parties has been entered in the registers kept. Member states may decide that producers who, during the wine year in question, do not produce more than 25 hL of wine or must themselves on their own premises are not required to withdraw their by-products. Finally, it has been established that producers may fulfill the obligation of disposal for a part or for the entireness of the by-products of wine making or any other processing of grapes, by delivering these to distillation (European Commission 1986–2007, 2003, 2006a,b, 2007a,b, 2007–2012; European Court of Auditors 2012).

5.6 South Africa's Issue

South Africa has undergone significant legislative change in virtually every sphere, notwithstanding the environment. The 1996 Constitution of South Africa determines that every person has the right to an environment, which is not harmful to one's own health or well-being, and to the conservation of the environment for current and future generations (Act 108 of 1996, Section 24). Historically, the Environmental Conservation Act (Act 73 of 1989) did not take into account how waste was produced, disposed of, or recycled. In this manner, therefore, accountability was largely absent in terms of waste emanating from production processes and the resultant impact on the environment. However, the National Environmental Management Act (Act 107 of 1998) addresses management of the environment through planned production processes. The act places full responsibility with the landowner for protecting and managing the environment through sustainable production processes. In addition, the 1998 National Water Act (No. 36) and the amended National Water Act of 1999 have introduced the management of water through the issue of water permits. Thanks to the latter act, users have the right to utilize water, unless they abuse it and thus the nature of ownership of water has changed from private to public. The protection of aquatic ecology that forms part of a water resource, and how abstraction and discharge affect the environment, are further important changes brought about by the amended Act of 1999. The Integrated Production of Wine Scheme (IPW) was introduced in 1998 by the South African

218 VALORIZATION OF WINE MAKING BY-PRODUCTS

wine industry under the guidance of the Wine & Spirits Board in response to changes in legislation (IPW 1998; SAWIS 2014). IPW guidelines and objectives address how the wine industry should comply with respect to production processes and the resultant impact on the natural environment. Preservation of the natural environment and prevention of soil erosion are the major requirements. Solid waste (including grape waste, lees and filter rests, but also glass, plastic, paper/carton, and metal) must be recycled or disposed of in an environmental friendly way and in accordance to legislative requirements. Management of wastewater appears to be a crucial topic. Monitoring its amount, quality, storing, and disposal results in a meticulous approach by program members. Noise from pumps, cooling apparatus, and vehicles may cause a disturbance to neighbors; for this reason, ambient noise should be limited and reduced.

5.7 Australia's Issue

The EPA's Guidelines provide information able to assist wineries and distilleries in developing an environmental monitoring program to comply with the Environment Protection Act and relevant Environment Protection Policies. The Environment Protection Act represents the key legislation addressing pollution in South Australia. Particularly, Section 25 of the Act imposes a general environmental duty on anyone who undertakes an activity that pollutes, or has the potential to pollute, to take all reasonable and practicable measures to prevent (or minimize) environmental harm. Environment protection legislation also includes environment protection policies (EPPs), which outline both recommendations and mandatory requirements to address environment protection matters (e.g., water quality, solid waste, air quality). Facilities, processing more than 50 tons of grape or grape product per annum (within the Mount Lofty Ranges Watershed Protection Area) or more than 500 tons elsewhere in the state, must have an Environment Protection Authority (EPA) license as declared under Part 8 of the Environment Protection Act 1993 (the Act). Moreover, licensed wineries and distilleries must develop and implement an environmental monitoring program and submit the data collected to the EPA annually. The main environmental impacts associated with wine production are pollution of water, degradation of soil, and damage to vegetation arising from liquid and solid waste disposal practices. At the same time, odors and air emissions resulting from the management of raw materials, and wastewater, solid and semisolid by-products from the wine making process represent a significant percentage. Finally, some considerations include noise from pumps, chillers, crushers, and other winery equipment, as well as vehicle noise, particularly during vintage. The different constituents of liquid and solid waste by-products from the wine making process are responsible for several effects on the environment (Figure 5.4). With respect to organic matter, three indicators are commonly used, such as BOD (biochemical oxygen demand), TOC (total organic carbon), and COD (chemical oxygen demand). Related effects include oxygen depletion when discharged into water, leading to the death of fish and other aquatic organisms, and odors generated by anaerobic decomposition. Several problems are related to alkalinity and acidity such as the death of aquatic organisms at extreme pH ranges. This issue affects both microbial activity in biological wastewater treatment processes, and the solubility of heavy metals in the soil and availability and/or toxicity in waters. Even nutrients, such as nitrogen (N), potassium (K), and phosphorus (P), show some effects on the environment, for example, eutrophication or algal bloom when

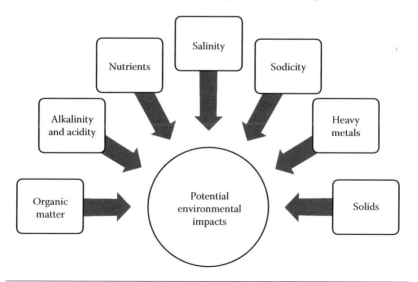

Figure 5.4 Winery waste constituents showing potential environmental impacts. (From South Australian EPA, Bunding and spill management, New South Wales Environment Protection Authority, Sydney, New South Wales, Australia, 2004.)

220 VALORIZATION OF WINE MAKING BY-PRODUCTS

discharged to water or stored in lagoons, thus causing undesirable odors. These compounds, in large amounts, can be rather toxic to crops. Nitrogen as nitrate in drinking water supply can be toxic to infants. Electrical conductivity (EC) and total dissolved salts (TDS) represent two indicators of salinity. The latter, in addition to being toxic to aquatic organisms, imparts an undesirable taste to water and affects water uptake by crops. Usually referred to as the sodium adsorption ratio (SAR) and exchangeable sodium percentage (ESP), sodicity affects soil structure, resulting in surface crusting, low infiltration and hydraulic conductivity, and hard and dense subsoil. Heavy metals represent a severe problem related to the environment mainly because of their high toxicity to plants and animals. Total suspended solids (TSS) reduce soil porosity, leading to reduced oxygen uptake. These can reduce light transmission in water, thus compromising ecosystem health and oppressing habitats. Environmental monitoring represents an effective tool that will assist the EPA and wineries to determine the load and effects of winery waste on the environment. Other objectives also include monitoring and maintaining the performance of waste management systems together with analyzing environmental management performance and comparing it with EPA standards. Wineries have to develop procedures to sample and monitor influent water, wastewater, soil, groundwater, and other receiving environments as required by the EPA license. In order to comply with both the Australian/New Zealand Standards (AS/NZS 5667 1998) and relevant EPA guidelines, influent water and wastewater sampling and monitoring procedures should be developed by the wineries (South Australian EPA 2002a,b, 2003a,b,c). Soil sampling, and related monitoring, should be also developed and take into account relevant schedules of the National Environment Protection (Assessment of Site Contamination) Measure 1999 (Site Contamination NEPM) to suit specific situations. A National Association of Testing Authorities (NATA) accredited laboratory must be responsible for the analysis of samples, by using NATA accredited procedures to ensure the integrity of the data. Before an eventual implementation, the EPA is solely responsible for approving the monitoring program. In fact, the EPA must be consulted before changes to the approved monitoring program are made. Data obtained from the monitoring requirements of the license have to be forwarded to the EPA, in order to be used to establish industry benchmarks and inform the public. In reporting the data, wineries have to use the EPA reporting format for ease in data management and consolidation and present the concentration of substances in water and wastewater two significant figures. The EPA also requires that, at intervals prescribed by the license, an independent qualified professional must verify the monitoring activity and resulting data. Before undertaking verification audits, the winery must contact the EPA to confirm that the independent verifier selected meets the required criteria. The quantity and types of waste produced by a winery vary due to waste management practices and the activities undertaken. Wineries must review and amend their monitoring programs regularly. The environmental monitoring program submitted to the EPA must include a schematic diagram showing the inputs (e.g., grapes, grape juice, chemicals, water) and outputs (e.g., various wastewater streams, grape marc, lees, filtered solids, stalks, wastewater sludge). A concise description of the processes adopted in the winery must be reported (e.g., crushing, fermentation, storage, maturation, bottling, and distillation). Moreover, wineries have to report details of annual processing inputs and outputs such as crush size, volume of grape juice produced, volume of processed grapes, percentage of wine, and sparkling wine and spirits. Through all this information, wineries are able to assist the EPA to consolidate and analyze the monitoring data and to understand the waste characteristics and generation patterns of each site.

References

Australian/New Zealand Standards (AS/NZS 5667:1998). Water quality—sampling. Standards Australia, New South Wales, Australia.

California Sustainable Winegrowing Alliance. 2004. California wine community sustainability report executive summary. http://www.sustaina blewinegrowing.org/docs/cswa_2004_report_executive_summary.pdf (accessed on May 2014).

Commission Decision 2000/532/EC of 3 May 2000 replacing Decision 94/3/EC establishing a list of wastes pursuant to Article 1(a) of Council Directive 75/442/EEC on waste and Council Decision 94/904/EC establishing a list of hazardous waste pursuant to Article 1(4) of Council Directive 91/689/EEC on hazardous waste.

222 VALORIZATION OF WINE MAKING BY-PRODUCTS

Commission Directive 98/15/EC of 27 February 1998 amending Council Directive 91/271/EEC with respect to certain requirements established in Annex I thereof.

Commission Regulation (EC) No. 555/2008 of 27 June 2008 laying down detailed rules for implementing Council Regulation (EC) No. 479/2008 on the common organization of the market in wine as regards support programs, trade with thirds countries, production potential and on controls in the wine sector.

Council Decision 2001/573/EC amending Commission Decision 2000/532/EC as regards the list of wastes.

Council Directive 76/464/EEC of 4 May 1976 on pollution caused by certain dangerous substances discharged into the aquatic environment of the community.

Council Directive 91/271/ECC of 21 May 1991 concerning urban wastewater treatment.

Council Directive 91/676/EEC of 12 December 1991 concerning the protection of waters against pollution caused by nitrates from agricultural sources.

Council Directive 1999/31/EC of 26 April of 1999 on the landfill of waste.

Council Directive 2000/60/EC of the European Parliament and of the council of 23 October 2000 establishing a framework for community action in the field of water policy.

Council Directive 2006/12/EC of the European Parliament and of the council of 5 April 2006 on waste.

Council Directive 2006/118/EC of the European Parliament and of the council of 12 December 2006 on the protection of groundwater against pollution and deterioration.

Council Directive 2008/1/EC of the European Parliament and of the council of 15 January 2008 concerning integrated pollution prevention and control.

Council Directive 2008/98/EEC of the European Parliament and of the council of 19 November 2008 on waste and repealing certain directives.

Council Regulation (EC) No. 479/2008 of 29 April 2008 on the common organization of the market in wine, amending Regulations (EC) Nos. 1493/1999, 1782/2003, 1290/2005, and 3/2008 and repealing Regulations (EEC) Nos. 2392/86 and 1493/1999. Official Journal of the European Union.

Council Regulation (EC) No. 491/2009 of 25 May 2009 amending Regulation (EC) No. 1234/2007 establishing a common organization of agricultural markets and on specific provisions for certain agricultural products (Single CMO Regulation). Official Journal of the European Union.

Council Regulation (EC) No. 607/2009 of 14 July 2009 laying down certain detailed rules for the implementation of Council Regulation (EC) No. 479/2008 as regards protected designations of origin and geographical indications, traditional terms, labelling and presentation of certain wine sector products. Official Journal of the European Union.

Council Regulation (EC) No. 1493/1999 of 17 May 1999 on the common organization of the market in wine. Official Journal of the European Union.

Council Regulation (EEC) No. 822/87 of 16 March 1987 on the common organization of the market in wine. Official Journal of the European Communities.

Council Regulation (EEC) No. 823/87 of 16 March 1987 laying down special provisions relating to quality wines produced in specified regions. Official Journal of the European Communities.

Council Resolution of 24 February 1997 on a community strategy for waste management (97/C 76/01). Council Resolution of 7 May 1990 on waste policy.

CSWA (2013). Sustainable winegrowing program. http://www.sustainablewinegrowing.org/ (accessed on May 2014).

European Commission (1986–2007). Agriculture and Rural Development budget: Guarantee Section of the European Agricultural Guidance and Guarantee Fund (EAGGF) financial reports, 1986–2007. Office for Official Publications of the European Communities, Luxembourg. http://aei.pitt.edu/view/eusubjects/eagus.html (accessed on June 2014).

European Commission (2003). Study on the use of the varieties of interspecific vines. Contract No. AGR 30881 of 30 December 2002. http://ec.europa.eu/agriculture/markets/wine/studies/vine_en.pdf (accessed on June 2014).

European Commission (2006a). Impact assessment—Annex to the Communication from the Commission to the Council and the European Parliament "Towards a Sustainable European Wine Sector." http://ec.europa.eu/agriculture/capreform/wine/fullimpact_en.pdf (accessed on June 2014).

European Commission (2006b). WINE-economy of the sector. European Commission. http://ec.europa.eu/agriculture/markets/wine/studies/rep_econ2006_en.pdf (accessed on May 2014).

European Commission (2007a). Accompanying document to the Proposal for a Council Regulation on the common organisation of the market in wine and amending certain regulations—Impact assessment. http://ec.europa.eu/agriculture/capreform/wine/impact072007/full_en.pdf (accessed on June 2014).

European Commission (2007b). Fact sheet—Towards a sustainable European Wine Sector. http://ec.europa.eu/agriculture/publi/fact/wine/072007_en.pdf (accessed on June 2014).

European Commission (2007–2012). Agriculture and Rural Development budget: European Agricultural Guarantee Fund (EAGF) financial reports, 2007–2012. Office for Official Publications of the European Communities, Luxembourg. http://ec.europa.eu/agriculture/capfunding/budget/index_en.htm (accessed on February 2014).

European Commission (2008). Reform of the EU wine market. http://ec.europa.eu/agriculture/capreform/wine/index_en.htm (accessed on February 2014).

224 VALORIZATION OF WINE MAKING BY-PRODUCTS

European Commission (2009–2012). Wine CMO: Financial execution of the national support program. Official Journal of the European Communities. Directorate-General for Agriculture and Rural Development. http://ec.europa.eu/agriculture/markets/wine/facts/index_en.htm (accessed on February 2014).

European Court of Auditors (2012). The reform of the common organization of the market in wine: Progress to date. Special Report No. 7. http://eca.europa.eu/portal/pls/portal/docs/1/14824739.pdf (accessed on May 2014).

European Landfill Directive (1999/31/EC). The landfill of waste. http://eurlex.europa.eu/legalcontent/EN/TXT/?uri=CELEX:31999L0031 (accessed on April 2014).

Eurostat (2013). Eurostat statistics database. Eurostat, Luxembourg. http://epp.eurostat.ec.europa.eu/portal/page/portal/education/data/database/ (accessed on February 2014).

FAO (2012). FAO Statistical Yearbook 2012. FAOSTAT. http://faostat.fao.org (accessed on June 2014).

International Organization for Standardization (2004). ISO 14000—Environmental management. http://www.iso.org/iso/iso14000 (accessed on March 2014).

IPW (1998). Integrated production of wine (IPW). http://www.ipw.co.za/ (accessed on June 2014).

Laufenberg G, Kunz B, and Nystroem M (2003). Transformation of vegetable waste into value added products: (A) the upgrading concept; (B) practical implementations. *Bioresource Technology* 87: 167–198.

New Zealand Wines (2014). Sustainability. http://www.nzwine.com/sustainability/ (accessed on May 2014).

OIV (2012–2014). Strategic plan. http://www.oiv.int/oiv/info/enplanstrategique (accessed on May 2014).

OIV Resolution CST 1/2004. Development of sustainable vitiviniculture. http://www.oiv.int/oiv/info/enresolution (accessed on January 2014).

OIV Resolution CST 1/2008. OIV guidelines for sustainable vitiviniculture: Production, processing and packaging of products. http://www.oiv.int/oiv/info/enresolution (accessed on January 2014).

Resolution OENO 1/2005. Distillates of vitivinicultural origin http://www.oiv.int/oiv/info/enresolution (accessed on January 2014).

Resolution VITI 2/2006. Measures used to prevent or limit the proliferation of wood diseases. http://www.oiv.int/oiv/info/enresolution (accessed on January 2014).

Resolution VITI 2/2003. Reasoned vine irrigation. http://www.oiv.int/oiv/info/enresolution (accessed on January 2014).

Saez L, Perez J, and Martinez J (1992). Low molecular weight phenolics attenuation during simulated treatment of wastewaters from olive oil mill in evaporation ponds. *Water Research* 26: 1261–1266.

SAWIS (2014). South African wine industry information and systems. http://www.sawis.co.za (accessed on June 2014).

SIP (2008). Sustainability in practice (SIP) certification program. http://www.sipcertified.org (accessed on March 2014).

South Australian EPA (2002a). Use of water treatment solids. New South Wales Environment Protection Authority, Sydney, New South Wales, Australia.

South Australian EPA (2002b). Wastewater lagoon construction. New South Wales Environment Protection Authority, Sydney, New South Wales, Australia.

South Australian EPA (2003a). Environment protection (water quality) policy. New South Wales Environment Protection Authority, Sydney, New South Wales, Australia.

South Australian EPA (2003b). Independent verification of monitoring programmes. New South Wales Environment Protection Authority, Sydney, New South Wales, Australia.

South Australian EPA (2003c). Odour assessment using odour source modelling. New South Wales Environment Protection Authority, Sydney, New South Wales, Australia.

South Australian EPA (2004). Bunding and spill management. New South Wales Environment Protection Authority, Sydney, New South Wales, Australia.

VIVA (2011). Sustainability in the Italian viticulture. National pilot project launched by the Italian Ministry for the Environment, Department for Sustainable Development, Climate Change and Energy. http://www.viticolturasostenibile.org/EN/Home.aspx (accessed on February 2014).

6

SUSTAINABILITY ISSUES

MATTEO BORDIGA

Contents

6.1	What Is the Definition of Sustainability?	228
6.2	Why Sustainable Winegrowing?	228
6.3	Environmental Impact of Wine Production	232
6.4	European Wine Regulations	234
6.5	OIV (Organization Internationale De La Vigne Et Du Vin)	235
	6.5.1 OIV—Mission (Laid Out in a Triennial Strategic Plan)	237
	6.5.2 Resolution CST 1/2004 Development of Sustainable Vitiviniculture	238
	6.5.3 Resolution CST 1/2008 Guidelines for Sustainable Vitiviniculture: Production, Processing, and Packaging of Products	239
6.6	Italian Wine Industry	246
6.7	Environmental Sustainability Assessment of Winegrowing Activity in France	248
6.8	The LIFE HAproWINE Project—Environmental Sustainability of the Wine Sector in Castilla y León (Spain)	249
6.9	Lodi Winegrowing Commission Sustainable Workbook and Lodi Rules	251
6.10	Sustainable Winegrowing in New Zealand	252
6.11	Sustainability in Practice and Low Input Viticulture and Enology	254
6.12	Integrated Production of Wine in South Africa	255
6.13	California's Sustainable Winegrowing Program	256
6.14	VineBalance, New York State's Sustainable Viticulture Program and Long Island's Sustainable Winegrowing	260
6.15	Wines of Chile Sustainability Program	261
6.16	McLaren Vale Sustainable Winegrowing Australia	262
6.17	United Kingdom Vineyard Association	264
6.18	General Conclusions and Discussion	265
References		266

228 VALORIZATION OF WINE MAKING BY-PRODUCTS

6.1 What Is the Definition of Sustainability?

Today, the central issue of international debate revolves around climate change and protection of the environment. Generally, environmental protection is considered an obstacle to the development and prosperity of enterprises, especially in the agricultural sector, and the wine industry is no exception. Vintners, constantly working to improve their products, tend to assume environmentally friendly actions as counterproductive for the quality of their wines as well as their incomes. Sustainability, a sound approach toward winegrowing and wine making, offers a solution to end this conflict by integrating environmental protection, profitability, and social issues.

In fact, considering sustainability just as an environmentally friendly approach is incorrect and very restrictive. First, it is necessary to distinguish sustainability from other two modern forms of winegrowing: biological and biodynamic approaches. Organic winegrowing focuses mainly on decreasing the environmental impact of its products and at the same time ensures the protection of wine from external ingredients, thus preserving its authentic nature. Biodynamic winegrowing strives to maintain the wholesomeness of their products in harmony with nature following the lunar phases. A general definition of organic farming is "an ecological production management system that promotes and enhances biodiversity, biological cycles, and soil biological activity. It is based on minimal use of off-farm inputs and on management practices that restore, maintain, and enhance ecological harmony" (U.S. National Organic Standards Board 1998). However, both approaches neglect two essential parts of sustainability: social issues and the cost-effectiveness of all the measures taken. The general idea is that sustainable viticulture integrates the following three main goals: environmental health, economic profitability, and social and economic equity. Therefore, sustainable winegrowing is not only relevant for the present but also critical for the future.

6.2 Why Sustainable Winegrowing?

Wine production, in parallel with the production of any good, leads to a number of by-products that might be dangerous for the environment as well as for human beings. Waste and especially wastewater are critical factors, considering that generally the production of a single bottle of red wine produces 0.5 kg of waste and emits 16 g of SO_2 (FAO 2012). Based on this, to develop and follow specific guidelines directed by sustainable winegrowing acquires a certain relevance (Figure 6.1). Water consumption in the wine making process appears to be crucial (on average wineries use 2 L of fresh water for each 0.75 L bottle of wine). Reducing water consumption in a winery is potentially a first step toward sustainability. However, the major challenge is rather to prevent marc, must, or wine to enter the drainage system. In addition to standard health and safety protection in the workplace, there is still a need for optimization of the entire process. All workers should be engaged in this constant optimization process while giving importance to long-term and strong cooperation with stakeholders, suppliers, and customers. However, sustainability, more than just being a recent trend, is a concept applied successfully by the international wine industry for several years that needs continuous improvement and dissemination. This approach is used not only to decrease costs and human influence on the environment but also to create a competitive advantage and public relations. Building wineries, oriented toward

Figure 6.1 Grape pomace (solid parts separated in skins, seeds, and stems).

230 VALORIZATION OF WINE MAKING BY-PRODUCTS

sustainability right from the beginning, seems to be an important opportunity especially for developing wine countries, taking recourse to numerous advantages and possibilities. Sustainable practices are also very significant when it comes to communications, advertising, or public relations. Countries may be keen to showcase efforts made by their wineries to adopt a sustainable practice. The promotion of green campaigns seems to be the main goal, not only in the production of wine but also to support other agricultural products. "Green" and "clean" are the two major branding aspects of the entire country. Winegrowers use sustainability as a sound marketing tool following a marked tourist-oriented approach. They take responsibility for the biodiversity by protecting local flora and fauna, which can be a tourist attraction. This philosophy shows two positive results: on the one hand, the overall health of the vineyard is supported and secured in the long term; and on the other hand, the biodiversity approach helps attract tourists to wineries. This principle certainly results in additional benefit and has a competitive advantage in today's overcrowded wine shelves leaving customers satisfied with this approach. Both development and execution of sustainability strategies in winegrowing prove to be the main objectives of the program. The business purpose, long-term goals, fundamental ideals, and actions represent some core elements responsible for the definition of business strategy. It appears crucial, starting from the small family-operated vineyard and winery to the corporate organization, to define and implement an accurate sustainability strategy. Providing winegrape growers and vintners with tools capable of guiding them to a proper sustainable development is the mission of the winegrowing program. Following these guidelines, growers and vintners themselves are able to apply this approach, assessing the sustainability of current practices, identifying areas of excellence and areas that require improvement, and developing action plans to increase an operation's sustainability. A long-term mission includes identification followed up by voluntarily disseminating best practices to maintain sustainable winegrowing to the wine community. Sustainable practices and how self-governing enhances the economic viability and future of the wine community is an important learning step. In addition, having an open dialog demonstrates how by working closely with neighbors and other stakeholders one can address concerns and enhance mutual respect.

SUSTAINABILITY ISSUES

The program defines sustainable winegrowing as winegrape growing and wine making practices that are sensitive to the environment, responsive to the needs of society, and feasible to implement and maintain. Sustainability is referred to as the combination of environmental, economic, and social principles (Figure 6.2). These three overarching principles provide a general direction for pursuing sustainability with a vision for both the present and future generations. However, these important principles are not easy to implement in the daily operations of winegrape growing and wine making. Striving to produce the best quality grapes and wine, providing leadership in protecting the environment, natural resources, and the long-term viability of agricultural lands, are a part of the numerous sustainability values. The economic and social wellbeing of vineyards and winery employees are supported, which enhance local communities through job creation, supporting local businesses, and actively working on important community issues. Furthermore, research and

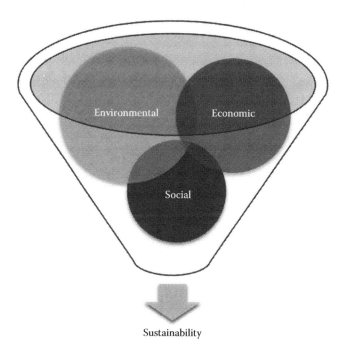

Figure 6.2 A combination of environmental, economic, and social principles is responsible for the general definition of sustainability. (From OIV Resolution CST 1/2008, OIV guidelines for sustainable vitiviniculture: Production, processing and packaging of products, http://www.oiv.int/oiv/info/enresolution, accessed on January 2014.)

232 VALORIZATION OF WINE MAKING BY-PRODUCTS

education are strongly supported as well as existing practices, which are continuously monitored and evaluated, in order to expedite continual improvements.

6.3 Environmental Impact of Wine Production

The main environmental impacts associated with wineries are related to pollution of water, degradation of soil and damage to vegetation arising from liquid and solid waste disposal practices. Another problem concerns both odors and air emissions resulting from the management of raw materials and wastewater, and solid and semisolid by-products from the wine making process. Even if with less impact, noise from pumps, chillers, crushers, and other winery equipment, as well as vehicle noise, particularly during vintage, represent some aspects that need further attention. Although the wine industry might be classified as less "dirty" than other sectors (e.g., the chemical sector), wine producers and vine growers have been increasingly engaged in sustainability, driven by different forces, mainly environmental concerns. The wine industry, indeed, has to face a number of environmental issues and challenges. The literature reports several environmental sustainable practices. Several relevant aspects often include soil management, water management, wastewater, biodiversity, solid waste energy use, air quality, and agrochemical use. Producers have to promote a sustainable use of chemicals, limiting their utilization, in order to preserve and enhance the level of biodiversity and soil fertility. The mandatory way to manage water is responsibly, optimizing its consumption and reducing runoff of contaminated wastewater. Furthermore, wineries must manage the landscape, minimizing their impact on the community to protect the health and safety of workers. Sustainability concerns have become increasingly important since the publication of the "Our Common Future" report by the United Nations Commission on Environment and Development in 1987 (World Commission on Environment and Development 1987). Since then, many countries have developed sustainability initiatives, which have in turn generated regulations, especially on the environmental and social aspects of sustainability. One of the common drivers of many agricultural sustainability initiatives is strictly linked to the harmful consequences of chemical inputs. Considering the high value of grapes, winegrape growing regions have developed some of the most

SUSTAINABILITY ISSUES

complex sustainability assessments and certifications, in some cases, incorporating a triple-bottom line approach, which evaluates the entire production system with regard to the interrelationship of economic, environmental, and social factors. Table 6.1 reports some of the main sustainability programs for winegrape growing worldwide. Considering the most important part of self-assessment, the Californian wineries definitely take the leading role (California Sustainable Winegrowing Alliance 2004; CSWA 2013). As early as 2001, the Californian wineries developed the "Code of Sustainable Winegrowing Practices" under the leadership of the Wine Institute and the "California Association of Wine Grape Growers." New Zealand is also one of the pioneers of sustainable winegrowing. "Clean" and "green" appear as the two indissoluble features representing the whole country and its products, and winegrowing has to meet the requirements stemming from this approach. "Sustainable Winegrowing New Zealand" is the organization responsible for the process focused on integrating sustainable practices into winegrowing. Similarly, Australia is very active about sustainability in winegrowing. The general strategy called "Sustaining Success," which has been established in 2002, integrates all sustainable activities. The State of Victoria in Australia results particularly active in terms of sustainable winegrowing, founding the "Environmental Protection

Table 6.1 Principal Programs from Different Countries

COUNTRY	PROGRAM
Italy	Tergeo
	V.I.V.A. Sustainable Wine
France	TerraVitis
Spain	LIFE HAproWINE project
California State	Lodi Winegrowing Commission Sustainable Workbook/Lodi Rules
California State	Vineyard Team/Sustainability in Practice (SIP)
Oregon State	Low Input Viticulture and Enology (LIVE)
California State	California Sustainable Winegrowing Alliance (CSWA)
	California Sustainable Winegrowing Program (SWP)
New York State	VineBalance
	Long Island Sustainable Winegrowing (LISW)
New Zealand	Sustainable Winegrowing (SWNZ)
South Africa	Integrated Production of Wine (IPW)
Chile	Sustainable Wine from Chile (SWC)
Australia	McLaren Vale Sustainable Winegrowing (MVSWGA)
United Kingdom	United Kingdom Vineyard Association (UKVA)

234 VALORIZATION OF WINE MAKING BY-PRODUCTS

Authority," which supports winegrowers and wineries in implementing sustainability schemes. The South African program for sustainable winegrowing, called "Integrated Production of Wine," was established as early as 1998. Different partners, also include research institutes and pesticide or fertilizer suppliers, have developed this system voluntarily. The World Bank financially supports another project, the Biodiversity & Wine Initiative established in 2004, focused on the protection of biodiversity, which is endangered by the monoculture vine plantings. However, analogous concepts also exist in European winegrowing countries. For example, at the University of Montpellier (France), there is a special course of study called "Integrated Wine Production," which solely deals with all aspects of sustainable measures in winegrowing, environmental laws, and marketing strategies toward sustainable wines. In Germany, international sustainability schemes such as ISO 14001 (also applied in some regional Australian environmental management schemes) and EMAS II (Eco-Management and Audit Scheme) have been successfully adopted. In addition to process optimization, resources preservation, and saving money, these companies clearly emphasize the potential effects of the communication of their certification. Showing the logo of the achieved certification enables them to communicate the value and reliability of their practices to customers.

6.4 European Wine Regulations

The European Union (EU) produces and consumes around 60% of the world's wine, having almost 50% of the world's vineyards (E-Bacchus 2012; Eurostat 2013). The EU is not only the largest global wine-producing region and the main importer and exporter of wine but it also has a highly regulated market (European Commission 1987–2007, 2003, 2006a,b, 2007a,b, 2007–2012, 2008, 2009–2012; European Court of Auditors 2012). Government intervention has taken many forms in EU wine markets, determining where (or not where) certain wines can be produced, the minimum spacing between vines, the typology of the vines that are allowed to be planted in certain regions, yield restrictions, and limiting imposition to new vineyard planting. However, wine regulation in the EU, characterized by several regulations related to different political pressures, presents a complex historical origin whose features are not always positive. General economic studies on the EU's wine markets have led to the conclusion that the policies have not been effective at solving the problems but conversely, causing some major distortions in the wine sector. Since 1962, more than 2000 regulations, directives, and decisions have been published in the EU concerning the wine sector, reforming the main wine framework law of 1962 at least five times (Council Regulation (EEC) No. 822/87, (EEC) No. 823/87, (EC) No. 479/2008; (EC) No. 491/2009; (EC) No. 607/2009). Many of the current EU regulations can be traced to French regulations of the late nineteenth and early twentieth centuries such as quality regulations like the "Appellations d'Origine Contrôlées" (AOC) system, which were introduced to protect producers of "quality wines," such as wealthy landowners of Bordeaux, from imitations and adulterations. At the same time, quantity regulations, such as planting restrictions, were introduced to protect French producers from cheap wine imports. After a period of general liberalization, however, surplus crises in the 1970s caused strong pressure from French producers to once again impose the regulations and extend them to the entire EU. Therefore, the actual integrated regulations, initially French and, to a lesser extent Italian, are now applied to approximately 60% of the world's wine production, resulting in a combination of economic, political, and institutional integration that may be responsible for partially inefficient institutions. Table 6.2 reports the chronology of principal European wine regulations. The extent of the regulatory interventions appears well represented if we consider that, for example, in the past three decades, every year on average, 20–40 million hectoliters (hL) of wine have been destroyed (through distillation). This volume represents 13%–22% of EU wine production, the equivalent of 3 to 6 billions bottles.

6.5 OIV (Organization Internationale De La Vigne Et Du Vin)

The International Organization of Vine and Wine (OIV), which replaces the International Vine and Wine Office, was established by the Agreement of April 3, 2001. The OIV is an intergovernmental organization of a scientific and technical nature of recognized competence for its works concerning vines, wine, wine-based beverages, table grapes, raisins, and other vine-based products. Thirty-five sovereign states signed this Agreement, subject to different internal procedures (approval, acceptance, or ratification). The Agreement went into force

236 VALORIZATION OF WINE MAKING BY-PRODUCTS

Table 6.2 Chronology of Principal European Wine Regulations

1962	Establishment of a viticultural land register.
	Harvest and stock declaration.
	Compilation of future estimates of resources and requirements (annually).
1970	Vines were classified into "recommended," "authorized," and "provisionally authorized" cv.
	Definition of different types of wine.
	Definition of the maximum amount for enrichment and alcohol strength in wine production.
	Common rules introduction on labeling and oenological practices.
	Introduction of distillation of excess production and obligatory distillation of the wine making by-products.
	Trade monitoring with non-member countries was established.
	Declaration of free movement of wine within the community.
1979	Compulsory distillation of wine obtained from table grape.
	Definition of oenological practices.
1987	Prohibition of new planting and temporary surpluses storage for table wines.
	Further subsidies introduction for vineyards conversion.
	New rules for quality wine production.
1999	Prohibition to planting new vines until 2010.
	Setting up of wine making processes and practices.
	Obligatory by-products, table wines, and crisis related event distillation is maintained.
	Introduction of restructuring and conversion measures for vineyards.

Source: From Council Regulation (EC) No. 1493/1999 of 17 May 1999 on the common organization of the market in wine, Official Journal of the European Union.

on January 1, 2004, following the deposit of the 31st Instrument of Ratification. As of June 13, 2013, the OIV was made up of 45 member states, which are as follows: Algeria, Argentina, Australia, Austria, Azerbaijan, Belgium, Bosnia-Herzegovina, Brazil, Bulgaria, Chile, Croatia, Cyprus, The Czech Republic, Finland, France, Georgia, Germany, Greece, Hungary, India, Israel, Italy, Lebanon, Luxembourg, FYR Macedonia, Malta, Moldavia, Montenegro, Morocco, the Netherlands, New Zealand, Norway, Peru, Portugal, Romania, Russia, Serbia, Slovakia, Slovenia, South Africa, Spain, Sweden, Switzerland, Turkey, and Uruguay (OIV 2012, 2013). The international treaty of April 3, 2001 also enables some territories and organizations to participate in the OIV work as observers: AIDV—International Wine Law Association; Amorim Academy; AREV—Assembly of Wine-Producing European Regions; AUIV—International University Association of Wine; CERVIM—Centre for Research, Environmental Sustainability and Advancement of Mountain Viticulture; FIVS—International Federation of Wines and Spirits; OENOPPIA—Oenological

Products and Practices International Association; UIOE—Union Internationale des Œnologues; VINOFED—World Federation of Major International Wine and Spirits Competitions; ASI—Association de la Sommellerie Internationale; Yantaï (China) prefecture-level municipality; and Ningxia Hui (China) autonomous region. Consensus is the normal method whereby the General Assembly shall adopt draft resolutions of a general, scientific, technical, economic, or legal nature. The new organization appears rationalized and has adapted to enable the organization to pursue its objectives, exercising its activities as an intergovernmental organization of a scientific and technical nature. Its competences contains these major topics as vines, wine, wine-based beverages, grapes, raisins, and other vine products.

6.5.1 OIV—Mission (Laid Out in a Triennial Strategic Plan)

The OIV provides information on actions to members whereby producers, consumers, and stakeholders show concerns requiring consideration. At the same time, the OIV provides assistance to other organizations mostly involved in carrying out standardization activities (both intergovernmental and nongovernmental). Its contribution to international practices and standards harmonization appears strategic and, moreover, the organization is also involved in the preparation of new international standards in order to improve the conditions for producing and marketing vine and wine products, still taking into account consumers' interests. To attain these objectives, the OIV's activities are focused on both promotion and guidance of scientific and technical research, monitoring the implementation of dedicated recommendations in collaboration with its members in different areas such as grape production, oenological practices, product definitions, labeling and marketing conditions, and analytical methods and assessing vine products. The submission of proposals is a crucial activity realized by the organization. Related topics range from guaranteeing the authenticity of vine products, especially with regard to consumers (label information), protecting geographical indications (vine- and winegrowing areas and the related appellations of origin), to improving scientific and technical criteria for recognizing and protecting new vitivinicultural plant varieties. Mutual recognition of practices is facilitated, contributing to the harmonization and adaptation of regulations by its members. The food safety area appears

238 VALORIZATION OF WINE MAKING BY-PRODUCTS

deeply investigated by specialist scientific monitoring, focusing on the estimation of the specific characteristics of vine products, by promoting and guiding research into appropriate nutritional and health aspects, followed by the dissemination of results to the medical and healthcare profession. The amount of biodegradable waste sent to landfills in member countries by 2016 must reach 35% of the levels reached in 1995 (European Landfill Directive 1999/31/EC). Therefore, European food processing industry operations have to comply with increasingly more stringent European Union environmental regulations related to the disposal or utilization of by-products and waste. These include growing restrictions on land, spraying with agro-industrial waste, and on disposal within landfill operations, and the requirements to produce final products that are stable and hygienic. Unless suitable technologies have been developed for the by-product processing, large numbers of food processing operations will be under threat. The importance of considering all dimensions is clearly reported in the definition of "sustainable viticulture" given by the OIV in its resolution of the Scientific and Technical Committee (CST) 1/2004. This resolution addresses the issue of sustainability in the production of grapes, wines, spirits, and other vine products, and lays the foundations for the guidelines for production, processing, and packaging of products further improved in the next resolution (Resolution 1/2008). Sustainable viticulture, defined as a "global strategy on the scale of the grape production and processing systems," incorporates at the same time the economic sustainability of structures and territories. This approach covers different areas of interest among which a product's quality, environmental risks, safety, and consumer health. In this manner the cultural and ecological aspect results are also therefore enhanced.

6.5.2 Resolution CST 1/2004 Development of Sustainable Vitiviniculture

Considering the guidelines for integrated production in viticulture defined by the International Organization for Biological Control (IOBC), the OIV General Assembly decided to adopt the following elements as general principles of sustainable development applied to vitiviniculture, observing the existence of different national regulations related to reasoned, integrated, and sustainable production. These general principles, observe the existence of different approaches and national regulations

SUSTAINABILITY ISSUES 239

(e.g., in particular reasoned, integrated, and sustainable production), have been adapted as needed depending on social, regulatory, and economic cultural aspects, and the natural climatic and soil conditions of each country and its regions. The program for the development of sustainable vitiviniculture has been developed as a priority in the context of the OIV strategic plan. Defined as a "global strategy" on the scale of grape production and processing systems, this program was aimed at certain objectives such as grape and wine production, able to meet consumer demands and safety, assuring its health but also at the same time protecting producers and staff associated with production. A main objective is the minimization of any environmental impacts linked to viticulture and the transformation process, thus promoting a sustainable vitiviniculture also from an ecological and economic point of view (e.g., rational use of energy). Efforts shall be made both to maintain biodiversity of viticulture and associated ecosystems, managing waste and effluents in an effective way, and preserving and developing viticultural landscapes. In this resolution, an implementation section reports some further goals such as the need to develop a sound strategy taking into account regional or national networks, undertaking an assessment of global production system. Some assessment criteria shall be developed to measure progress made with this strategy, subsequently adapting the latter to local and territorial specificities. A technical pathway evaluation shall be required based on qualitative economic constraints, consumer safety, and environmental aspects, and developing practices related to precision techniques. Starting from an initial assessment, the need to establish an improvement plan together with a regular progress report based on adapted environmental indicators appeard valid even because these criteria are being potentially used by producers in their communications with consumers.

6.5.3 Resolution CST 1/2008 Guidelines for Sustainable Vitiviniculture: Production, Processing, and Packaging of Products

Following the proposal of the Scientific and Technical Committee, where Resolution CST 1/2004 establishes guidelines for the production of grapes, wines, spirits, and other vine products in accordance with the principles of sustainable development applied to vitiviniculture, the OIV General Assembly decided to adopt the following guidelines for implementing environmental sustainability in the

240 VALORIZATION OF WINE MAKING BY-PRODUCTS

vitivinicultural sector. Member states are recommended to refer to this guidelines as a basis for the development, updating, and review of national or regional procedures for environmentally sustainable vitiviniculture, continuing the programs related to the development of sustainable vitiviniculture, as indicated in the Strategic Plan, within the OIV. Resolution 1/2008 represents an implementation guide for environmentally sustainable production in the world vitiviniculture sector. Table 6.3 shows some aspects of the 2008 reform of European wine policy. Protection and preservation of natural resources (solar energy, climate, water, soil) assets through environmentally sustainable practices are mandatory for the long-term viability of vitivinicultural activities. Economic, environmental, and social sustainability are the three aspects that an appropriate program should satisfy. The environmental risk assessment results as the base to the development of sustainable activities. In this regard, several aspects should be taken into consideration (e.g., site selection; biodiversity; variety selection; solid waste; soil and water management; energy use; air quality; wastewater; neighboring land use; human resource management; and agrochemical use). "Self-assessment" and other forms of evaluation are able to measure the lack of and improvement in environmental performance that should be incorporated into a sound sustainability program,

Table 6.3 The 2008 Reform of European Wine Policy

Support measures	Each country is entitled with a dedicated finding budget. Support programs contain one or more of the following measures as single payment scheme support; promotion; conversion of vineyards; green harvesting; mutual funds; harvest insurance; investments; by-product distillation.
Trade with third world countries	According to the World Trade Organization, market intervention measures followed this policy such as distillation and public storage items.
Regulatory measures	The Commission can approve or change wine making practices.
	Wines are now divided into "wine with a Geographical Indication (GI)" and "wine without a GI." Geographical Indication category is divided into two subcategories: Protected Designation of Origin (PDO) wines (highest quality level) and Protected Geographical Indication (PGI).
	National quality labeling schemes are maintained for PDO wines, while wines without a GI can be labeled with grape variety and vintage.
Production potential	The planting right regime will finish at community level starting from 2016. Each member state has the right to extend the limit until 2018.

Source: From OIV Resolution CST 1/2008, OIV guidelines for sustainable vitiviniculture: Production, processing, and packaging of products, http://www.oiv.int/oiv/info/enresolution, accessed on January 2014.

SUSTAINABILITY ISSUES   241

ensuring continuous control. Intra- and inter-sectorial cooperation for natural resource management for a comprehensive ecological and social management is recognized to be important to increase the overall awareness within the vine and wine sectors. The identification of requiring protection areas (environmental and landscaping interests) is a major topic that the management of vineyards and wineries, in compliance with regional, national, and international regulations, must carefully supervise. Work must be adapted in order to optimize energy use, also focusing on a management plan related to effluents and waste for their reduction, recycling, or reusing. Constant training for personnel (responsible for sustainable activities) must be achieved together with regular updating on techniques, which contribute to sustainable development. The selection of the infrastructure, equipment, and services is realized in accordance with continual improvement principles, taking into account the environmental performance of the supplier, the energy and water use, its durability, and in addition, recycling possibilities. During soil preparation/cultivation, the damage and harmful effects caused to the landscape and environment should be limited in so far as possible. Fundamental principle of environmentally sustainable production results inputs reduction. The main goal is restricting the use of inputs as additives, processing aids, and packaging materials (mentioned in the Oenological CODEX), favoring renewable resources. Using optimized infrastructure, equipment, and efficient processes, water and energy consumption (required for cultivation, wine production operations, and packaging) can be reduced. In environmentally sustainable wine making, a fundamental consideration should be directed on by-products and waste management, comprising their reduction at the source, useful recovering, and recycling. The end use of effluents should determine the treatment and the choice of chemicals to be used as disinfectants and cleaning agents, minimizing their impact on the environment and the local community. The characterization of effluents should be based on analytical parameters such as biological oxygen demand (BOD) or chemical oxygen demand (COD), pH, thus enabling the identification of the treatment required and its optimization. At the same time, limiting the presence of solid matter and optimizing by-product separation (stalks, skins, seeds, and yeast lees) appear to be important requirements related to sustainability together with avoiding mobile equipment (tractors, harvesting

Figure 6.3 Grape skins (Muscat).

machines) wash near a watercourse or sampling site (Figure 6.3). Both the storage and treatment of effluents and solid waste might be carried out, avoiding alteration or contamination risks, in properly located areas to minimize their impact and pollution potential with respect to the landscape and community. The ideal treatment systems should be adapted both to assist liquid waste separation (contaminated and uncontaminated) and to minimize potential airborne contamination, supporting agronomical or biological processes with an efficient use of energy. Application of treated waste in vineyards, orchards, and fields should take into account the characteristics of the soil and crops, based notably on the following criteria: BOD or COD and pH. Sustainable production applied to viticulture production operations is certainly achievable on condition that some principles are fulfilled. For example, establishing the vineyard appears a fundamental step, including the determination of the viticultural aptitude and potential of the site (soil study, water availability, and water protection requirements). Biodiversity maintenance must be properly ensured and attention must be paid to use plant material (vine type and rootstock) free from serious viral diseases and suitable for the local conditions and the

SUSTAINABILITY ISSUES

required type of production. Compatible with sustainable production, the vine training system might be realized, taking into account different items such as water requirements, grape quality, soil potential and protection, vine vigor, density and layout of the vines, and phytosanitary product applications, while of course not neglecting landscape protection and risks of disease reduction. Fertilizer input should be compatible with quality grape production, vine health, and soil fertility maintenance, minimizing its quantity where possible in relation to plant needs, soil type, and risks of leaching. However, preference should be given to recycling organic nutrients. Soil maintenance, designed to create optimum conditions for the plants, preventing erosion, compaction, and nutrient leaching, promotes biological diversity. All the appropriate measures to protect the soil against erosion should be taken, such as green covering, cover cropping, ground coverage or mulches (straw, compost), site adaptation, and terrace maintenance. The strategy must be evaluated considering the precipitation levels, soil water reserves, erosion, and frost risks. Moreover, other items appear to be important to consider especially those related to the vine and its age, the training system used, and the grape quality and output, in particular the nitrogen content of musts. Water input should be related to production objectives (winegrapes, table grapes, raisins) for the vine at the various stages of its development, the specific nature of the grape, and the wine requirements, taking into account the water balance of each vineyard (Resolution VITI 2/2003). Irrigation techniques designed to optimize water efficiency, such as micro or drip irrigation, might be preferred, ensuring a sustainable viticulture which is able to limit the risks of environmental impact, in particular, in terms of soil salinity and water reserves. The winter pruning period should be selected according to local climatologic conditions in order to limit both contamination risk and wounds, thus reducing the liability of wood rotting diseases (Resolution VITI 02/2006). The basic purpose of phytosanitary protection is to protect the vine against pests and diseases, while respecting the environment. Preventative treatments should be evaluated according to the potential risk of developing diseases and pests (VITI-OENO 1/2005). Annual and regional information documents, as well as fungal disease forecasting models should be used as the basis for a protection strategy. Prevention is feasible using suitable vine types and rootstocks, appropriate vine

244 VALORIZATION OF WINE MAKING BY-PRODUCTS

training systems, proper cultivation methods, soil maintenance (green covering, soil cultivation period), and preserving beneficial organisms. However, if phytosanitary protection is required, the quantities used should be compliant with legal restrictions taking into account important issues such as the phenological stage and the surface area of the plant, accidental effects on beneficial fauna and nontarget organisms, and the beneficial organism's toxicity (particularly bees). In addition, several risks should be carefully taken into consideration, especially those related to developing resistance, water or soil pollution, and grape and wine residues affecting the vinification. During the handling and application of phytosanitary products, it is important to pay attention to the proper techniques and weather conditions, also having available a dedicated area with a system that both avoids possible network contamination and limits the risk of accidental overflows or spillage. The spray operator has to utilize suitable techniques (maintenance and calibration of the spray plant apparatus is recommended) and protective equipment, thus avoiding any risk of intoxication/contamination associated with the preparation of the mixture and with spraying. Also, the storage of phytosanitary products appears a crucial step that should be methodically realized and with high security. Phytosanitary products (in their original packaging with the label and safety factsheets) have to be stored into a specific ventilated and locked area, avoiding any contamination or accidents. During harvesting, some specific challenges emerge regarding inputs, pollution management of by-products and effluents as reported (VITI-OENO 1/2005). An additional vigilance is required because vintage results as a period of intense physical activities (operation of machinery, mainly into confined spaces and chemical handling). Harvesting operations such as picking temperature and transport period must take into account limiting energy consumption for harvest transport, heating or cooling. At the same time, solid and liquid by-products, resulting from vintage operations, should be properly stored minimizing their risks of alteration, contamination, and environmental impact. A sustainable approach, including hygiene, energy use, and by-product management, have to be applied to production operations, involving physical processes, such as centrifugation, filtration, and heating/cooling or oenological processes. Wherever possible, solid or liquid residues (from clarification or stabilization operations) should be reprocessed to

SUSTAINABILITY ISSUES

recover active materials but disposing any residues, unable to be remanaged, in a proper manner in order to minimize their impact to the environment and the local community. Consideration should be given also to inert containers or wooden barrels (generally used for wine maturation and aging) principally toward their durability, integrity, and opportunity to recycling. Wooden containers demand particular vigilance with respect to hygiene due to their porous nature, so preserving the products that meet the barrel surface. Also in this case, hot water or steam should be preferred to clean and sterilize equipment rather than chemical agents. The possibility to recycle packaging elements (even materials when they are no longer usable) should be the first option. Efforts should be made to efficiently manage waste as packaging containers (glass, plastic, or metal), corks, capsules, labels, and cartons. Minimization of packaging materials results appropriated but still permitting an optimal conservation and presentation of the product. Physical treatments, such as hot water or steam, rather than chemical cleaning or sterilizing agents, are mostly favored to clean and sterilize packaging equipment surfaces (which come into contact with products), but taking into account the energy consumption and water availability. Table 6.4 summarizes the main points of the OIV Strategic Plan (2012–2014).

Table 6.4 Strategic Axes of the Plan 2012–2014

1	Statistical analysis of the sector
2	Economic analysis of the sector
3	Vitiviniculture biophysical, economic, and social environments
4	Sustainable vitiviniculture, integrated production, and organic production
5	Climate change and vitiviniculture
6	Greenhouse effect
7	Biodiversity and genetic resources
8	Regulation and impact of biotechnologies
9	Oenological practices and techniques
10	Identification and analysis methods
11	Safety and quality
12	Nutrition and health—individual and societal aspects
13	Designation and labeling
14	Collection, processing, and dissemination of information
15	International cooperation

Source: From the OIV, Strategic Plan, 2012–2014, http://www.oiv.int/oiv/info/enplanstrategique, accessed on May 2014.

246 VALORIZATION OF WINE MAKING BY-PRODUCTS

6.6 Italian Wine Industry

The Italian wine industry is deeply committed to sustainability, and the attention for this area is constantly growing between the sector workers. In recent years, both private businesses and consortiums have launched several sustainability programs, so pointing out the commitment of farmers and wine producers to the implementation of sustainability principles in viticulture and wine production. However, the manifold design of the sustainability initiatives and the differences related to the objectives, methodologies, and proposed tools, risks to create confusion and undermine the positive aspects of these initiatives. The growing interest from consumers, together with pressures from governments and environmental groups, for green products and the higher commitment to export in countries with a strong attention for sustainable products are among the institutional process to sustainability. However, in the wine sector, environmental preservation is an instinctive approach for winemakers, concerned with maintaining proper environmental conditions and in preserving the natural resources in order to maintain land productivity. Considering this, the wine industry receives an important contribution toward sustainability directly from the inside, especially because small-medium companies mainly form the sector, and frequently the ownership coincides with the management. At the international level, developing wine countries have been pioneers in introducing sustainability in the wine industry (vine growing and wine production). In 2008, following the proposal of the Scientific and Technical Committee, whereas Resolution CST 1/2004 established guidelines for the vitivinicultural sector, the OIV General Assembly decided to implement the guidelines for environmental sustainability deliberating a new resolution, CST 1/2008. Since then, several guidelines and programs for sustainable winegrowing and production have been defined. The most consistent sign of interest for sustainability in the wine sector in Italy can be considered the wide range of sustainability programs launched in recent years by private producers and consortiums. However, if on the one hand this is a positive signal regarding the sustainability issues in viticulture, on the other hand, confusion can arise among vine growers and wine producers considering the large number of

SUSTAINABILITY ISSUES

programs and their fragmentariness. The great number of different strategies and guidelines make their comparison extremely arduous, with the risk that producers do not have a clear understanding of the opportunities arising from the implementation of a specific sustainability program. Moreover, it is evident that a lack of clarity related the growing range of sustainability-sounding names and adjectives reported on the bottles. Terms like *sustainable, organic, natural*, or *eco-friendly*, often not adequately explained, influence consumers. Sustainability initiatives spreading appears to be a great opportunity for the overall wine sector, but confusion and overlapping of initiatives, methodologies, and results must be avoided. In the Italian wine sector, the promotion of a common notion of sustainability is mandatory together with a broader industry-wide sustainability strategy. "Tergeo" and "V.I.V.A. Sustainable Wine" represent two examples of this approach (VIVA 2011; Unione Italiana Vini 2011). The Tergeo program, realized by the UIV (Unione Italiana Vini) an Italian wine trade association, aims to support environmental, social, and economic sustainability in the Italian wine sector. The main objective is to enhance the "knowledge and technology transfer" from companies and researchers to farmers and wine producers by acting as a sustainability promoter. The Tergeo Scientific Committee, composed of academic professors, researchers, and experts of the wine sector, is responsible for evaluation of different applications proposed by the partners (wineries not included). Once accepted, the application is proposed to the association members (producers and farmers). In 2011, the Italian Ministry for the Environment, Land and Sea launched a project named "V.I.V.A. Sustainable Wine" aiming to evaluate the wine-sector sustainability performance (based on Water & Carbon Footprint calculation). Participants of this program are some large Italian wine-producing companies together with universities and research institutes. Similar to Tergeo's program, its main objective is to establish a common methodology for the environmental, social, and economic sustainability assessment in the wine sector. Air, Water, Vineyard, and Territory represent the four indicators applied in the V.I.V.A. Sustainable Wine approach. The creation of a common understanding of sustainability is crucial for producers as well as for the entire sector both resulting in a more effective communication toward consumers and reducing the uncertainty linked to the presence of a wide range of certifications and sustainability labels on the market. It is evident that a common language and framework is required. The large number of sustainability initiatives and programs generates a situation of uncertainty that affects companies. This condition induces the risk that both the real characteristics of each program and benefits do not appear clearly understood by the management and ownership. The competitiveness of Italian wine, particularly on foreign markets promoting sustainable products, is somehow resized in the absence of a single and unique sustainability framework.

6.7 Environmental Sustainability Assessment of Winegrowing Activity in France

Also in France, sustainability has become a public matter, raising governmental interest for sustainable wine production and viticulture practices. The concept of integrated farming (*agriculture raisonnée*) aims to reduce the negative impacts from agriculture and the use of chemicals. In France, the system is supported by overall specifications, including waste and water management, landscape, and biodiversity. Integrated farming is a first level of sustainable agriculture, characterized by an enhanced practices control. TerraVitis®, framework created in 1998 and based on integrated farming protocol, publicizes and recognizes the concept of sustainable wine production, offers serious guarantees by the application of technical specifications, and proponing to consumers' wines from sustainable production (TerraVitis 1998). TerraVitis interestingly focuses on agricultural practices, expanding the original scope through the association's action, which tends to respect local conditions (terroir approach). It is possible to effort to reduce the environmental impact generated by the agricultural production process, while also searching for alternative solutions toward disease control. The National Federation TerraVitis (FNTV) includes five local associations, affecting the winegrowers from different departments: Burgundy Beaujolais, Loire, Bordeaux, Champagne Vineyard, and Rhone—Mediterranean. In 2009, TerraVitis wine producers improved further by including in their approach and their technical reference, some new elements such as water resources and air management, biodiversity, soil resources, and energy resources. This framework represents the best answer to a competitive modern agriculture, offering quality products and resulting in an example of successful cooperation between technicians and chambers of agriculture. In 2000, another innovative integrated farming system was developed to meet the assessment tool demand of vine growers by adapting to the vine production, the INDIGO® method. INDIGO indicators, developed initially for arable farming, have been tested to answer the need of the viticulture industry in Alsace, Champagne, Burgundy, Jura, and Loire Valley vineyards. Six indicators were tested in vineyards including pesticides, energy, nitrogen, organic-matter, soil-cover, and frost. The results assessed the feasibility and the robustness of the INDIGO indicators (multicriteria method of environmental evaluation), also showing large variations in rain intensity, fungi attacks, and winegrower practices amongst vineyards. This positive example of adaptation of indicators (initially from arable farming) for different vintages, vineyard conditions, and winegrower practices fits perfectly with today's key factors of the environmental impact of viticulture, particularly in the context of the global wine markets.

6.8 The LIFE HAproWINE Project—Environmental Sustainability of the Wine Sector in Castilla y León (Spain)

The project (co-financed by the European Union) goal is favoring a sustainable development in the wine sector in Castilla y León (an autonomous community in northwestern Spain) (LIFE HAproWINE Project 2013). The strategic lines of action include six items as waste, sustainable management, environmental technologies, oenology-tourism-heritage, water, and energy. Challenges in managing waste, not only obtained during the production process but also in distribution and consumption, are becoming ever more important for the wine industry. Waste logistic appears to be crucial for the sector with respect to establishing collaborative programs between companies (a specialized database) to promoting research and development projects focused on waste revaluation. New product designs, derived from grapes and wine, must be improved, generating a by-products market applying feasible waste recycling processes. Sustainable management aims to minimize the economic, social, and environmental impacts over the life cycle of a product. To do this, an appropriate and effective system of sustainable management is required that includes also a proper

250 VALORIZATION OF WINE MAKING BY-PRODUCTS

control of inputs and supply planning, an effective impact detection systems, planned measures to minimize negative impacts, a monitoring system, and communication. Collaboration of all stakeholders at all stages, able to launch mechanisms that promote sustainability through the value-chain, is required to implement environmental management systems, certified (ISO 14001/EMAS) or not, or energy management systems (International Organization of Standardization 2004). This strategy includes also elaborating certification protocols according to current regulations and integrating the different aspects of sustainability within the legal framework. The wine sector must invest in adapting its production to new processes and suitable technology systems, constantly updating about innovations related to winegrowing practices as well as winery processes (e.g., CO_2 emissions capture). The use of renewable energies should be increased and attention paid on designing new and less polluting containers and packaging. The impact that climate change exerts on vineyards must be monitored together with process development, techniques, and technologies to the sector's adaptation to these fluctuations. Promoting wine as part of rural tourism and associated with other values such as the natural, historical and cultural heritage results a fundamental business opportunity. This concept is strictly linked to the protection of vineyards landscape (territorial and urban planning), as well as to the conservation of their biodiversity. A major environmental impact related to wine making is sewage water (cleaning and disinfection of equipment in the winery). Considering that water is becoming an even more scarce and expensive resource, its consumption results in a key point. Water management in the land and wineries must be improved, favoring technology integration for water reuse that is able to minimize underperforming irrigation practices and promote more efficient systems (dripping). The main areas for energy improvement focus on refrigeration, heating, air conditioning, pumping, air compression, and lighting. In order to minimize energy consumption at all stages of the wine making process, it is necessary to improve the efficiency of facilities with best practices that are able to increase competitiveness, reducing energy and environmental costs. Systems that are able to obtain bio-energy from by-products (seed compost, skins, and stems) appear suitable in this direction but also a sustainable building approach (winery design and construction) could improve energy efficiency.

6.9 Lodi Winegrowing Commission Sustainable Workbook and Lodi Rules

In 1991, Lodi Winegrape Commission (LWC) has been created with the main objective of promoting the Lodi wine region (California) and its wines. The most important issues for winegrowers resulted in the Integrated Pest Management (IPM) program. In 1995, the program implemented its objectives by tracking the results of a series of sustainable winegrowing projects. In 2005, the Lodi Rules program (a third-party certification scheme) was established to respond to the growers' demand for a marketing application for the self-assessment workbook (Lodi Winegrape Commission 2005; Ohmart 2008). The program has grown vigorously since its creation as of 2012, nearly 26,000 winegrape acres were "Certified Green" in California. The certification process encompasses two components: the Lodi Rules (practice standards; last detailed revision was accredited in 2013), and a Pesticide Environmental Assessment System (PEAS), a risk assessment tool that measures the total impact of all organic and synthetic pesticides used during the year. This certification is designed to lead to measurable improvements in the health of the surrounding ecosystem, society-at-large, and wine quality. The Lodi Rules standards are the backbone of the program, and are based on six items which include business, human resources, the ecosystem, soil, water, and pest management. The second key component (PEAS) represents a model used to quantify the total environmental and human impact of pesticides applied to vineyards annually by generating an Environmental Impact Unit (EIU) for each pesticide. The Lodi Rules takes a comprehensive approach to farming, promoting practices that enhance biodiversity, water and air quality, soil health, and employee and community well-being. Before accreditation by Protected Harvest, standards are peer-reviewed by scientists, members of the academic community, and environmental organizations. Through this approach, this accreditation has received the Consumers Union's highest rating as an eco-label certifier. Integrated pest management is a main goal by limiting crop protection to only essential measures in vineyards and preserving native grasses and trees (wildlife habitat protection). Air and water quality must be monitored constantly. Efforts must be done to reduce air pollution and conserve energy, and to optimize water application through

252 VALORIZATION OF WINE MAKING BY-PRODUCTS

careful program able to maintain an efficient irrigation systems. Soil health results fundamental to obtain excellent grape. Organic matter, by planting cover crops and using compost, appears a sound item in this direction. Finally, the people involved are the foundation for sustainable winegrowing success. Workers must receive proper training that enables them to perform their jobs safely, maximizing their ability.

6.10 Sustainable Winegrowing in New Zealand

Sustainable Winegrowing New Zealand (SWNZ) has its origins in 1995 starting from a pilot project of the Hawke's Bay Winegrowers Association (New Zealand Wines 2014). Following the international demands and constraints, the original inspiration was to assess vineyard chemical use. This was a successful approach that other winegrape regions rapidly began, thus developing a sustainability program able to aim to produce high-quality wine by employing environmentally responsible and economically viable processes in vineyards and wineries. The New Zealand Winegrowers Sustainability policy states that wine must be produced from 100% certified grapes in fully certified wine making facilities. Moreover, certification must be through an independently audited program in which members self-assess their operations online (annually) and provide supporting documentation for their responses in order to become part of the plan. The self-assessment consists of three sets of questions: major, minor, and best practices. The majors are mandatory, minors are generally relevant practices, and best practices are the future step. Winegrowers made vineyard and wine accreditation to the SWNZ a prerequisite to participation in promotional events, resulting that approximately 90% of the wines produced became part of the program. The program is continuously improved adhering to standards that reflect seven key points which include biodiversity, soil, water and air, energy, chemicals, by-products, and people and business practices. Biological diversity (essential for health, vitality, and stability) must be conserved, encouraging benign native flora and fauna. Program members should create an ecologically balanced vineyard environment with a diverse ecosystem of plants and animals, favoring biological control of pests, diseases as a substitute for chemical items. Responsible management must focus on improving soil health, avoiding erosion, and limiting waste and harmful discharges. Soil preservation and vine nutrition, including strategies to maintain or increase organic matter, must be enhanced by suitable practices also applying a sound management of waste disposal systems. Ground cover must be preserved, proving to have a positive effect toward soil structure, erosion, and nutrients also improving biodiversity. Generally, preserving water quality is of critical importance to the industry and wine sector is not an exception. Sustainable water management includes using water efficiently, while avoiding harmful environmental effects. Apply the optimum amount of water to vines to ensure balanced growth, periodically monitoring the irrigation system's performance. Where possible, efforts must be done to recycle winery water also minimizing water wastage. Similar approaches must be adopted to air quality. Program members are required to reduce harmful emissions, and minimize noise and light. New actions to improve energy efficiency should be initiated because, as well as being of comprehensive environmental benefit, energy-saving initiatives greatly reduce the running costs. Vineyards and wineries are encouraged to monitor their energy use steadily by adopting efficient energy management practices. SWNZ aims to a minimal use of pesticides, insecticides, and herbicides. Quality production of grapes and wine can be obtained by applying a sound integrated plant protection giving priority to natural, cultural, and biological methods. Approved agrichemicals should only be used when justified, based on monitoring programs ensuring safety to both environment and human health. The program focuses on initiatives that promote the minimization, reuse, and recycling of waste/by-product, in addition to its responsible disposal. Training staff in the management and handling of these materials appears a crucial point. According to different government, regional, and industry initiatives, responsible waste management practices must be developed and properly adopted, thus reducing the risk of spills of by-products, chemicals, and petrochemicals. There is a similar approach with regard to wastewater management and disposal. A wastewater treatment system has to be designed and managed in order to minimize any environmental impacts. Desirably, wine sector workers should have an exhaustive knowledge of the aims and principles of the program by understanding the effects on the environment, human health, and safety. This goal is achievable by protecting the well-being of workers, encouraging training, promoting workplace quality, and positive contributions to the local community. Constant growth can be reached by adopting a sustainable

254 VALORIZATION OF WINE MAKING BY-PRODUCTS

business models and prudent financial management able to increase value and reduce costs. These guidelines inevitably allow precious ongoing reinvestment in people, businesses, and the community.

6.11 Sustainability in Practice and Low Input Viticulture and Enology

In 1994, the Central Coast Vineyard Team (CCVT) was created by a group of growers, wineries, and service providers, followed 2 years later, by a self-assessment on sustainable vineyard practices called the Positive Points System (PPS) (Central Coast Vineyard Team [CCVT] 1994). In 2004, through a strategy focused on meetings and viticultural educational initiatives within the community, the CCVT began the development of a third-party certification program called Sustainability in Practice (SIP) (SIP 2008). Certification (now extended to the entire State of California) aimed to be a distinguishing program able to authenticate vineyard practices by distinguishing their wines within the sector market. The program's standards are annually peer-reviewed by universities, government departments, and industry associations. The program not only assesses vineyards, also wineries can certify their wines, allowing them to use a SIP certified seal on wine bottles. Similarly, in 1997, a group of Oregon winegrowers established a voluntary organization called Low Input Viticulture and Enology (LIVE) program (Low Input Viticulture and Enology [LIVE] 1999). The pilot project started with a voluntary inspection through a partnership developed with the Oregon State University, aiming to understand their level of compliance with the guidelines. In 1999, LIVE was annexed and certified by the International Office of Biological Control (IOBC) to certify individual farmers (also including growers from Washington State since 2006). Program strategy mainly focuses on an item as transparency, thereby engaging growers and consumers through an assessment system available online on the LIVE website. The program aims to promote viticulture in conjunction with environment preservation by supporting related social, cultural, and recreational aspects. High-quality grape results in strongly sustained minimized pesticide use, by encouraging biological diversity and exploitation of natural regulating mechanics. Exclusively, after the completion of 2 years of farming under LIVE standards, it is possible to achieve certification, which must be renewed every 3 years.

6.12 Integrated Production of Wine in South Africa

In 1998, a South African governmental Act on Liquor Products (Act No. 60 of 1989) promulgated the Integrated Production of Wine (IPW) scheme (IPW 1998; SAWIS 2014). The IPW scheme emerged out of the need for identification of environmental standards in the production of wine in South Africa. Its success has induced international adoption of the principles of the scheme (OIV has taken the IPW scheme forward for recognition by all member countries as accepted global guidelines). IPW appears as a system of sustainable viticultural and practices, which promotes environmentally friendly and profitable production, encouraging wine producers to voluntarily adopt a set of ecologically friendly minimum standards (from vineyard establishing to packaging). The concept of integrated production is well known to international agriculture even if unofficially adopted for several years before the formalization of the IPW scheme. The Wine of Origin (WO) scheme was mandatory, highly regulated since IPW started. The IPW program is structured on two main documents as the guidelines and the manual. The guidelines present recommendations, as well as minimum standards. A committee of industry experts updates guidelines bi-annually, by including any new research and up-to-date legislation. In South Africa, two other schemes, created during the last decade, are suited to winegrape sustainability: the Biodiversity and Wine Initiative (BWI; related to the conservation of the Cape Floral Kingdom) and the Wine and Agricultural Industry Ethical Trade Association (WIETA; related to fair labor practices) Code. The IPW guidelines include items both for the vineyard and for the cellar. One of the most important principles of IPW is that production should proceed in harmony with nature, requiring the natural environment to be improved and conserved. Soil erosion should be avoided at all costs through regular maintenance and environmentally friendly methods. Where possible, it is recommended to preserve areas of indigenous vegetation improving the occurrence of the natural enemies of pests.

To ensure the effective implementation of the scheme, it is compulsory that at least one representative of a winery is required to attend an IPW training course, incorporating integrated pest management (IPM). Preservation of the natural environment and prevention of soil erosion are the major requirements. Cultivars selection should be

256 VALORIZATION OF WINE MAKING BY-PRODUCTS

directed toward those most suitable to the soil, natural environment, and disease resistant. Concerning vineyard layout, row direction must be planned to avoid soil erosion and to allow for maximum aeration in order to reduce disease incidence. Limiting mechanical cultivation is a recommended cultivation practice, together with a strictly necessary use of chemical herbicides (authorized by the IPW program only).

Excessive fertilization for vine nutrition is discouraged causing water and environment pollution (susceptible to pests and disease). Connected to this, every feasible natural method (pest and disease control) must be used to minimize the need for chemical control. Similarly, irrigation should be scheduled according to the water properties of the soil and moisture content measurements. An up-to-date chemical and microbial analysis of the water used in the winery should be available. Climate change is probably the most important environmental aspect currently under investigation in the world. Wineries contribute to greenhouse gas emissions/CO_2 equivalents during different activities in the winery (mainly combustion of fossil fuels). Emissions records should be kept to each winery before setting objectives, ensuring that wineries continuously decrease CO_2 equivalent. An equipment and infrastructure maintenance schedule must be in place to ensure the integrity of all equipment so avoiding any food safety hazards. Management of wastewater (defined as all water used and generated in the winery during processes like the cleaning of tanks, winery equipment and floors, as well as wine making processes) appears to be a crucial topic. Monitoring its amount, quality, storing, and disposal results in a meticulous approach by program members. Solid waste (including grape waste, lees, and filter rests but also glass, plastic, paper/carton, and metal) must be recycled or disposed of in an environmentally friendly way and in accordance to legislative requirements (Figures 6.4 and 6.5). Noise from pumps, cooling apparatus, and vehicles may cause a disturbance to neighbors, for this reason, ambient noise should be limited.

6.13 California's Sustainable Winegrowing Program

In 2001, the California Sustainable Winegrowing Program (CSWP) was established through a partnership of the Wine Institute and the California Association of Winegrape Growers (CAWG), by developing

SUSTAINABILITY ISSUES 257

Figure 6.4 Lees from Nebbiolo after fermentation.

Figure 6.5 Lees from Nebbiolo after fermentation (overview from the top).

the Code of Sustainable Winegrowing Practices workbook. The latter serves as the basis for the program, providing a tool for vintners and growers to assess their practices and learn how to improve their overall sustainability. In 2003, the Wine Institute and CAWG established a nonprofit organization, named California Sustainable Winegrowing Alliance (CSWA), in order to promote the adoption of sustainable winegrowing practices through the SWP. In 2010, the Certified California Sustainable Winegrowing program (CCSW-Certified) was launched as a voluntary third-party certification to verify the adoption of sustainable practices and improvement. The long-term sustainability of the California wine community represents the SWP vision. The program defines practices sensitive to the environment, responsive to the needs of society-at-large, and economically feasible to implement and maintain (this combination of three principles is referred to as the three "E's" of sustainability) (Figure 6.6). The third (and current) edition of the voluntary California Code of Sustainable Winegrowing Workbook was released in 2013. For each of the areas, the workbook describes four categories in order of increasing sustainability,

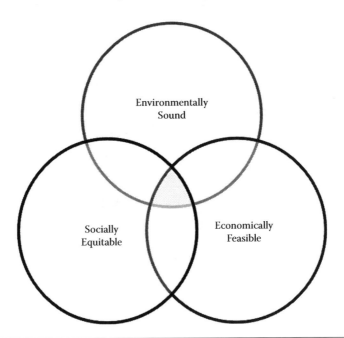

Figure 6.6 Sustainability as defined by the three overlapping principles of Environmentally Sound, Economically Feasible, and Socially Equitable. (From CSWA, Sustainable winegrowing program, 2013, http://www.sustainablewinegrowing.org/, accessed on May 2014.)

and each category has specific measures to assess the current state of the vineyard or winery. A key desired outcome appears to be the widespread development and execution of sustainability strategies in the California winegrowing community. Several potential benefits arise from the different sustainable practices adopted as a long-term viability of both land and business (e.g., cost savings). Looking at future International Trade Certification such as ISO 14001, attention will be required in preparing adequately, however, always committed in improving wine quality. The market value of wine produced in California needs to be maintained and improved steadily thereby enhancing relations with other markets (e.g., Europe). Long-term viability of land and conservation of natural resources results crucial environmental benefits together with a performing management of a unique and specific territory. Important benefits from a social perspective will be the strengthening of relations with neighbors, communities, consumers, tourists but also with regulators and public policy institutions (government, media, and educators). Efforts might be made to ensure the proper health and well-being of farm and winery employees and neighbors. Developing a statewide Code of Sustainable Winegrowing Practices appears one of the major goal by the Wine Institute and the California Association of Winegrape Growers. In this direction, the program is focused on the realization of dedicated and up-to-date accompanying workbooks intended to winegrower and vintner together with the build of a credible measurement system to document and communicate statewide adoption of sustainable practices. The voluntary adoption by the entire wine community of the high standards for sustainable practices presented in the code results one of the desired outcomes that this program shall aim. Moreover, this approach plans to promote the widespread use of the winegrowing and wine making workbooks directly in business planning, training, execution, and analysis, without neglecting to apply the measurement system to evaluate progress in the adoption of sustainable practices within the community. Finally, it shall be fundamental to prove that self-governance, education, and an open dialog be truly able to improve the economic viability and future of the wine community. In order to obtain all the desired outcomes, the SWP is projected to stimulate a "cycle of continuous improvement." Growers, vintners, and the industry are all involved in making this constant

260 VALORIZATION OF WINE MAKING BY-PRODUCTS

progression by enabling them to evaluate their operations, learn about new approaches, and develop action plans in order to increase their adoption of sustainable practices. The implementation program cycle consists of some phases that participants are encouraged to re-assess themselves in order to consolidate this process of evaluation, learning, and improvement. Some crucial steps of this approach include operations, such as providing participants with practical self-assessment workshops; tracking and measuring the related results; offering information about sustainable practices (on areas needing improvement); facilitating information exchange; and motivating participants to implement effective changes.

6.14 VineBalance, New York State's Sustainable Viticulture Program and Long Island's Sustainable Winegrowing

In 2004, regional Cooperative Extension grape programs started developing an education program able to promote the adoption of sustainable viticultural practices in New York vineyards (VineBalance 2004). New York State's sustainable viticulture program represents the result of a cooperative project between industry groups, the Finger Lakes Grape Program, Lake Erie Regional Grape Program, and Long Island's Grape Extension Program. The grower self-assessment workbook represents the base of the program that covers all the management decisions made by New York State grape growers. The workbook sections represent an evolution obtained improving two earlier programs in New York: NYS Agricultural Environmental Management (AEM) worksheets and the Long Island Sustainable Practices Workbook. In 2007, the VineBalance's New York Guide to Sustainable Viticulture Practices Grower Self-Assessment Workbook was launched, representing the tool by which growers are able to evaluate their viticultural practices, addressing the diversity of the state's sector industry with a broad range of questions. VineBalance does not have a certification scheme, as its main objective is to promote the adoption of sustainable practices. Since 2012, two different groups, Welch's (grape juice) and Long Island Sustainable Winegrowing (LISW) started a certification process based on the VineBalance workbook. LISW represents a not-for-profit organization that provides education and certification for

Long Island vineyards by using international standards of sustainable practices in quality winegrape production. This tool is based on an independent third-party-verified checklist system consisting of recommended and prohibited practices and materials, planning and ecological options. Implementing cultural practices able to reduce the use of chemicals and fertilizers results one of the major goals of the program together with the protection of the farmer, the environment, and society. Vineyard biodiversity must be promoted and maintained using proper practices. In order to both produce and conserve the highest level of quality grapes, soil conservation appears crucial and this should be preserved healthy and fertile. Particularly, practices that aim to preserve the integrity of the maritime ecosystem, counteracting its runoff and leeching, are encouraging.

6.15 Wines of Chile Sustainability Program

In 2009, the Wines of Chile, a nonprofit organization representing 95% of the bottled wine exported from Chile, released the Wines of Chile Strategic Plan 2020 (sustainability represents one of its key principles) (WOC 2010). The Consorcio Tecnológico (Technological Consortium), representing the technical tool of the sector industry, has been appointed to develop a sustainability program. Then, a joint project between industry representatives and the University of Talca started the development of the related Sustainability Code. The Code covers three main areas: Vineyards (Green Area), Winery and Bottling Plants (Red Area), and Social (Orange Area), by providing a checklist of control points and a compliance standard. The green area includes issues on natural resources, pests and disease, agrochemicals, and job safety. The red area contains chapters about energy, water management, contamination prevention, and waste. The orange chapter includes all social issues as relationships with the workers, community, environment, and clients. Wines of Chile has adopted the principles of the code in the development of an official accreditation system and in 2012 the first wineries has been officially certified. There is even more of an increase in consumers' preference for products that can demonstrate some level of environmental sustainability and social responsibility. This consumer trend has created a growing market for products produced in a sustainable manner,

262 VALORIZATION OF WINE MAKING BY-PRODUCTS

resulting in a great opportunity for Chile. The program focused on the idea that sustainability means recognizing wine as part of a complete system that is not limited to the vineyard but that encompasses the winery, the employees, and the local community. The Wines of Chile Sustainability Program includes a series of initiatives and projects intended to establish the sustainability of wine industry. The program is structured applying production processes environmentally friendly, socially equitable, and economically viable. The program includes several issues that include the National Sustainability Code, geology, climate changes, pesticides, biodiversity, water management, energy, and social responsibility. The Sustainability Code is a voluntary tool focused on incorporating sustainable practices in winegrowing companies, following the basis of the area requirements: vineyards (natural resources, handling agrochemicals, and industrial safety); wineries and bottling plants (energy efficiency, water management, waste, recycling, and pollution prevention); and social (ethics, environment, working conditions, community relationships, marketing, and consumers). Its objective is to encourage producers and winemakers to improve their management by meeting the requirements stipulated in the standard. The Sustainability Code includes a series of requirements, which fall within a long-term vision through a combination of environmental, social, and economic principles. To comply with the requirements, the vineyards must have an environmental and social management system that, at least, meets the current national legislation.

6.16 McLaren Vale Sustainable Winegrowing Australia

In the early 2000s, the McLaren Vale Grape Wine and Tourism Association (MVGWTA) has established the McLaren Vale Sustainable Winegrowing Australia (MVSWGA) program by developing a series of viticultural initiatives directed to the improvement of sector practices, fruit quality, and financial viability in the region (MVSWGA 2012, 2013). Moreover, the Association also released a Pest and Disease Code of Conduct in 2006 (voluntarily endorsed by the growers the next year). In 2007, the Soil Management, Water Management, and Preservation of Biodiversity Codes were also established (South Australian EPA 2002a,b, 2003a,b,c, 2004).

SUSTAINABILITY ISSUES 263

Similar to other programs, the program assesses sustainability through the triple bottom line approach (environment, economics, and social), focusing on continuous improvement of the results over time. The method of assessment is similar to that of the Lodi and CSWA workbooks, reporting questions ranging from zero (explicitly unsustainable) to four (most sustainable) as well as nonapplicable (NA). This methodology differs from the other two workbook methods with the addition of having a "zero" scoring option. The MVSWGA places growers into four certification categories: category 1 (red, needs attention); category 2 (yellow, good); category 3 (green, very good); and category 4 (blue, excellent). The program's content also changes annually (peer-reviewed by experts primarily from universities and governmental departments) to incorporate any relevant findings to the assessment, also in line with FIVS and OIV. Land preservation (followed by a constant degree of improvement) results in a major objective in a sustainable approach like this one. Together with the soil, the quality of different products must also be increased, always satisfying the requirements of the wineries. However, in this constant enhancing approach, efforts must be made to minimize the potential negative impacts on the environment. The program also aims to promote social development and well-being of the people who live and work in the community. The six assessment areas involved in the program correspond to: soil health; pest and disease; biodiversity; water; waste and social. Herbicides and pesticides should be reduced to a minimum. On the other hand, biodiversity must be favored (pest management) as the application of organic compost (soil management). The optimization of the use of energy represents another crucial step together with smart waste management by reducing its production but favoring its reuse and recycling. Overall, even if the approach resulting from the program is comparable to the previously reported ones, it should be noted that a fundamental aspect arising from the sustainable program results the theme of biodiversity. Vineyards should adopt a conservation friendly approach, reducing any harm to native vegetation or habitats for native wildlife. A proper environmental management focused to the biodiversity preservation of the vineyard represents the key point to allow for the reduction of chemical inputs into the environment. Biodiversity conservation must be intended as a part of core business and this message needs to be transmitted

264 VALORIZATION OF WINE MAKING BY-PRODUCTS

to the consumers, while benefiting from the many advantages this approach provides. Recently, the Grape and Wine Research and Development Corporation (GWRDC) gave a grant to the McLaren Vale Grape Wine & Tourism Association to explore the most effective and efficient ways a wine region can receive International Standard Organization (ISO) certification. ISO provides practical tools (more than 18,000 standards) for all the three dimensions of sustainable development: economic, environmental, and societal. Applying ISO 14001:2004 (criteria for an environmental management system) can provide assurance to company management, employees, and external stakeholders that environmental impact is being measured and improved. Accreditation of the McLaren Vale region appears a key priority able to allow the community to receive the environmental and economic benefits associated with the implementation of the ISO 14001 standard, thus resulting in the first certified region in the world (reinforcing the innovation, leadership, and environmental credentials of McLaren Vale). Several benefits can emerge from this certification such as an improved image (among regulators, customers, and the public) and higher technical credibility. Undoubtedly, the international exposure results have improved, enhancing investor confidence and promotional opportunities.

6.17 United Kingdom Vineyard Association

In 2009, the United Kingdom Vineyard Association (UKVA) set up the Sustainability Working Group able to establish the Policy Statement for Sustainability addressing three specific areas (economic, environmental, and social) (United Kingdom Vineyard Association [UKVA] 2009). Economic objectives include monitoring and promoting wine quality, following closely national and international trends in order to build a sound understanding of aspects that affect the future viability of English wine production. Both education and training represent the fundamental tools which are also able to promote product quality through the UK quality wine schemes and annual competition. Vineyards and wineries shall be supported by realizing a handbook with basic information (business planning, marketing, sales strategies, accessing grants, and quality assurance including HACCP). Consistent with other similar programs, environmental issues appear to be a core topic, including carbon management (greenhouse gas reduction plans), integrated pest management (reducing pesticide use), agrochemicals, soil management (improving its structure and organic content), biodiversity (reducing herbicide use), energy efficiency, water and waste management (achieving reducing, reusing, recycling). Staff training is essential in this perspective of development together with recruiting and maintaining a productive workforce (recruitment, retention, health, safety, staff welfare, and employee reward). The community must be involved in these activities by collecting and sharing ideas, always favoring good neighbor relations. One aspect, that has been taken into serious consideration, is alcohol misuse. In this direction, efforts should be taken to support the government's strategy to tackle the harm of alcohol.

6.18 General Conclusions and Discussion

If we consider the most widely used definition of sustainable development, coined by the Brundtland Commission, as the "development that meets the needs of the present without compromising the ability of future generations to meet their own need," it appears evident that the direction taken by the different programs described is the right one. Even though there is a great emphasis on the environmental aspects, this item, certainly important, is not the main driver for the conception of sustainability assessment programs for viticulture. Programs have been realized to increase the overall sustainability through both operations improvement and education (in this approach universities play a crucial role). However, it is evident how the states that belong to the "old continent" may be placed at a lower level concerning relevant sustainability assessment programs for viticulture. Closing the gap represents the current challenge toward the "new world." Even though it may appear obvious, the success of a sound program is largely due to the people driving the programs as managers and innovative growers. Moreover, the way these people communicate and engage with their stakeholders is fundamental together with the usefulness of the developed program to improve sustainability. Dedicated programs only make sense if they are useful to help growers to improve their sustainability in the context of

266 VALORIZATION OF WINE MAKING BY-PRODUCTS

the community and environment in which they are located. "Old continent" states, even if the understood this message clearly (also possessing excellent managers and growers), still lack in defining a common sustainability strategy such as the positive examples of the United States, Australia, and New Zealand.

References

California Sustainable Winegrowing Alliance (2004). California wine community sustainability report executive summary. http://www.sustainablewinegrowing.org/docs/cswa_2004_report_executive_summary.pdf (accessed on May 2014).

Central Coast Vineyard Team (CCVT) (1994). Promoting Sustainable Vineyard Practices. http://www.vineyardteam.org (accessed on February 2014).

Council Regulation (EC) No. 479/2008 of 29 April 2008 on the common organization of the market in wine, amending Regulations (EC) Nos. 1493/1999, 1782/2003, 1290/2005, and 3/2008 and repealing Regulations (EEC) Nos. 2392/86 and 1493/1999. Official Journal of the European Union.

Council Regulation (EC) No. 491/2009 of 25 May 2009 amending Regulation (EC) No. 1234/2007 establishing a common organization of agricultural markets and on specific provisions for certain agricultural products (Single CMO Regulation). Official Journal of the European Union.

Council Regulation (EC) No. 607/2009 of 14 July 2009 laying down certain detailed rules for the implementation of Council Regulation (EC) No. 479/2008 as regards protected designations of origin and geographical indications, traditional terms, labelling and presentation of certain wine sector products. Official Journal of the European Union.

Council Regulation (EEC) No. 822/87 of 16 March 1987 on the common organization of the market in wine. Official Journal of the European Communities.

Council Regulation (EEC) No. 823/87 of 16 March 1987 laying down special provisions relating to quality wines produced in specified regions. Official Journal of the European Communities.

Council Regulation (EC) No. 1493/1999 of 17 May 1999 on the common organization of the market in wine. Official Journal of the European Union.

CSWA (2013). Sustainable winegrowing program. http://www.sustainablewinegrowing.org/ (accessed on May 2014).

E-Bacchus (2012). Wine register of designations of origin and geographical indications protected in the EU. European Commission, online database. http://ec.europa.eu/agriculture/markets/wine/e-bacchus/ (accessed on March 2014).

SUSTAINABILITY ISSUES   **267**

European Commission (1986–2007). Agriculture and Rural Development budget: Guarantee Section of the European Agricultural Guidance and Guarantee Fund (EAGGF) financial reports, 1986–2007. Office for Official Publications of the European Communities, Luxembourg. http://aei.pitt.edu/view/eusubjects/eagus.html (accessed on June 2014).

European Commission (2003). Study on the use of the varieties of interspecific vines. Contract No. AGR 30881 of 30 December 2002. http://ec.europa.eu/agriculture/markets/wine/studies/vine_en.pdf (accessed on June 2014).

European Commission (2006a). Impact assessment—Annex to the Communication from the Commission to the Council and the European Parliament "Towards a Sustainable European Wine Sector." http://ec.europa.eu/agriculture/capreform/wine/fullimpact_en.pdf (accessed on June 2014).

European Commission (2006b). WINE—Economy of the sector. European Commission. Official Journal of the European Communities. http://ec.europa.eu/agriculture/markets/wine/studies/rep_econ2006_en.pdf (accessed on May 2014).

European Commission (2007a). Accompanying document to the proposal for a Council Regulation on the common organisation of the market in wine and amending certain regulations—Impact assessment. http://ec.europa.eu/agriculture/capreform/wine/impact072007/full_en.pdf (accessed on June 2014).

European Commission (2007b). Fact sheet—Towards a sustainable European Wine Sector. http://ec.europa.eu/agriculture/publi/fact/wine/072007_en.pdf (accessed on June 2014).

European Commission (2007–2012). Agriculture and Rural Development budget: European Agricultural Guarantee Fund (EAGF) financial reports, 2007–2012. Office for Official Publications of the European Communities, Luxembourg. http://ec.europa.eu/agriculture/capfunding/budget/index_en.htm (accessed on February 2014).

European Commission (2008). Reform of the EU wine market. http://ec.europa.eu/agriculture/capreform/wine/index_en.htm (accessed on February 2014).

European Commission (2009–2012). Wine CMO: Financial execution of the national support program. Directorate-General for Agriculture and Rural Development. http://ec.europa.eu/agriculture/markets/wine/facts/index_en.htm (accessed on February 2014).

European Court of Auditors (2012). The reform of the common organization of the market in wine: Progress to date. Special Report No. 7. http://eca.europa.eu/portal/pls/portal/docs/1/14824739.pdf (accessed on May 2014).

European Landfill Directive (1999/31/EC). The landfill of waste. http://eurlex.europa.eu/legalcontent/EN/TXT/?uri=CELEX:31999L0031 (accessed on April 2014).

Eurostat (2013). Eurostat statistics database. Eurostat, Luxembourg. http://epp.eurostat.ec.europa.eu/portal/page/portal/education/data/database/ (accessed on February 2014).

268 VALORIZATION OF WINE MAKING BY-PRODUCTS

FAO (2012). Food and Agriculture Organization of the United Nations. FAOSTAT. http://faostat.fao.org (accessed on June 2014).

International Organization for Standardization (2004). ISO 14000—Environmental management. http://www.iso.org/iso/iso14000 (accessed on March 2014).

IPW (1998). Integrated production of wine scheme (IPW). http://www.ipw.co.za/ (accessed on June 2014).

LIFE HAproWINE Project (2013). Integrated waste management and life cycle assessment in the wine industry: From waste to high-value products. http://www.haprowine.eu/introduccion_e.php (accessed on May 2014).

Low Input Viticulture and Enology (LIVE). (1999). Project focused on Integrated Solutions for Sustainable Winegrowing in the Pacific Northwest. http://liveinc.org (accessed on June 2014).

Lodi Winegrape Commission (2005). Lodi Rules for sustainable winegrowing. http://www.lodiwine.com/certified-green (accessed on March 2014).

MVSWGA (2012). Sustainability report McLaren Vale 2012. http://www.mclarenvale.info (accessed on June 2014).

MVSWGA (2013). McLaren Vale Sustainable winegrowing Australia system (MVSWGA). http://www.sustainablewinegrowing.com.au (accessed on June 2014).

New Zealand Wines (2014). Sustainability. http://www.nzwine.com/sustainability/ (accessed on May 2014).

Ohmart, C. 2008. Lodi Rules certified wines enter the marketplace. *Practical Winery Vineyard* 29:32–42.

OIV (2012). Statistical report on world vitiviniculture. http://www.oiv.int/oiv/info/enizmiroivreport/ (accessed on May 2014).

OIV (2012–2014). Strategic plan. http://www.oiv.int/oiv/info/enplanstrategique (accessed on May 2014).

OIV (2013). State of the Vitiviniculture World Market. http://www.oiv.int/oiv/info/enconjoncture/ (accessed on May 2014).

OIV Resolution CST 1/2004. Development of sustainable vitiviniculture. http://www.oiv.int/oiv/info/enresolution (accessed on January 2014).

OIV Resolution CST 1/2008. OIV guidelines for sustainable vitiviniculture: Production, processing and packaging of products. http://www.oiv.int/oiv/info/enresolution (accessed on January 2014).

Resolution OENO 1/2005. Distillates of vitivinicultural origin. http://www.oiv.int/oiv/info/enresolution (accessed on January 2014).

Resolution VITI 2/2003. Reasoned vine irrigation. http://www.oiv.int/oiv/info/enresolution (accessed on January 2014).

Resolution VITI 02/2006. Measures used to prevent or limit the proliferation of wood diseases. http://www.oiv.int/oiv/info/enresolution (accessed on January 2014).

SAWIS (2014). South African wine industry information and systems. http://www.sawis.co.za (accessed on June 2014).

SIP (2008). Sustainability in practice (SIP) certification program. http://www.sipcertified.org (accessed on March 2014).

South Australian EPA (2002a). Use of water treatment solids. New South Wales Environment Protection Authority, Sydney, New South Wales, Australia.

South Australian EPA (2002b). Wastewater lagoon construction, New South Wales Environment Protection Authority, Sydney, New South Wales, Australia.

South Australian EPA (2003a). Environment protection (water quality) policy. New South Wales Environment Protection Authority, Sydney, New South Wales, Australia.

South Australian EPA (2003b). Independent verification of monitoring programmes. New South Wales Environment Protection Authority, Sydney, New South Wales, Australia.

South Australian EPA (2003c). Odour assessment using odour source modelling. New South Wales Environment Protection Authority, Sydney, New South Wales, Australia.

South Australian EPA (2004). Bunding and spill management. New South Wales Environment Protection Authority, Sydney, New South Wales, Australia.

VIVA. 2011. Sustainability in the Italian viticulture. National pilot project launched by the Italian Ministry for the Environment, Department for Sustainable Development, Climate Change and Energy. http://www.viticolturasostenibile.org/ EN/Home.aspx (accessed on February 2014).

TerraVitis®. (1998). Project based on integrated farming protocol, recognizing the concept of sustainable wine production. http://www.terravitis.com/ (accessed on April 2014).

Unione Italiana Vini. (2011). Tergeo. http://www.uiv.it/progettotergeo/ (accessed on February 2014).

United Kingdom Vineyard Association (UKVA) (2009). Sustainability. http://www.ukva.org.uk/ (accessed on June 2014).

U.S. National Organic Standards Board (1998). http://www.ams.usda.gov/ AMSv1.0/nosb (accessed on May 2014).

VineBalance (2004). New York guide to sustainable viticultural practices. http://www.vinebalance.com (accessed on June 2014).

WOC (2010). Wines of Chile sustainability program. http://www.winesofchile.org (accessed on May 2014).

World Commission on Environment and Development (1987). Report "Our Common Future". United Nations, Tokyo, Japan.

7

MARKETING POTENTIAL

JORGE A. CARDONA AND
THELMA F. CALIX

Contents

7.1 Introduction	271
7.2 Grapes	273
7.2.1 Composition	273
7.2.2 Production	274
7.2.3 Grape Prices	276
7.3 Grape By-Products	277
7.3.1 Facts	277
7.3.2 Pomace	277
7.3.2.1 Seeds	278
7.3.2.2 Skins	282
7.3.2.3 Stems	284
7.3.3 Lees	284
7.4 Closing Remarks	286
References	287

7.1 Introduction

Grapes and their products are among the most important horticultural products worldwide and are listed as one of the five most important fruits with 13% of the total fruit production (FAO 2015; Moulton and Possingham 1998). There are vestiges of grape cultivation for more than 4000 years originating in the region of the Mediterranean Sea and Eastern Europe. *Vitis vinifera* is the main variety cultivated but there are various grapevines grown that have adapted to various regions in the world which include *Vitis labrusca*, *Vitis rotundifolia*, *Vitis mustangensis*, and *Vitis aestivalis* (Martín-Belloso and Marsellés-Fontanet 2007). Consumption of grape includes fresh fruit, raisins, jam, jelly, juice, spirits, and principally wine, which is the most important product manufactured from this commodity (Moulton and Possingham 1998). The use of grapes in the wine industry depends on regions as Europe and Oceania destined more than 60% of their grape production to wine manufacture over the past 10 years, whereas Asia only utilized less than 10% for the same purpose during the same period (FAO 2015). Although yield of wine from grapes vary depending on the grape variety, agricultural practices (nutrition, irrigation, harvest), vineyard density, weather conditions, and extraction method (press), taking all these details into account an average yield from grape to wine would be 45%–60% (Sousa et al. 2014). This means that about half of the total fruit volume is considered by-product which needs further processing to increase the value of this crop. The composition of the by-product includes water, proteins, lipids, carbohydrates, vitamins, minerals, and compounds with high biological importance such as fiber, organic acids, and phenolic compounds (Sousa et al. 2014). Recently, various efforts have been made to utilize this by-product as it contains various phytochemicals of importance to the food, cosmetic, agricultural, and medical industries. The consumption of functional foods and beverages has been growing due to a major consumer trend toward health consciousness (Childs 1999; Milo 2005). Therefore, natural products have gained substantial attention in the market. By the end of the last century, a quarter of the food antioxidant market was occupied by natural antioxidants with an annual growth rate of 6%–7%. The main sources of natural antioxidants include vitamin C, tocopherols, polyphenolics, and organic acids which are present in fruits, vegetables, spices, and herbs (Meyer et al. 2002; Peschel 2006). This growing demand for natural products has also created an opportunity to substitute synthetic antioxidants which have been associated with potential toxicity. Furthermore, the manufacturing costs of these compounds are more expensive than natural antioxidants from fruits and vegetables (Moure et al. 2001). This chapter aims to review the different parts of the fruit, their main components, and the potential to be used further as ingredients for human, animal, and plant consumption.

7.2 Grapes

7.2.1 Composition

As the grape berry develops, modifications in its size, composition, color, texture, flavor, and pathogen susceptibility occur (Conde et al. 2007). Mature grapes are physiologically composed of pulp or flesh (75%–85%), skin (15%–20%), and less than 5% of seeds (Figure 7.1) (Martín-Belloso and Marsellés-Fontanet 2007). Within the grape berry, the distribution of compounds important to wine quality has denoted that skin contains significant amounts of anthocyanins, procyanidins, and flavor compounds; pulp provides water, organic acids, sugars, and flavor compounds to wine; and seeds are rich sources of procyanidins (Kennedy 2008), among other significant compounds which are considered later in this chapter as by-products. Chemically, the most prevalent compounds in the fruit are tartaric (3–7 g/L) and malic acids (1–3 g/L), and it has been reported that they start to accumulate at the initial stages of berry development (Conde et al. 2007). Tartaric acid is the only

Figure 7.1 Physiology and composition of grapes. (From Ruiz-Villareal, M., Illustration of grape physiology and composition, 2008, reviewed February 2015, available at: http://en.wikipedia.org/wiki/File:Wine_grape_diagram_en.svg.)

274 VALORIZATION OF WINE MAKING BY-PRODUCTS

compound allowed to be added to wines to regulate acidity and to prevent oxidation (Marchitan et al. 2010; Palma and Barroso 2002). It is important to consider that its chemical synthesis is more expensive than its recovery from grapes or grape by-products from the wine industry; therefore, its recovery is mainly obtained from grape seeds and red grape skins (Palma and Barroso 2002). Additionally, extraction of tartaric acid is preferably obtained from grape seeds because of its biological stability in comparison to the fruit itself, which suffers biological degradations such as fermentation of sugars (Palma and Barroso 2002). The presence of grape phenolics is of great significance to the wine industry. The grape phenolic compounds contribute to the color, taste, and feel of wines and their concentration increases throughout berry development (Kennedy 2008). These phenolic compounds are originated as secondary plant metabolites and are essential for plant reproduction, stability, and growth processes in plants (Croft 1999; Shahidi and Naczck 2003). Depending on the amount of phenolic rings, these phenolic compounds are divided into four classes at a molecular structure level: cinnamic acids and benzoic acids are made of one phenolic ring; stilbenes are made of two phenolic rings; anthocyanins, flavonols, and flavan-3-ols are made of three rings; and ellagic acids, which are made of a complex ring structure (Zhu et al. 2012).

7.2.2 Production

Data from 2013 indicated a world grape production of 77.2 million tons yearly (FAO 2015). Europe is the biggest grape producer (38%) followed by Asia (32%) and United States (21%) (Figure 7.2). Grape production has grown 15% in the last decade due to the increase in production in China which doubled from 6 million tons per year to almost 12 million tons in 2013 which positioned this country as the biggest producer of grape in the world (Figure 7.3). The biggest grape producers other than China include Italy, Spain, and the United States, which produced around 8 million tons per year (2013), and France (5.5 million tons). Grapes are mainly used for wine, juice, distilled liquors, dried fruit, and fresh consumption (Williams et al. 1994). There are close to 800 varieties of grapes destined for the production

MARKETING POTENTIAL

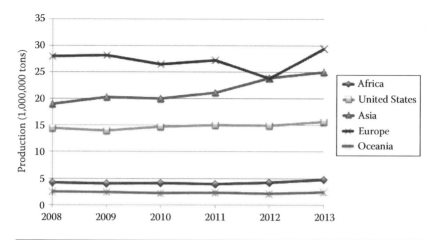

Figure 7.2 World grape production: 2008–2013. (From Food and Agriculture Organization [FAO], Grape production statistics, United Nations, 2015, available at: http://faostat3.fao.org/compare/E.)

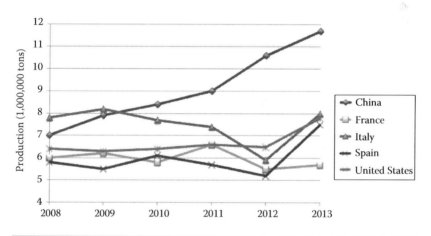

Figure 7.3 Most important grape producers: 2008–2013. (From Food and Agriculture Organization [FAO], Grape production statistics, United Nations, 2015 available at: http://faostat3.fao.org/compare/E.)

of wine and spirits. However, the use of this fruit for the wine industry depends on regions. France, Germany, Portugal, Australia, and New Zealand destine close to three quarters of their grape production to wine manufacture while countries like Turkey and Iran destined less than 1% to this product as they are the biggest producers of raisins worldwide together with the United States (FAO 2015). There has not been an important increase in wine production in the last decade

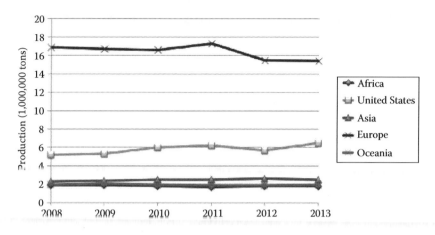

Figure 7.4 World wine production 2008–2013. (From Food and Agriculture Organization [FAO], Grape production statistics, United Nations, 2015, available at: http://faostat3.fao.org/compare/E.)

with a steady average production of 27.5 million tons of wine per year (Figure 7.4). Europe is the main wine producer with almost 60% of the world's production followed by United States with 30%.

7.2.3 Grape Prices

Grape prices have been fluctuating affecting the grape and wine industry. This effect was mainly due to the global economic crisis and overproduction of grapes. A report in Australia indicated a 45% reduction from U.S. $657/ton (2000) to an estimated price of U.S. $453/ton for 2005 (PIRSA 2005). In parallel, Blake (2012) stated that prices were as low as a quarter of current prices between 1998 and 2008 in California. Reports suggested grape prices from U.S. $300/ton for generic grapes and up to U.S. $600/ton for specific varieties like Merlot and Cabernet Sauvignon (Blake 2012). These prices have been increasing as there is a shortage of grape production worldwide, which increases the price of the raw materials, and the wines produced (McMillan 2015). This issue has generated more pressure on wineries to assess the possibilities to process by-products as their revenue is lowered by the increase in grape prices while wine consumption has not been increased significantly. In addition, it is important to stress the shift of markets as China has increased its the production of grapes and wines, and the United States became the leading consumer of wines displacing European countries as the leaders in grape production and wine consumption.

7.3 Grape By-Products

7.3.1 Facts

The processing of food and beverages generates large amounts of by-products. Almost 50% of the grape fruit is used in the process (fermentable juice) and the remaining half portion is considered to be by-products (stems, skins, and seeds). The by-product composition includes water, proteins, lipids, carbohydrates, vitamins, minerals, and compounds with high biological importance such as fiber, organic acids, and phenolic compounds (tannins, phenolic acids, procyanidins, anthocyanins, and resveratrol) (Arvanitoyannis et al. 2006; Sousa et al. 2014). By-products, mostly considered as a disposal problem, could represent a significant cost if they are not managed properly. Therefore, by-products are normally used as organic fertilizers or carbon sources following composting, and as livestock feed (Ferrer et al. 2001). Recently, various efforts have been made to provide more opportunities to use this by-product as it contains various phytochemicals of importance to the food, cosmetic, agricultural, and medical industries, which can also provide a substantial increase in income to the wine industry. Additionally, global competitiveness in the wine industry is forcing processors to use all the materials produced while fermenting and aging wine to reduce cost and environmental damage. Examples of by-product uses includes extracts utilized as nutraceuticals, antioxidants, colorants, antimicrobials, animal feed, medical remedies, and natural ingredients to improve the nutritional value of food matrixes such as milk (Jing and Giusti 2005), model muffins (Mildner-Szkudlarz et al. 2015), and sourdough mixed rye bread (Mildner-Szkudlarz et al. 2011).

7.3.2 Pomace

Polyphenolics can be widely found in pomace of various fruits and vegetables (Bonilla et al. 1999; Koplotek et al. 2005; Pastrana-Bonilla et al. 2003; Visioli et al. 1999), and these materials signify an inexpensive source to generate food ingredients (Cardona et al. 2010). Therefore, an interest in natural antioxidants and its existence in fruit by-products has motivated fruit and vegetable processors to consider the extraction of polyphenolics, sterols, and organic acids to increase

278 VALORIZATION OF WINE MAKING BY-PRODUCTS

profitability of their operations. In grapes, pomace constitutes around 40% of the total fruit and it is an important source of polyphenolics (Cardona et al. 2009; Morris and Brady 2004; Pastrana-Bonilla et al. 2003), such as anthocyanins (about 131 ± 0.4 mg/100 g of flour) (Sousa et al. 2014). Additionally, it is a source of other phenolic compounds as they share characteristics with anthocyanins in terms of solubility in water and stability to temperature and oxygen. Research has focused on converting muscadine and blueberry pomace into nutraceutical products that could be sold for as much as U.S. $45/kg (Cardona 2007; Lee and Wrolstad 2004; Phillips 2006). According to Sousa et al. (2014), the microbiological quality of grape pomace is acceptable for human consumption due to its unfavorable conditions of low moisture and pH lower than 4. Furthermore, it has been proved that this extract is considered nontoxic by the bioassay established by Meyer et al. (1982) against larvae of *Artemia salina* sp.

7.3.2.1 Seeds Grape seeds only represent less than 5% of the total fruit's weight, though they comprise about 38% (w/w) of the dry weight of pomace (Beveridge et al. 2005). Grape seeds are complex matrixes mainly composed of approximately 40% fiber, 16% oil, 11% protein, and 7% phenolic compounds and tannins; such as procyanidins (Dos Santos Freitas et al. 2008; Murga et al. 2000) (Figure 7.5). Procyanidins are the major constituents of de-fatted grape-seed extracts, which consist of chains of flavan-3-ol units, (+)-catechin, and (-)-epicatechin, linked through C4–C6 and C4–C8 interflavan bonds (Weber et al. 2007). Extracts from grape seeds containing these complex phenols and tannins could inhibit lipid peroxidation since these compounds are efficient free radical scavengers (Lau and King 2003; Murga et al. 2000). Therefore, the dietary intake of grape-seed extracts has been inversely linked to the risk of coronary heart disease (Murga et al. 2000) and confirmed to provide anticancer and cancer chemopreventive potential (Kaur et al. 2009), by their intact bioavailability in some parts of the body, such as the colon (Choy et al. 2013). It has been shown that grape-seed extracts also have antimicrobial activity against a wide range of microbial pathogens, including *S. aureus*, *S. coagulans niger*, *C. freundii*, *E. cloacae*, *E. coli*, and *C. echinulatum* (Palma and Taylor 1999). Grape-seed oil is mainly extracted by using the supercritical carbon dioxide method (about 6%–14% w/w yield;

Figure 7.5 Grape seeds and related powdered extract.

Beveridge et al. 2005); however, it has been proved that the yielded extraction and the kinetics velocity extraction are increased by using propane as a solvent (Dos Santos Freitas et al. 2008). Quantitative analyses of grape-seed oil have demonstrated that it contains high quantities of polyunsaturated fatty acids (linoleic and oleic acids) and sterols, such as campesterol (26.3–94.4 mg/100 g), stigmasterol (28.5–96.1 mg/100 g), and β-sitosterol (172.3–823.5 mg/100 g) (Crews et al. 2006; Dos Santos Freitas et al. 2008; Palma and Taylor 1999). Linoleic acid is present in amounts ranging 61.3%–74.6% and oleic acid level ranges 14.0%–20.9% of total fatty acid content (Crews et al. 2006). Beveridge et al. (2005) calculated that the seed oil by-product from wine manufacture in 2001 could be about 201 tons of oil (~CDN $1.6–$2.1 million—retail) to be consumed as a dressing, specialty salad, or cooking oil. Olive oil is the best comparison to grape-seed oil as specialty oil and it has experienced a 20% price increase between 2010 and 2015. Reports from 2013 estimated extra virgin olive oil prices at about U.S. $3812/ton (Indexmundi 2015). If this information could be used for estimation, an average income of almost U.S. $140 million could have been generated on grape seed

280 VALORIZATION OF WINE MAKING BY-PRODUCTS

oil in Europe alone in 2013. This result was based on the assumption that 29.1 million tons of grapes were produced that season in Europe from which 63% were destined to wine production (18.3 million tons). From this volume of fruit, 1%–4% represented seeds (364,155 tons) from which close to 36,416 tons of grape-seed oil could be extracted (10% yield). Global results using the same estimations for seed content (2%) and oil recovery (10%) from grape production destined for wine production are presented in Table 7.1. This oil production could have a yearly marketing potential higher than U.S. $200 million except for 2012 where global production of grapes was reduced (Table 7.2). This estimation was very conservative as income from this operation could raise up to U.S. $336 million if a higher percentage of the fruit was considered as seeds (4%) and the oil yield extracted from those seeds (12%). Additionally, this amount would increase as the price for olive oil has increased 17.5% from 2013 to prices reviewed at the beginning of 2015 (U.S. $4480/ton). In parallel, a production of de-fatted seed

Table 7.1 Potential World Grape-Seed Oil Production (Ton)

	YEAR				
REGION	2009	2010	2011	2012	2013
Africa	2,223	2,257	2,147	2,316	2,403
United States	10,972	11,757	12,280	11,888	13,359
Asia	3,782	3,696	3,884	4,404	4,600
Europe	35,020	33,545	34,674	29,655	36,416
Oceania	2,873	2,617	2,799	2,679	2,860
Total	**54,869**	**53,872**	**55,783**	**50,943**	**59,639**

Source: Food and Agriculture Organization (FAO), Grape production statistics, United Nations, 2015, available at: http://faostat3.fao.org/compare/E.

Table 7.2 Potential Income (Million U.S.$) from Grape-Seed Oil Commercialization

	YEAR				
REGION	2009	2010	2011	2012	2013
Africa	8.47	8.60	8.18	8.83	9.16
United States	41.83	44.82	46.81	45.32	50.93
Asia	14.42	14.09	14.81	16.79	17.54
Europe	133.49	127.87	132.18	113.05	138.82
Oceania	10.95	9.98	10.67	10.21	10.90
Total	**209.16**	**205.36**	**212.64**	**194.20**	**227.34**

Source: Food and Agriculture Organization (FAO), Grape production statistics, United Nations, 2015, available at: http://faostat3.fao.org/compare/E.

extract could be produced and sold due to its high phenolic and fiber content. Grape-seed extracts are common in the nutraceutical market and could cost up to U.S. $25 for a 100 capsule flask (Dwyer et al. 2014). Taking into account various considerations, over 10 billion flasks of grape-seed extracts could have been produced worldwide between 2009 and 2013 (Table 7.3). Assumptions included grape production of each continent destined to wine production; average seed content in the fruit (2%) and a remnant of 85% after oil extraction; and 75% recovery after milling and production of 300 mg capsules. With the price stated earlier in this paragraph, a gross income of more than U.S. $193 billion could have been generated in 2013 in Europe alone (Table 7.4). It is important to note that the manufacture of these capsules requires direct compression and packaging, thus the cost need to be examined prior to project implementation. Finally, the residual de-fatted and treated extract could be used for compost or animal feed as it still contains important concentration of fiber and protein.

Table 7.3 Potential World Production (Million Flasks) of Grape-Seed Extract Rich in Polyphenols

	YEAR				
REGION	2009	2010	2011	2012	2013
Africa	472	480	456	492	511
United States	2,332	2,498	2,609	2,526	2,839
Asia	804	785	825	936	978
Europe	7,442	7,128	7,368	6,302	7,738
Oceania	610	556	595	569	608
Total	**11,660**	**11,448**	**11,854**	**10,825**	**12,673**

Source: Food and Agriculture Organization (FAO), Grape production statistics, United Nations, 2015, available at: http://faostat3.fao.org/compare/E.

Table 7.4 Potential Income (Million U.S.$) from Commercialization of Grape-Seed Extract (Capsules)

	YEAR				
REGION	2009	2010	2011	2012	2013
Africa	11,808	11,989	11,405	12,305	12,767
United States	58,291	62,458	65,235	63,158	70,972
Asia	20,091	19,636	20,633	23,398	24,439
Europe	186,042	178,209	184,206	157,543	193,457
Oceania	15,262	13,902	14,867	14,232	15,196
Total	**291,493**	**286,194**	**296,346**	**270,635**	**316,831**

Source: Food and Agriculture Organization (FAO), Grape production statistics, United Nations, 2015, available at: http://faostat3.fao.org/compare/E.

282 VALORIZATION OF WINE MAKING BY-PRODUCTS

7.3.2.2 Skins Grape skins, like grape seeds are good sources of phytochemicals such as catechin, epicatechin, and gallic acid (Yilmaz and Toledo 2004). Chemically, the composition of grape skin is approximately of 5.8% protein, 4.1% fat, 13.3% cellulose, 4.1% ash, 54.2% water, 9.2% sucrose, 2.4% flavonoids, and 0.75% titratable acids (acetic acid) (Sadovoy et al. 2011). In Georgia alone, around 700,000 kg of dry muscadine skins are produced yearly and could be transformed into a very profitable product (Phillips 2006). An economic analysis was conducted based on profitability, sensitivity, and economic return of three alternatives of polyphenolic isolation following fermentation (Cardona et al. 2010). Operations evaluated included spraying, freezing, and vacuum drying after a fermentation procedure. Results showed an average skin production of 33,000 kg (fresh) which yielded skin that was intended to be mixed with hot water for a production of 66,000 kg of liquid extract and 33,000 kg of residual skins. It was estimated that the production of 114 kg of dried phenolic could be marketed for at least U.S. $45/kg. From the remaining skins after fermentation, 30% were considered solids (Phillips 2006) and were proposed to be sold as animal feed as part of the by-product operation. From the remaining skins obtained, a total of 10,000 kg of dry skin could have been produced in a season in Georgia alone. If these results were to be extrapolated to the production of skins from the global wine industry, 636 million flasks of phenolic extract could have been produced in 2013 (Table 7.5), taking into account 40% of the fruit to be considered as skin. 1600 mg/kg of polyphenols in the skin, a concentration of 1600 mg/kg of phenolic compounds and similar assumptions as the ones discussed for grape-seed extracts for capsule production. With similar prices as a

Table 7.5 Potential World Production (Million Flasks) of Grape Skin Extract Rich in Polyphenols

	YEAR				
REGION	2009	2010	2011	2012	2013
Africa	24	24	23	25	26
United States	117	125	131	127	142
Asia	40	39	41	47	49
Europe	374	358	370	316	388
Oceania	31	28	30	29	31
Total	**585**	**575**	**595**	**543**	**636**

Source: Food and Agriculture Organization (FAO), Grape production statistics, United Nations, 2015, available at: http://faostat3.fao.org/compare/E.

grape-seed extract (U.S. $25 per flask), a gross income of almost U.S. $15 billion could have been generated worldwide and U.S. $10 billion in Europe alone (Table 7.6). It is important to mention that these values could be very conservative as the product created is an isolation of phenolic compounds whereas the grape-seed extract mentioned in the previous section was a mixture of fiber, protein, and phenolic compounds. Therefore, the price for these capsules could increase representing a growth of the marketing potential of this isolated product. Other alternative for the skin is the extraction of anthocyanins to be used as natural colorants. With an estimated 1000 mg/kg of total anthocyanins and 40% skin in the fruit, a world production of more than 10,000 tons could be generated yearly (Table 7.7). This product could be priced at $50/kg to yield a marketing potential of U.S. $500 million per year from which 70% represents Europe (Table 7.8). As mentioned with grape-seed extracts, there is a remnant after phenolic isolation

Table 7.6 Potential Income (Million U.S.$) from World Grape Skin Extract Commercialization

	YEAR				
REGION	2009	2010	2011	2012	2013
Africa	593	602	572	618	641
United States	2,926	3,135	3,275	3,170	3,562
Asia	1,008	986	1,036	1,174	1,227
Europe	9,339	8,945	9,246	7,908	9,711
Oceania	766	698	746	714	763
Total	**14,632**	**14,366**	**14,875**	**13,585**	**15,904**

Source: Food and Agriculture Organization (FAO), Grape production statistics, United Nations, 2015, available at: http://faostat3.fao.org/compare/E.

Table 7.7 Potential World Production (Tons) of Grape Skin Extract Rich in Polyphenols as a Natural Colorant (Anthocyanins)

	YEAR				
REGION	2009	2010	2011	2012	2013
Africa	445	451	429	463	481
United States	2,194	2,351	2,456	2,378	2,672
Asia	756	739	777	881	920
Europe	7,004	6,709	6,935	5,931	7,283
Oceania	575	523	560	536	572
Total	**10,974**	**10,774**	**11,157**	**10,189**	**11,928**

Source: Food and Agriculture Organization (FAO), Grape production statistics, United Nations, 2015, available at: http://faostat3.fao.org/compare/E.

284 VALORIZATION OF WINE MAKING BY-PRODUCTS

Table 7.8 Potential Income (Million U.S.$) from Commercialization of Grape Skin Extract (Natural Colorant)

	YEAR				
REGION	2009	2010	2011	2012	2013
Africa	22	23	21	23	24
United States	110	118	123	119	134
Asia	38	37	39	44	46
Europe	350	335	347	297	364
Oceania	29	26	28	27	29
Total	**549**	**539**	**558**	**509**	**596**

Source: Food and Agriculture Organization (FAO), Grape production statistics, United Nations, 2015, available at: http://faostat3.fao.org/compare/E.

that could be used for human and animal consumption as it contains soluble and insoluble fiber.

7.3.2.3 Stems Grapes are fruit berries attached to the stem and make up the cluster or bunch of grapes (Hellman 2003). During harvest, the stem is cut from the plant and is discarded when the juice of whole grapes is processed. This by-product contains tannins and fiber that could be processed to extract those compounds for animal feed (Llobera and Cañellas 2007).

7.3.3 Lees

During fermentation, yeast strains are responsible for the conversion of sugars into alcohol in addition to the creation of other compounds that exert aroma and flavor in the final product (Bisson 1993; Fleet and Heard 1993). This action is instrumental in the development of wine. However, yeast and other components (tannins, polyphenols, and crystallized acids) are commonly removed through filtration and flocculation to obtain a clarified wine, resulting in a by-product called lees (Lempereur and Penavayre 2014; Vlyssides et al. 2005). Hwang et al. (2009) used wine lees to improve the rheological and antioxidant properties of ice cream (Figure 7.6). Additionally, lees is known to provide high amounts of tartaric acid (~77.5 g/L), which is used in cream of tartar in European cuisine, and, as an alternative, to provide nutritional support to growth lactic acid bacteria cultures, such as *Lactobacillus pentosus* (Rivas et al. 2006). Aside for the food industry

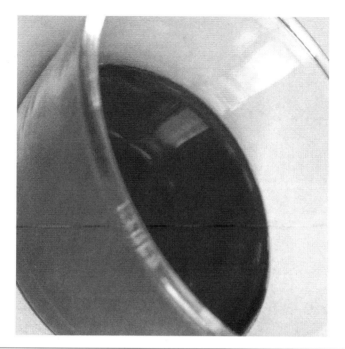

Figure 7.6 Lees.

Table 7.9 Potential World Production (Tons) of Tartaric Acid from Lees

	YEAR				
REGION	2009	2010	2011	2012	2013
Africa	389	395	376	405	421
United States	1920	2057	2149	2080	2,338
Asia	662	647	680	771	805
Europe	6128	5870	6068	5190	6,373
Oceania	503	458	490	469	501
Total	**9602**	**9428**	**9762**	**8915**	**10,437**

Source: Food and Agriculture Organization (FAO), Grape production statistics, United Nations, 2015, available at: http://faostat3.fao.org/compare/E.

applications, tartaric acid is used in the pharmaceutical and cosmetic industries and in construction as a fluidizer. To estimate the world production of tartaric acid (Table 7.9) from lees and wine fermentation, a production of 0.5 tons of lees per 100 tons of fruit was used as explained by Bacic (2003) where various wineries were assessed in Australia to get that proportion. In addition to that value, the fraction of tartaric acid in the lees ranges from 65 to 77 g/L (Rivas et al. 2006) and for this calculation a concentration of 70 g/L was used yielding

286 VALORIZATION OF WINE MAKING BY-PRODUCTS

Table 7.10 Potential Income (Thousand U.S.$) from the Commercialization of Tartaric Acid Extracted from Lees

	YEAR				
REGION	2009	2010	2011	2012	2013
Africa	778	790	751	811	841
United States	3,840	4,115	4,298	4,161	4,676
Asia	1,324	1,294	1,359	1,542	1,610
Europe	12,257	11,741	12,136	10,379	12,745
Oceania	1,005	916	979	938	1,001
Total	**19,204**	**18,855**	**19,524**	**17,830**	**20,874**

Source: Food and Agriculture Organization (FAO), Grape production statistics, United Nations, 2015, available at http://faostat3.fao.org/compare/E.

close to 10,000 tons of tartaric acid that could have been produced (2009–2013). The price of this product was estimated at U.S. $2000/ton which could have yielded U.S. $18– $21 million worldwide with the biggest income produced in Europe (63%) (Table 7.10). Lees are also a high source of amino acids from dead yeast and the nutrients that were used by yeast, containing nitrogen at about 2 g/100 g of dried lees, and carbon at ranges around 31.1–37.4 g/100 g of dried lees (Rivas et al. 2006). The lees can be crushed to obtain alcohol which can be isolated through distillation and tartaric acid can be further processed to obtain calcium tartrate. Ethanol is the main product of fermentation and the ingredient in wine and spirits (Fleet and Heard 1993). However, ethanol can be produced from by-products as they still contain residual sugars. The concentration of ethanol in lees can range from 55% to 80% (Bayrak 2013). Research has indicated that the production of ethanol from by-products was not only advantageous to use as a biofuel but also improved the extraction of anthocyanins which otherwise could interact with sugars creating undesirable reactions and color change in the product (Cardona et al. 2009). A rapid fermentation converted 4%–5% of sugar into ethanol which was then evaporated during drying to obtain a powder. However, this ethanol could be distilled and used as biofuel.

7.4 Closing Remarks

Arvanitoyannis et al. (2006) have presented a thorough analysis of potential uses and applications of wine industry waste showing many of the discussed opportunities in this chapter. Additionally, these authors have proposed the use of these by-products as soil conditioners, fertilizers, and adsorbent of heavy metals. Products such as pullulans could also be extracted from grape skins and used as low-calorie thickeners for animal and human consumption. Other alternatives include the combustion of by-products to generate electricity (Celma et al. 2007), composting of the pomace or performing anaerobic digestion to produce biogas (Lempereur and Penavayre 2014), which could be performed in waste that has already been de-fatted (in case of grape seeds) and processed for phenolic extraction (seeds and skins). Other authors illustrated the marketing potential of various by-products from Canadian wine industry used for nutraceuticals (dried pomace and grape-seed oil) and fertilizers (Dwyer et al. 2014). The possibilities exposed in this chapter illustrate the importance of processing the by-products generated by the wine industry, not only from an economical perspective but also from an environmental perspective as the waste production gets reduced significantly. Additionally, more laws are being considered regarding gas emissions and waste management which will force industries to further process their by-products regardless of the economic benefit this could generate.

References

Arvanitoyannis IS, Ladas D, Mavromatis A. 2006. Potential uses and applications of treated wine waste: A review. *Int. J. Food Sci. Technol.* 41:475–487.

Bacic T. 2003. Recovery of valuable products from lees and integrated approach to minimize waste and add value to wine production. Final Report to Grape and Wine Research and Development Corporation. University of Melbourne, Melbourne, Victoria, Australia, 43pp.

Bayrak E. 2013. Utilization of wine waste for fermentative processes (thesis). Izmir Institute of Technology, İzmir, Turkey, 62pp.

Beveridge THJ, Girard B, Kopp T, Drover JCG. 2005. Yield and composition of grape seed oils extracted by supercritical carbon dioxide and petroleum ether: Varietal effects. *J. Agric. Food Chem.* 53:1799–1804.

Bisson LF. 1993. Yeast—Metabolism of sugars. In Fleet GH (ed.), *Wine Microbiology and Biotechnology*. Boca Raton, FL: CRC Press, pp. 55–76.

Blake C. 2012. Higher California wine grape prices expected to continue. *Western Farm Press*. Available at: http://westernfarmpress.com/grapes/higher-california-wine-grape-prices-expected-continue?page=1. Last accessed March, 2015.

288 VALORIZATION OF WINE MAKING BY-PRODUCTS

Bonilla F, Mayen M, Merida J, Medina M. 1999. Extraction of phenolic compounds from red grape marc for use as food lipid antioxidants. *Food Chem.* 66:209–215.

Cardona JA. 2007. Chemical and economic analysis of a value-added product from muscadine grape pomace (thesis). University of Florida, Gainesville, FL, 106pp.

Cardona JA, Lee JH, Talcott ST. 2009. Color and phenolic stability in extracts produced from muscadine grape (*Vitis rotundifolia*) pomace. *J. Agric. Food Chem.* 57:8421–8425.

Cardona JA, Wysocki A, Talcott ST. 2010. Economic analysis of an isolated product obtained from muscadine grape pomace. *Hort. Technol.* 20(1):160–168.

Celma AR, Rojas S, López- Rodríguez F. 2007. Waste-to-energy possibilities for industrial olive and grape by products in Extromadura. *Biomass Bioenergy* 31:522–534.

Childs NM. 1999. Nutraceutical industry trends. *J. Nutraceut. Funct. Med. Foods* 2(1):73–85.

Choy YY, Jaggers GK, Oteiza PI, Waterhouse AL. 2013. Bioavailability of intact proanthocyanidins in the rat colon after ingestion of grape seed extract. *J. Agric. Food Chem.* 61:121–127.

Conde C, Silva P, Fontes N, Dias ACP, Tavares RM, Sousa MJ, Agasse A, Delrot S, Gerós H. 2007. Biochemical changes throughout grape berry development and fruit and wine quality. *Food* 1(1):1–22.

Crews C, Hough P, Godward J, Brereton P, Lees M, Guiet S, Winkelmann W. 2006. Quantification of the main constituents of some authentic grape-seed oils of different origin. *J. Agric. Food Chem.* 54:6261–6265.

Croft KD. 1999. Antioxidant effects of plant phenolic compounds. In Basu TK, Temple NJ, Garg ML (eds.), *Antioxidant in Human Health and Disease*. New York: CABI Pub., pp. 109–202.

Dos Santos Freitas L, De Oliveira JV, Dariva C, Assis Jacques R, Bastos Caramao E. 2008. Extraction of grape seed oil using compressed carbon dioxide and propane: Extraction yields and characterization of free glycerol compounds. *J. Agric. Food Chem.* 56:2558–2564.

Dwyer K, Hosseinian F, Rod M. 2014. The market potential of grape waste alternatives. *J. Food Res.* 3(2):91–106.

Ferrer J, Páez G, Mármol Z, Ramones E, Chandler C, Marín M, Ferrer A. 2001. Agronomic use of biotechnologically processed grape wastes. *Bioresour. Technol.* 7:39–44.

Fleet GH, Heard GM. 1993. Yeast—Growth during fermentation. In Fleet GH (ed.), *Wine Microbiology and Biotechnology*. Boca Raton, FL: CRC Press, pp. 27–54.

Food and Agriculture Organization (FAO). 2015. Grape production statistics. United Nations. Available at: http://faostat3.fao.org/download/Q/QC/E. Last accessed February, 2015.

Hellman EW. 2003. Grapevine structure and function. In Hellman EW (ed.), *Oregon Viticulture*. Corvallis, OR: Oregon State University Press, pp. 5–19.

MARKETING POTENTIAL 289

Hwang JY, Shyu YS, Hsu CK. 2009. Grape wine lees improves the rheological and adds antioxidant properties to ice cream. *Food Sci. Technol.* 42(1):312–318.

Indexmundi. 2015. Historic olive oil prices (2010–2015). Available at: http://www.indexmundi.com. Last accessed March, 2015.

Jing P, Giusti MM. 2005. Characterization of anthocyanin-rich waste from purple corncobs (*Zea mays* L.) and its application to color milk. *J. Agric. Food Chem.* 53:8775–8781.

Kaur M, Agarwal C, Agarwal R. 2009. Anticancer and cancer chemopreventive potential of grape seed extract and other grape-based products. *J. Nutr.* 139:1806–1812.

Kennedy JA. 2008. Grape and wine phenolics: Observations and recent findings. *Cienc. Invest. Agraria.* 35(2):107–120.

Koplotek Y, Otto K, Bohm V. 2005. Processing strawberries to different products alters contents of vitamin C, total phenolics, total anthocyanins and antioxidant capacity. *J. Agric. Food Chem.* 53:5640–5646.

Lau DW, King AJ. 2003. Pre- and post-mortem use of grape seed extract in dark poultry meat to inhibit development of thiobarbituric acid reactive substances. *J. Agric. Food Chem.* 51:1602–1607.

Lee J, Wrolstad RE. 2004. Extraction of anthocyanins and polyphenolics from blueberry processing waste. *J. Food Sci.* 69(7):C564–C573.

Lempereur V, Penavayre S. 2014. Grape marc, wine lees an deposit of the must: How to manage oenological by-products? *BIO Web Conf.* 3(01011):1–6.

Llobera A, Cañellas J. 2007. Dietary fibre content and antioxidant activity of Manto Negro red grape (*Vitis vinifera*): Pomace and stem. *Food Chem.* 101:659–666.

Marchitan N, Cojocaru C, Mereuta A, Duca Gh, Cretescu I, Gonta M. 2010. Modeling and optimization of tartaric acid reactive extractin from aqueous solutions: A comparison between response surface methodology and artificial neural network. *Sep. Purif. Technol.* 75:273–285.

Martín-Belloso O, Marsellés-Fontanet AR. 2007. Grape juice. In Hui YH (ed.), *Handbook of Fruits and Fruit Processing*. Ames, IA: Blackwell Publishing, 697pp.

McMillan R. 2015. Wine report. State of the Wine Industry. Silicon Valley Bank. St. Helena, CA. Available at: http://www.svb.com/uploadedFiles/Content/Blogs/Wine_Report/2015_Wine_Report/wine-report-2015-pdf.pdf. Last accessed March, 2015.

Meyer AS, Suhr KI, Nielsen P, Lyngby, Holm F. 2002. Natural food preservatives. In Ohlsson T, Bengtsson N (eds.), *Minimal Processing Technologies in the Food Industry*. Boca Raton, FL: CRC Press, pp. 124–176.

Meyer BN, Ferrigni NR, Putnam LB, Jacobsen LB, Nichols DE, Mclaughlin JL. 1982. Brine shrimp: A convenient general bioassay for active plant constituents. *J. Med. Plant Res.* 45(5):31–34.

Mildner-Szkudlarz S, Siger A, Szwengiel A, Bajerska J. 2015. Natural compounds from grape by-products enhance nutritive value and reduce formation of CML in model muffins. *Food Chem.* 172:78–85.

290 VALORIZATION OF WINE MAKING BY-PRODUCTS

Mildner-Szkudlarz S, Zawirska-Wojtasiak R, Szwengiel A, Pacyński M. 2011. Use of grape by-product as a source of dietary fibre and phenolic compounds in sourdough mixed rye bread. *Int. J. Food Sci. Technol.* 46:1485–1493.

Milo L. 2005. Nutraceuticals and functional foods. *Food Technol.* 59(5):65–67.

Morris JR, Brady PL. 2004. The muscadine experience: Adding value to enhance profits. Arkansas Agricultural Experiment Station, Institute of Food Science and Engineering, University of Arkansas, Fayetteville, AR.

Moulton K, Possingham J. 1998. Research needs and expectations in the grape and grape products sector. *World Conference on Horticultural Research International Society for Horticultural Science*, Rome, Italy.

Moure A, Cruz JM, Franco D, Domínguez JM, Sineiro J, Dominguez H, Núñez MJ, Parajó JC. 2001. Natural antioxidants from residual sources. *Food Chem.* 72:145–171.

Murga R, Ruiz R, Beltrán S, Cabezas JL. 2000. Extraction of natural complex phenols and tannins from grape seeds by using supercritical mixtures of carbon dioxide and alcohol. *J. Agric. Food Chem.* 48:3408–3412.

Palma M, Barroso CG. 2002. Ultrasound-assisted extraction and determination of tartaric and malic acids from grapes and winemaking by-products. Departamento de Química Analítica, Facultad de Ciencias, Universidad de Cádiz, Cádiz, Spain.

Palma M, Taylor LT. 1999. Fractional extraction of compounds from grape seeds by supercritical fluid extraction and analysis for antimicrobial and agrochemical activities. *J. Agric. Food Chem.* 47:5044–5048.

Pastrana-Bonilla E, Akoh C, Sellapan S, Krewer G. 2003. Phenolic content and antioxidant capacity of muscadine grapes. *J. Agric. Food Chem.* 51:5497–5503.

Peschel W, Sánchez-Rabaneda F, Diekmann W, Plescher A, Gartzia I, Jimenez D, Lamuela-Raventos R, Buxaderas S, Codina C. 2006. An industrial approach in the search of natural antioxidants from vegetable and fruit wastes. *Food Chem.* 97(1):137–150.

Phillips RD. 2006. Pilot-scale, pre-commercial production of nutraceuticals from Georgia commodities. In Fiscal Year 2005–2006 Report to industry. Georgia's Traditional Industries Program for Food Processing, Food Processing Advisory Council, Athens, GA, p. 10.

Primary Industries and Resources South Australia (PIRSA). 2005. A report on the impact of current grape-pricing trends on the Riverland Region. Government of South Australia, Canberra, Australia, 69pp.

Rivas B, Torrado A, Moldes AB, Domínguez JM. 2006. Tartaric acid recovery from distilled lees and use of the residual solid as an economic nutrient for *Lactobacillus. J. Agric. Food Chem.* 54:7904–7911.

Sadovoy V, Silantyev A, Selimov M, Shchdrina T. 2011. An examination of chemicals and molecular properties of grape berry skin flavonoids. *Food Nutr. Sci.* 2:1121–1127.

Shahidi F, Naczk M. 2003. *Phenolics in Food Nutraceuticals.* Boca Raton, FL: CRC Press, 558pp.

Sousa EC, Uchoa-Thomaz AMA, Carioca JOB, Morais SM, Lima A, Martins CG, Alexandrino CD et al. 2014. Chemical composition and bioactive compounds of grape pomace (*Vitis vinifera* L.), Benitaka variety, grown in the semiarid region of Northeast Brazil. *Food Sci. Technol. (Campinas)* 34(1):135–142.

Visioli F, Romani A, Mulinacci N, Zarini S, Conte D, Vincieri FF, Galli C. 1999. Antioxidant and other biological activities of olive mill waste waters. *J. Agric. Food Chem.* 47:3397–3401.

Vlyssides AG, Barampouti EM, Mai S. 2005. Wastewater characteristics from Greek wineries and distilleries. *Water Sci. Technol.* 51:53–61.

Weber HA, Hodges AE, Guthrie JR, O'Brien BM, Robaugh D, Clark AP. 2007. Comparison of proanthocyanidins in commercial antioxidants: Grape seed and pine bark extracts. *J. Agric. Food Chem.* 55:148–156.

Williams LE, Dokoozlian NK, Wample R. 1994. Grape. In Schaffer B, Anderson PC (eds.), *Handbook of Environmental Physiology of Fruit Crops*. Boca Raton, FL: CRC Press Inc., pp. 85–133.

Yilmaz Y, Toledo RT. 2004. Major flavonoids in grape seeds and skins: Antioxidant capacity of catechin, epicatechin, and galic acid. *J. Agric. Food Chem.* 52:255–260.

Zhu L, Zhang Y, Lu J. 2012. Phenolic contents and compositions in skins of red wine grape cultivars among various genetic backgrounds and originations. *Int. J. Mol. Sci.* 13:3492–3510.

8

FUTURE PERSPECTIVES

MATTEO BORDIGA

Contents

8.1	Oligosaccharides	294
8.2	Oligosaccharide Production	299
8.3	Extraction of Polysaccharides from Natural Sources	300
8.4	Production of Oligosaccharides by Acid Hydrolysis	301
8.5	Production of Oligosaccharides by Enzymatic Hydrolysis	302
8.6	Production of Oligosaccharides by Physical Hydrolysis	303
8.7	Purification of Oligosaccharides	303
8.8	Potential Case Study I	304
	8.8.1 Prebiotic Oligosaccharides: A Biorefinery Approach to Valorize Vine and Wine Making By-Products	304
	8.8.2 Primary Objectives	307
	8.8.3 Secondary Objectives	308
	8.8.4 Methods	309
	8.8.5 Schematic Scheduling Activities	314
	8.8.6 Potential Applications, Impact	316
8.9	Spent Grape Pomace	317
	8.9.1 Potential of By-Products	317
	8.9.2 Potential Exploitation of Residues from Vineyards and Wine Production	318
	8.9.2.1 Vine Shoots	318
	8.9.2.2 Vine Leaves	319
	8.9.2.3 Grape Skins	319
	8.9.2.4 Grape Seeds	319
	8.9.2.5 Vinification Lees	320
	8.9.2.6 Residue from Residues	320
8.10	Potential Case Study II	320
	8.10.1 Spent Grape Pomace: Still a Potential By-Product	320
	8.10.2 Distillation Process	322
	8.10.3 Extraction of Fatty Acids and Phenolic Compounds	323

294 VALORIZATION OF WINE MAKING BY-PRODUCTS

8.10.4	Phenolic Compound Analysis	323
8.10.5	Proanthocyanidin Analysis	324
8.10.6	Fatty Acid Analysis	325
8.10.7	Antiradical Activity	326
8.10.8	Statistical Analysis	326
8.10.9	Results	326
	8.10.9.1 Phenolic Compounds	326
	8.10.9.2 Hydroxybenzoic Acids	328
	8.10.9.3 Hydroxycinnamic Acids and Tartaric Esters	328
	8.10.9.4 Flavan-3-Ols	328
	8.10.9.5 Flavonols	329
	8.10.9.6 Anthocyanins	329
	8.10.9.7 RP-HPLC-Phloroglucinolysis	330
	8.10.9.8 Fatty Acid Content	332
	8.10.9.9 Antiradical Activity	334
8.11 Discussion		334
8.12 Conclusions		337
References		338

8.1 Oligosaccharides

The macromolecules of wines include polyphenols, proteins, and polysaccharides. Polysaccharides have been thoroughly studied because of their important technological and sensory properties in wine. They have the ability to interact and aggregate with tannins (Riou et al. 2002), to decrease astringency in wine-like model solutions (Vidal et al. 2004), to inhibit hydrogen tartrate crystallization (Gerbaud et al. 1996), to interact with wine aroma compounds (Chalier et al. 2007), and to form specific coordination complexes with Pb^{2+} ions (O'Neill et al. 1996; Pellerin et al. 1997). The structure and amount of polysaccharides released into wine depend on the wine making process and can be modified by enzymatic treatment (Ayestaran et al. 2004; Doco et al. 2007). Unlike wine polysaccharides, which have been the subject of many studies, oligosaccharides have only been partially studied, although their presence in wine is known (Figures 8.1 and 8.2). It is thus necessary to determine their exact composition in wine and to analyze their molecular structures in order to better understand the technological and organoleptic properties associated with them. The study of oligosaccharides has led

FUTURE PERSPECTIVES

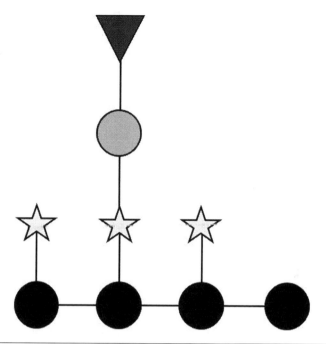

Figure 8.1 The structure suggested for xyloglucans reported in wine. Glucose (dark circle), galactose (light circle), fructose (triangle), and xylose/arabinose (star). (From Bordiga, M. et al., *J. Agric. Food Chem.*, 60, 3700, 2012.)

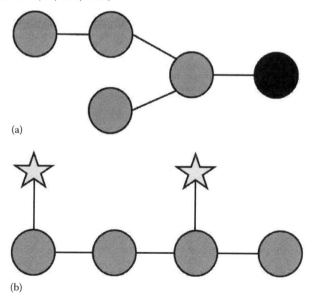

Figure 8.2 The structures suggested for (a) Galacto-oligosaccharides and (b) Arabinogalactans reported in wine. Glucose (dark circle), galactose (light circle), and xylose/arabinose (star). (From Bordiga, M. et al., *J. Agric. Food Chem.*, 60, 3700, 2012.)

296 VALORIZATION OF WINE MAKING BY-PRODUCTS

to considerable confidence that complex carbohydrates provide health benefits (Kunz et al. 2000; Urashima et al. 2001). Scientists are beginning to realize that sugar, oligosaccharides, and polysaccharides exhibit both high structural specificity, much as do proteins and polynucleotides, and complexity of structure and function. The structure–function properties of oligosaccharides are being studied, much the same as are those of proteins, as bioactive components that aid intestinal functions in humans (Hopkins and Macfarlane 2003; Shoaf et al. 2006). Oligosaccharides, structurally related to plant cell wall polysaccharides, are involved in plant physiology and in particular in plant defense responses, and they can play an important role in medicinal, food, and agricultural applications. The study of oligosaccharides (Newburg et al. 2004; Newburg 2005) has thus far been focused almost exclusively on animal milks, with hundreds of articles published just on the topic of human milk oligosaccharides. One of the most studied and well-demonstrated actions of oligosaccharides is its prebiotic activity. It is thus necessary to determine the exact composition of oligosaccharides in wine and to analyze their molecular structures in order to understand their organoleptic and bioactive properties. The definition of prebiotics is nondigestible food ingredients that beneficially affect the host by selectively stimulating the growth and/or activity of one or a limited number of bacteria in the colon that can improve the host health. Oligosaccharides fit this definition because they are neither digested nor absorbed in the upper intestinal tract of humans but are delivered intact into the colon, where they can act as nutrients for colonic microflora. There is evidence that neutral oligosaccharides present in human milk act as growth enhancers for bacteria of the genus *Bifidobacterium* in infants. A comparison of the fecal flora of breast-fed infants with that of bottle-fed infants showed that the levels of bifidobacteria were significantly higher in the former than in the latter (Boehm and Stahl 2007). Briefly, it can be concluded that oligosaccharides and glycoconjugates are important bioactive molecules, present not as nutrients for energy but to provide protection in the first few days in the life of a newborn. They may have a role as micronutrients to meet the specific need of a rapidly growing infant, for example, a source of sialic acid which has a role in the development of the nervous system. The classical structural characterization of oligosaccharides has been obtained by glycosyl and linkage analyses, by NMR or by mass spectrometry techniques

FUTURE PERSPECTIVES 297

(Macfarlane et al. 2008). The analysis of oligosaccharides by MS has been made possible by the development of soft ionization techniques such as the matrix-assisted laser desorption ionization (MALDI). The matrix-assisted laser desorption ionization/Fourier transform ion cyclotron resonance (MALDI–FTICR) method is a sensitive and robust analytical method with high performance capability, and it allows rapid and unambiguous assignment of oligosaccharide signals (Ninonuevo et al. 2006; Penn et al. 1997). Oligosaccharide production is of great interest for the food industry due to the beneficial effects on human health. However, the production of prebiotics in large amounts and of good quality is still a challenge to food science and technology. Several methods have been developed, modified, and adapted to optimize the quality of oligosaccharides and enable their production. The food industry needs more efficient, simple, and less expensive processes for their application on a large scale. Trends in oligosaccharide production involve the improvement of existing processes by both hydrolysis and synthesis of oligosaccharides. On the other hand, the polysaccharide hydrolysis by chemical, enzymatic, or physical processes is simpler and has a greater yield than that of synthesis. Depolymerization of polysaccharides by chemical or enzymatic processes has shown good results, but requires an improvement. Physical processes emerge as innovative alternatives due to their quickness, convenience, and quality of the end product. Although good results have been observed, their applicability in the food industry is still limited, and further tests are necessary. There is no doubt that the physical process is the better choice when sustainable and long-term economies are taken into account; however, the production of oligosaccharides by this way needs to be improved to obtain better results. Therefore, each process has advantages and disadvantages, which must be evaluated before determining the best process. Given the complexity of the production of oligosaccharides and the features of the source, the process used will determine the characteristics of the end product and the applicability of the process on an industrial scale. According to their molecular weight or to the degree of polymerization, carbohydrates are classified into monosaccharides, oligosaccharides, or polysaccharides. Oligosaccharides are defined as saccharide compounds containing 2–15 sugar units, including also some nondigestible disaccharides (Mussatto and Mancilha 2007). Besides this classification, carbohydrates can also be classified into digestible or nondigestible.

298 VALORIZATION OF WINE MAKING BY-PRODUCTS

Nondigestible carbohydrates include some polysaccharides and oligosaccharides delivered intact to the large intestine. In the colon, these compounds show the ability to be selectively fermented stimulating the intestinal microbiota (several beneficial systemic effects). Oligosaccharides with prebiotic properties have received great interest from both researchers and food industry due to their potential as ingredients for the formulation of functional foods (Barreteau et al. 2006; Mussatto and Mancilha 2007). Among the most cited, physiological benefits of these compounds are the stimulation of the intestinal microbiota; production of short-chain fatty acids; reduction of intestinal pH; inhibition of the development of pathogenic microorganisms; decrease of gastrointestinal infections; decreased insulin response and glucose uptake; improvement of the lipoprotein profile; reduction of cancer development risk; and stimulation of the uptake of minerals, such as calcium, magnesium, and iron (Gobinath et al. 2010; Moure et al. 2006; Qiang et al. 2009). The introduction of oligosaccharides as functional ingredients in the food industry has increased, especially in beverages (fruit juices, coffees, chocolates, and teas) and dairy products (yoghurt, powdered milk, and ice cream). Fructooligosaccharides (FOS) and galactooligosaccharides (GOS) are among the main commercially available oligosaccharides with bifidogenic properties. A novel generation of oligosaccharides, namely, xylooligosaccharides (XOS), pecticoligosaccharides (POS), and agarooligosaccharides has received attention in recent years (Chen et al. 2013; Kang et al. 2014; Moure et al. 2006; Qiang et al. 2009). The properties of oligosaccharides differ with their composition, molecular weight, and structural linkages. These characteristics are directly related to the natural source from which they are extracted and to the extraction process applied (Kang et al. 2014; Nabarlatz et al. 2007). In general, these oligomers are substantially composed of fructose, galactose, glucose, and/or xylose monomers. Their degree of purity is also very important with respect to the physiological effects. Some methods for oligosaccharides production can generate products with the presence of salts or other nonsaccharide compounds, unhydrolyzed polysaccharides, or monosaccharides. In this case, the elimination of these compounds appears mandatory. The reduction of the amount of nonsaccharide compounds present in the medium is very important to increase the quality of the final product for prebiotic applications (Gullon et al. 2011). Although there are some

FUTURE PERSPECTIVES 299

oligosaccharides with prebiotic properties commercially available (FOS and GOS), there is an increasing interest in the development of new prebiotics with added functionality. Production of this functional ingredient in large scale requires constantly the developing of new techniques but also improving the existing ones (Gullon et al. 2011; Kang et al. 2014). The synthesis of oligosaccharides is more complex than that of other polymers due to the several combinations of monomer units. Moreover, specific enzymes and highly controlled reaction conditions are necessary, usually resulting in a lower yield. Therefore, the synthesis process is more expensive and difficult to reproduce on a large scale. Obtaining oligosaccharides from the hydrolysis of polysaccharides is often simpler, has lower costs, and is more reproducible on an industrial scale (Barreteau et al. 2006). However, the peculiarities inherent to each hydrolysis method proposed in the literature can yield different qualitative and productive efficiencies.

8.2 Oligosaccharide Production

Nondigestible oligosaccharides are available from several foods and other natural sources. Also, oligosaccharides can be developed from substrates extracted from natural sources. When oligosaccharides are naturally available in the raw material, only the extraction process is necessary. This process can be accomplished by substrate solubilization in water, methanol, or other alcoholic medium. Other than available oligosaccharides, hydrolytic or synthetic methods also allow for obtaining low-molecular-weight carbohydrates. Presently, most oligosaccharides used as prebiotics are produced by extraction from natural sources, are obtained by the enzymatic or acid hydrolysis of polysaccharides, or produced by enzymatic transglycosylation (Du et al. 2011). Low-molecular-weight FOS can be enzymatically synthesized from sucrose. Lactulose is also produced by synthesis, through an isomerization reaction. GOS and glycosylsucrose are produced by transglycosylation—the former from lactose and the latter from maltose or sucrose (Mussatto and Mancilha 2007). Lactosucrose is produced by an enzymatic reaction that transfers the final fructosyl fraction of sucrose to lactose. Isomaltulose naturally occurs in some foods (honey and sugarcane juice) or can be produced from sucrose, through an enzymatic reaction that alters the glycosidic bonds. The production of oligosaccharides by hydrolysis usually requires

300 VALORIZATION OF WINE MAKING BY-PRODUCTS

the extraction of polysaccharides from the raw material, followed by depolymerization (Du et al. 2011). FOS and XOS are most commonly produced by hydrolysis of polysaccharides, which can be caused by chemical, enzymatic, or physical agents. Inulin from chicory is a good source for the enzymatic production of FOS, and some alternative sources have been investigated, such as agave fructans (Avila-Fernandez et al. 2011). Xylooligosaccharides are produced by the depolymerization of lignocellulosic materials, which are abundant in xylans. For this purpose, some agricultural by-products can be used, including husks, straw, and corncobs (Moure et al. 2006). The enzymatic or acid hydrolysis of starch results in maltooligosaccharides (MOS), isomaltooligosaccharides, gentiooligosaccharides, and cyclodextrins. The latter two still undergo transglycosylation (Warrand and Janssen 2007). Emerging chitinoligosaccharides (COS) are produced from deacetylated chitin and POS are obtained from pectin, both by depolymerization (Chen et al. 2013). Usually, oligosaccharides for food applications are not pure and often contain polysaccharides from the original material, monosaccharides, and a small fraction of nonsaccharide compounds. A pure final product facilitates their inclusion in food formulations, although there is evidence that mixtures of oligosaccharides may have interesting physiological effects, such as the decrease of the discomfort caused by excessive gas production by the intestinal microbiota. Successive purification processes for the elimination of undesirable compounds, besides detailed molecular characterization and investigations of impurities, are very important to prove the final quality of oligosaccharides. The physiological effects of oligosaccharides are related to their monosaccharide composition, degree of polymerization, molecular weight, and the presence of impurities. These characteristics depend on the natural source and on the chosen process. Thus, structural studies can provide valuable information to access oligosaccharide structure–function relationships, and a series of *in vitro* and *in vivo* studies should be performed, particularly when there are new ingredients or production processes, before their approval for commercialization and consumption (Kang et al. 2014).

8.3 Extraction of Polysaccharides from Natural Sources

Oligosaccharides can be obtained by direct extraction from raw materials (Coelho et al. 2014). Extraction can also be necessary to solubilize

FUTURE PERSPECTIVES 301

and to isolate polysaccharides for subsequent depolymerization. Once extracted from raw materials, polysaccharides are more accessible to the acid, enzymatic, or physical depolymerization processes. In this case, it is easier to control the reaction and the characteristics of the final product. POS and FOS are produced by this method: pectin is extracted from raw materials, such as citrus peel, apple pomace, or sugar beet, while inulin and fructans are extracted from chicory or agave, with the subsequent hydrolysis of these polysaccharides. Water, ethanol, acids, or enzymes are applied for the extraction of polysaccharides from raw materials (Doco et al. 2003). Extraction of polysaccharides in aqueous solution can occur at different temperatures (from room temperature to boiling temperature) and, in most cases, some hours are required for the extraction. The complexity of the extraction process depends on the chemical composition of the source and on the specificity desired for the final product. For the extraction of xyloglucans from grape berry cell walls, ethanol washes were performed for precipitation of soluble polysaccharides, removal of lipids with methanol/chloroform, and removal of phenols with ethanol and acetone. For higher purification, depectinization was performed with some enzymes. Polysaccharides from *Nerium indicum* were isolated with a successive process of extraction with ethanol and boiling water. This process was carried out for 6 days, with an average yield of 2.8% of crude high-molecular-weight polysaccharides. The extraction of polysaccharides is not always a step for oligosaccharide production. Autohydrolysis is a physical process in which the whole raw material is used. Proteins, lignin, and minerals present in the complex matrix of raw materials do not interrupt autohydrolysis because the reaction is not performed by an external agent. When acids or enzymes are added, these substances present in the raw material can disrupt the reaction.

8.4 Production of Oligosaccharides by Acid Hydrolysis

Acid hydrolysis is a relatively simple, inexpensive, and easy-to-control process because the reaction is interrupted with neutralization of the medium. Sulfuric, hydrochloric, and trifluoroacetic acids are the most used ones for polysaccharide depolymerization. Generally, the process is optimized at higher temperatures, above 60°C, and the time

302 VALORIZATION OF WINE MAKING BY-PRODUCTS

often varies, 2–6 h on average. However, acid hydrolysis may have disadvantages, such as degradation of monosaccharides with sequential formation of toxic substances (furfural and 5-hydroxymethylfurfural) and low yield of oligosaccharides. Moreover, the length of chains of the resulting oligosaccharides can vary. These facts indicate the necessity of constantly monitoring the reaction conditions (acid concentration, temperature, and time) and of conducting preliminary tests. Some polysaccharides are more susceptible to the acid or temperature employed. In fructans, for example, just few minutes are sufficient for FOS production. Warrand and Janssen (2007) hydrolyzed pure amylose under dilute acidic conditions (0.45 M HCl, 90°C) with two types of heating: microwave radiation and conventional heating. Microwave treatment proved to be more efficient. A similar range of oligosaccharides was observed, as with the conventional heating procedure, without any degradation compounds (brown products), and a faster production rate, leading to very short production time (no more than 15 min).

8.5 Production of Oligosaccharides by Enzymatic Hydrolysis

Enzymatic depolymerization of polysaccharides is the best choice for larger oligosaccharide production (Barreteau et al. 2006). Oligosaccharides with desired molecular weights and minimum adverse chemical modifications in the final products are possible by enzymatic hydrolysis (highly region and stereo selective). Some prebiotics were prepared with enzymes: MOS, by the action of a-amylases on starch; isomaltooligosaccharides, by the action of α- and β-amylases and α-glycosidases on the starch; FOS production, by the action of inulinase on inulin and fructans; XOS, by the action of β-1,4-xylanases on xylans; and oligogalacturonides, by the action of pectinases, pectolyases, and other polygalacturonases on the pectic substrate (Samanta et al. 2014). Enzymes for oligosaccharide production are produced by microorganisms (bacteria and fungi), and they may be endoenzymes or exoenzymes. Although efficient, these enzyme productions should undergo concentration/isolation to eliminate the microbiological contaminants. For this reason, obtaining oligosaccharides by enzymatic hydrolysis can be more expensive. Another disadvantage is the necessity of a buffer, hindering the purity of the final product. Moreover,

FUTURE PERSPECTIVES **303**

the enzyme yield and quality are highly associated with the capacity of the microorganism to adapt to the substrate, which depends on pH, temperature, and composition of the culture medium.

8.6 Production of Oligosaccharides by Physical Hydrolysis

Autohydrolysis occurs in water and at high temperatures, between 130°C and 230°C. Autohydrolysis is widely used for XOS production from lignocellulosic materials (Nabarlatz et al. 2007). This physical process also proved efficient for POS production from sugar beet pulp. One of the advantages of this method is that it does not require the use of chemical products, which increase cost, hinder the purification of the final product, and generate environmental waste. In addition to sugar decomposition, a series of other simultaneous processes occur, including the removal of solid-phase extractives, partial dissolution of lignin (acid-soluble lignin) and ashes, solubilization of proteins, and reaction of proteins and amino acids with sugars, producing melanoidins. To minimize the production of undesirable compounds, pretreatment can be performed. Rapid preheating of the biomass in a microwave at 120°C–150°C and washing remove some undesirable compounds. Microwave, gamma radiation, ultrasonication, ultraviolet light, and dynamic high-pressure microfluidization are also applied for oligosaccharide production (Chen et al. 2013). Physical processes represent a fast and clean way to produce oligosaccharides. However, they are under development, and their application to the industry is still very limited. In addition, some issues need to be solved such as the optimization of reaction conditions for obtaining a better yield and the basic kinetics of polysaccharide hydrolysis. Higher temperatures promote depolymerization, debranching, de-esterification, and formation of brown products.

8.7 Purification of Oligosaccharides

In oligosaccharide production, a purification step is necessary. Undesirable compounds, such as monosaccharides, acids, or salts, are eliminated in this stage. For that purpose, several methods can be applied, such as fungal cultures, and filtration and separation processes (Gullon et al. 2011). For example, *Saccharomyces cerevisiae*

304 VALORIZATION OF WINE MAKING BY-PRODUCTS

eliminated glucose and fructose from the medium in oligosaccharides extracted from pitaya after 48 h of incubation. Each culture is suitable for a given substrate; thus, its energetic necessities and specific development conditions must be considered. Biological purification is not always viable and sufficiently effective. In some cases, more complex and complete processes are necessary. A variety of strategies have been reported with this purpose, including successive physicochemical processes. Vacuum evaporation could be one option for the removal of volatile compounds and the increase of oligosaccharide concentration. Extraction with solvents can be used to remove nonsaccharide fractions. Adsorption and ion exchange resins can be used to remove salts and other undesirable compounds. Chromatographic separation methods are efficient for molecular weight classification and separation of oligosaccharides from other compounds (Hu et al. 2009). Membrane and chromatographic separation allow verifying specific degrees of polymerization in physiological functions and preclude the use of solvents. However, they are difficult to reproduce on a large scale and also very expensive.

8.8 Potential Case Study I

8.8.1 Prebiotic Oligosaccharides: A Biorefinery Approach to Valorize Vine and Wine Making By-Products

Although food by-products may be reused for animal feeding, they usually represent an environmental problem and they add disposal costs to the food industry (Barile et al. 2009; British Nutrition Foundation 1990). Many studies have assessed the potential reutilization of several plant-derived streams for their inclusion in human diet using green biotechnological sustainable approaches. This approach, in general, could reduce industrial costs and justify new investments in equipment, enabling new sustainable bio-based industry. Generally, plant cell wall is primarily composed of polysaccharides, which can be classified as cellulose and cell wall matrix components (namely pectin), hemicellulose and xyloglucans (Vicens et al. 2009). Cell wall polysaccharides are largely used by the food industry as thickeners, stabilizers, gelling agents and, in some cases, emulsifiers. Dietary fiber can be divided into two categories: insoluble and soluble. Insoluble dietary fiber is mostly included into plant cell

FUTURE PERSPECTIVES

wall and comprises cellulose, part of hemicelluloses, and lignin. In addition to not being digested in the small intestine, most insoluble fiber is not digested by bacteria in the gut either. In contrast, soluble dietary fiber, digested by gut bacteria, includes pentosans, pectins, gums, and mucilage, and potentially prebiotic compounds. Prebiotics are not digestible food ingredients that beneficially affect the host by selectively stimulating the growth and/or activity of one or a limited number of bacteria in the colon, thus improving host health. One broad definition accepted for biorefinery is the sustainable processing of biomass into a spectrum of marketable products and energy. The lignocellulose biorefinery is based on the fractionation of lignocellulosic-rich biomass sources into three major intermediate output streams: cellulose, hemicelluloses and lignin, which can be further processed into a portfolio of bio-based products.

Considering both the previously reported research focused on the identification and characterization of complex bioactive oligosaccharides in wine and further preliminary results showing a significant prebiotic activity of these oligosaccharidic fractions, to improve the knowledge of this class of compounds appears sound. Moreover, the topic of this project denotes environmental and biotechnological impact especially since their extraction/production will be optimized starting from by-products. Over the next few years, areas of food processing waste management will expand rapidly. For example, according to Directive 2008/98/EC of the European Parliament, future legislation regarding industrial waste including those of wineries will become even more demanding, thus increasing the cost of waste management. The recovery of value-added products can help in that direction, being part of a new philosophy of sustainable agriculture. By-products of plant-based food processing, which represent a major disposal problem for the industry concerned, appear as very promising sources of high value-added substances. Vine cultivation and wine making process generate a significant number of by-products with low economic value but still characterized by a certain activity. These materials mainly include skins and seeds (pomace) but also pruning. Generally, only a very small portion of these materials is used (e.g., fertilizer, animal feed, etc.), so appearing suitable raw materials to be industrially valorized. Recent research in the area of carbohydrate-based food ingredients has shown the efficiency of oligosaccharides

306 VALORIZATION OF WINE MAKING BY-PRODUCTS

as prebiotic, but also as gelling, thickening, and emulsifier agents. Oligosaccharides, with degrees of polymerization between 3 and 15 covalently linked through glycosidic bonds, are neither digested nor absorbed in the upper intestinal tract of humans and delivered intact into the colon, where they can stimulate the growth and development of gut flora described as probiotic bacteria. To improve the knowledge of this class of compounds, also confirming this bioactive attitude, appears sound especially since their extraction/production will be optimized starting from by-products. Evaluating new strategies to enhance extraction yield appears mandatory if we consider that pruning and pomace are waste/by-products still characterized by a not well-defined oligosaccharidic compounds. Different procedures could be evaluated in order to obtain both the starting matrices and the different extracts (lipids and polyphenols will be removed during the extractive steps; avoiding the polyphenols interference could be a strategy to improve the prebiotic effect). Considering the multistep biorefinery approach, every new by-products achieved (from cellulose, hemicelluloses and lignin) will be further characterized and analyzed, so evaluating a possible novel valorization, including these into a portfolio of bio-based products (also with different technological properties such as texturizing and stabilizing). Different extractive techniques, such as ultrasound and microwave assisted and enzymatic-driven digestion could be developed. Enzymatic extraction could be deepened regarding both the typology and the different potentially applicable methodologies (e.g., immobilization). The mechanism for enzyme-assisted extraction of oligosaccharides from residual sources is based on the cell wall degrading enzymes that can weaken or break down the cell wall, leaving the intracellular materials more exposed for extraction. At the same time, the enzymes may be used to produce new smaller oligosaccharidic molecules degrading complex polymers. The different oligosaccharidic extracts obtained could be characterized using appropriate techniques (HPLC-MS, GC-MS), considering some aspects related to their bioactivity. Particularly important will be to assess the impact that the extracted oligosaccharides will produce on the probiotic bacteria activity (growth promotion) once considered some of their aspects such as chemical structure and bioactivity, as well as improving their shelf life after lyophilization (cryoprotection) (Guggenbichler et al. 1997; Harmsen et al. 2000; Ruiz-Palacios et al.

FUTURE PERSPECTIVES 307

2003). Finally, the potential outcomes will be the release of innovative solutions both at process and product level, allowing to new perspectives in food and nutraceutical market.

8.8.2 Primary Objectives

The primary objectives as listed here, are … to valorize the main vine cultivation and wine making by-products (seeds, skins, and pruning) and to evaluate the technological impact that distillation process exerts on pomace (also exhausted pomace). These are different matrices with a low economic value but still characterized by a certain activity (economically exploitable), considering that only a very small portion of these materials is reused. To improve knowledge of oligosaccharidic compounds related to grape by-products. To evaluate new strategies to enhance oligosaccharides extraction yield (multistep biorefinery approach), investigating different extractive methods. To evaluate the proper procedures in order to obtain the more suitable starting matrices (e.g., milling and defatting). To compare innovative extractive technologies (microwaves, ultrasounds, and enzymes), thus optimizing the best operational conditions and their synergies. To evaluate the proper procedures required in the utilization of a professional multimode oven (microwaves/ultrasounds), resulting the right compromise in "green" technology extraction. This technology allows milder reaction conditions, low production costs, formation of cleaner products with higher yields, and minor waste when compared with the use of traditional reagents. To evaluate the more suitable and performing enzymatic combination (mono- or multistep), optimizing the operational parameters related to mechanism of this assisted extraction. This is based on the cell wall degradation (to weaken or to break down the cell wall), both leaving the intracellular materials more exposed to the extraction, modifying polysaccharidic polymers too. To evaluate different methodologies potentially applicable to the enzymatic-driven digestion process such as immobilization, also comparing diverse parameters and suitable support. In this manner, it will be improved the knowledge of a certainly more performant process if compared to the free enzyme one (also economical). To obtain new scalable processes, based on the "white" biotechnologies with a "green" footprint (as requested by a sustainable modern industrial process) for isolation and preparation of oligosaccharides from natural

308 VALORIZATION OF WINE MAKING BY-PRODUCTS

sources such as by-products. To characterize chemical structures and biological activities of oligosaccharidic compounds aiming to provide fundamental information about their potential future applications. Their characterization and quantification will be performed by accurate mass spectrometry combined with gas chromatography techniques. Considering that the prebiotic properties of oligosaccharides appear to be influenced by their degree of polymerization, monosaccharide composition, and by the glycosidic linkages, to improve knowledge of the composition and structure of oligosaccharides with different extraction methods could result in the development of novel prebiotics (Bradburn et al. 1993; Egert and Beuscher 1992; Englyst et al. 1987; Gibson and Roberfroid 1995). To avoid the polyphenols interference toward the potential prebiotic effect of the extracts. Residual polyphenols (still present in the matrices and characterized by an antimicrobial effect such as oligomeric and polymeric proanthocyanidinic classes) will be removed during the extractive steps. To confirm the prebiotic attitude of oligosaccharides achieved from grape by-products in relation to previously published researches carried out by the principal investigator focused on the identification and characterization of complex bioactive oligosaccharides in white and red wine (Bordiga et al. 2012; Belleville et al. 1993; Doco and Brillouet 1993; Ducasse et al. 2010). These oligosaccharidic fractions extracted from wine have shown a significant prebiotic activity on probiotic bacteria such as *Bifidobacterium infantis*.

To evaluate the different oligosaccharidic extracts as potential functional ingredients with prebiotic activity toward other well-known probiotic bacteria (e.g., *Lactobacillus plantarum* and *Lactobacillus crispatus*), allowing them to improve the growth during *in vitro* fermentation. To evaluate the different oligosaccharidic extracts as potential novel cryoprotective agents in the lyophilization process of probiotics. To describe the effect that stabilization processes and different extraction techniques may have on the content and availability of oligosaccharides.

8.8.3 Secondary Objectives

The secondary objectives as listed here, are ... to further characterize and analyze every new by-product achieved (from cellulose, hemicelluloses, and lignin), considering the multistep biorefinery approach, thus evaluating a possible novel valorization (bio-based products also with different technological properties such as texturizing and stabilizing). Alternatively, for example, the evaluation of different yield of vanillin biotechnological production, obtained following enzymatic treatment of the lignin removed from the matrices during the oligosaccharides purification. To improve the expertise and the ability on developing the most suitable enzymatic-driven digestion process toward plant-based matrices (eventually also integrated with microwaves, ultrasounds) in order to obtain bioactive compounds. To improve the efficient utilization of wine chain by-products (very promising sources of value-added substances), which represent a major disposal problem for the industry concerned.

8.8.4 Methods

The selection of an appropriate number of vegetable matrix by-products to be examined (pomace, exhausted pomace after distillation and pruning) will be performed in collaboration with sector companies like wineries, distilleries, and specialized oil mills, also considering the seasonality of the different matrices. The characterization of the wine chain by-products in relation to their content of residual oligosaccharidic compounds will be performed. These by-products prove to be available at a low cost and in large quantities and are therefore usable because of their nutritional value, such as potential ingredients. Evaluating new strategies to enhance extraction yield appears mandatory if we consider that pomace and pruning are by-products/waste which are still characterized by a certain amount of oligosaccharidic compounds even after the first step of extraction. Different extractive methods, suitable for the recovery of the different oligosaccharides present in by-products, will be investigated. The task of the extractive step will be the comparison and the optimization of different procedures that will include innovative technologies (microwaves, ultrasounds, and enzymes) also taking into consideration the description of the effect that different techniques may have on the content and availability of oligosaccharides. Different procedures will be evaluated in order to obtain both the starting matrices (e.g., conditioning, milling) and different extracts. Seeds and skins will be separately analyzed from pomace in order to evaluate in detail the effective oligosaccharidic distribution and related extractive yield. Seeds will be de-fatted before extraction process. For a better evaluation of

310 VALORIZATION OF WINE MAKING BY-PRODUCTS

the innovative technologies and their possible synergism, a professional multimode oven (microwaves and ultrasounds) appears to be the right compromise and an alternative and feasible method to extract oligosaccharides. This technology allows milder reaction conditions, low production costs, formation of cleaner products with higher yields, and minor waste when compared with the use of traditional reagents (e.g., alkali). Temperature is one of the most important factors contributing to the recovery yield when using this assisted extraction. Generally, the higher temperature applied, the higher recovery yield is obtained. However, high temperatures may cause degradation of products, dependent on the structure of the polysaccharides. Furthermore, the solid/solvent ratio is also an important parameter when performing this extraction technology, yielding higher recoveries when using more diluted conditions. All these aspects need to be certainly investigated in order to obtain the better compromise, considering that this technology allows obtaining oligosaccharides in proportions that are dependent on the operating conditions. The oligosaccharides corresponded to hexose-oligosaccharides, xyloglucans, and arabinogalactans may be the natural by-products of the degradation of cell wall polysaccharides. The permeability of the different parts of grape cell walls can be increased by enzymes, which can help the partial hydrolysis of cell wall polysaccharides. Enzyme treatments have been shown to modify the wine polysaccharide composition. The mechanism for enzyme-assisted extraction is based on the cell wall degrading enzymes that can weaken or break down the cell wall, both leaving the intracellular materials more exposed to the extraction and modifying polysaccharidic polymers. Especially enzymatic extraction will be evaluated mainly regarding the different methodologies potentially applicable such as immobilization. The immobilization of enzymes offers several improvements for enzyme applications because the storage and operational stabilities result enhanced and their reusability represents a great advantage if compared to free enzymes (also economic). The immobilization of enzymes will be performed via the selection of an appropriate number of methods using different supports. The recovery of enzymatic activity after the immobilization process will be evaluated together with the improvements obtained concerning the stability of the enzyme to temperature, pH, solvents, storage, and operation. Similar to the free ones, immobilized enzymes may be applied in a huge number of industrial processes, especially in environmental applications. To ensure

FUTURE PERSPECTIVES 311

proper oligosaccharide identification by MS, an appropriate multistep solid-phase extraction will be performed (e.g., using C-18 cartridge to eliminate substances such as residual polyphenols followed by carbograph cartridge to remove residual salts and monosaccharides that would interfere with MS analysis). The chemical characterization of the different oligosaccharidic extracts obtained using appropriate techniques (mainly HPLC-MS and GC-MS, but if required, also by MALDI-FTICR MS and nano LC-Chip-Q-ToF) will be achieved (Figures 8.3 and 8.4). In parallel, some other aspects related to their bioactivity will be considered (e.g., radical scavenging activity). Moreover, both the analysis of glycosyl composition after methanolysis and derivatization (trimethylsilyl groups), and then a methylation to obtain the glycosidic linkages, will be performed. Considering that the prebiotic properties of oligosaccharides appear influenced by their degree of polymerization, by the monosaccharide composition and by the glycosidic linkages, the use of optimal hydrolysis conditions may lead to a more effective process for their production. Additionally, improved knowledge of the composition and structure of oligosaccharides with different extraction methods could result in the development of novel prebiotics. The next phase of this project will be focused on the evaluation of the different extracts with potential prebiotic activity able to promote the growth of well-known probiotic bacteria, such as *Lactobacillus plantarum* and *L. crispatus*, as well as improving their shelf life after lyophilization. The different oligosaccharidic extracts will be evaluated as potential novel cryoprotective ingredients. The optimization of the industrial production of probiotics strongly depends on the technologies. All the production phases (preparation and sterilization of the medium, fermentation, concentration stabilization/preservation, and formulation) are critical in order to obtain a long shelf life, as well as improving viability. Freeze-drying is commonly used to preserve probiotics, but it could cause cell damage and undesirable effects, such as denaturation of sensitive proteins and decreased viability of cells. These effects are caused by low temperatures, freezing, osmotic and desiccation stress. The cryoprotectant agents play a fundamental role in the conservation of viability during freeze-drying, improving quality in shelf life. Generally, nonreducing disaccharides, polyalcohols, and polysaccharides were suggested and used as cryoprotectors. Considering the multistep extractive approach, every new by-products achieved from cellulose, hemicelluloses, and lignin will be further characterized and

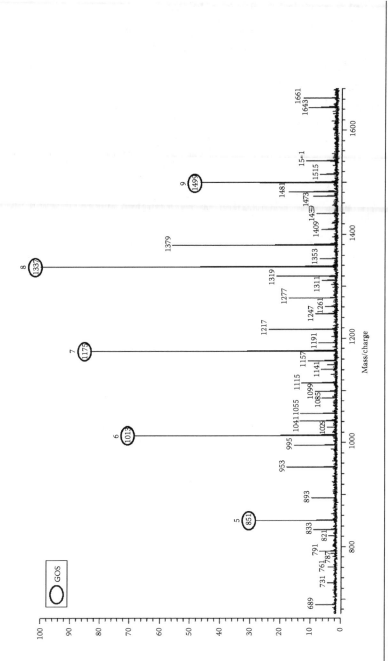

Figure 8.3 Positive-mode MALDI-FTICR spectra of an oligosaccharide fraction of Chardonnay wine. Major peaks at m/z 851, 1013, 1175, 1337, and 1499 represent sodium-coordinated ([M-Na]+) GOS with DP ranging from 5 to 9. (From Bordiga, M. et al., *J. Agric. Food Chem.*, 60, 3700, 2012.)

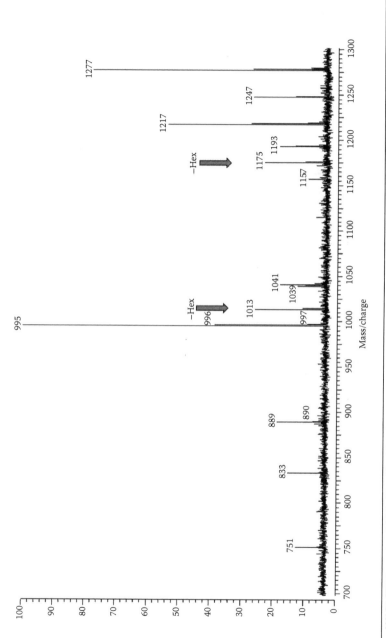

Figure 8.4 MALDI-FTICR spectra of GOS with a DP of 8 (m/z = 1337). Fragment ions corresponding to glycosidic bond cleavages (Hex) and cross-ring cleavages (60, 90, and 120) were obtained. (From Bordiga, M. et al., *J. Agric. Food Chem.*, 60, 3700, 2012.)

314 VALORIZATION OF WINE MAKING BY-PRODUCTS

analyzed, thus evaluating a possible novel valorization (bio-based products also with different technological properties such as texturizing and stabilizing). Alternatively, for example, different yield of vanillin production, obtained after enzymatic treatment of the lignin present in the matrices, by means of microbiologic fermentation/enzymatic conversion will be also evaluated.

8.8.5 Schematic Scheduling Activities

Step 1: Appropriate matrices selection. Starting matrix preparation (milling, de fatting, seed, and skin separation). Different extractive procedures evaluation. Setting operational parameters of multimode ovens (microwaves and ultrasounds). Selection of the best performing enzymes. Setting operational parameters of enzymatic-driven digestion. Evaluation of the potentially suitable enzyme/s for immobilization methodologies and different related support. Setting the best performing operational parameters for immobilized enzyme/s.

Step 2: Proper oligosaccharide purification using an appropriate multistep solid-phase extraction (e.g., using a C-18 cartridge followed by a carbograph cartridge). Chemical characterization of the different oligosaccharidic extracts achieved. Prebiotic activity evaluation of different oligosaccharidic extracts obtained.

Step 3: Fine chemical characterization of more performing (prebiotic activity) oligosaccharidic extracts (MALDI-FTICR MS and nano LC-Chip-Q-ToF) (Figure 8.5). Fine glycosidic linkage characterizations of the previous extracts.

Step 4: Evaluation of the different oligosaccharidic extracts as potential novel cryoprotective ingredients in the lyophilization process of probiotics.

Further characterization of the new by-products achieved (from cellulose, hemicellulose, and lignins), considering the multistep biorefinery approach include the ... evaluation of a potentially novel valorization also with different technological properties such as texturizing and stabilizing. Evaluation of the extractive yield of vanillin, obtained after enzymatic treatment of the lignin removed from the matrices.

FUTURE PERSPECTIVES

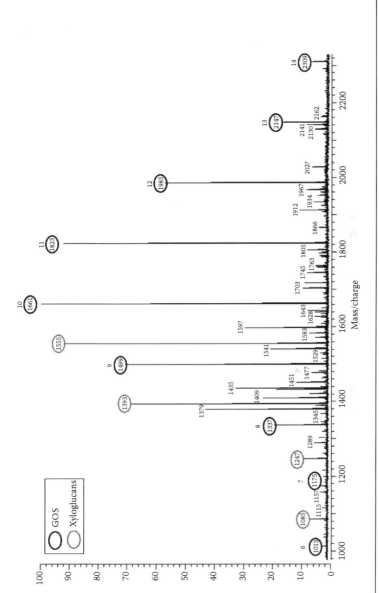

Figure 8.5 Positive-mode MALDI-FTICR spectra of an oligosaccharide fraction of Chardonnay wine. Major peaks at m/z 1337, 1499, 1661, 1823, 1985, and 2147 represent sodium-coordinated ([M-Na]+) GOS with DP ranging from 8 to 13; 1393, and 1555 sodium-coordinated ([M-Na]+) xyloglucans. (From Bordiga, M. et al., *J. Agric. Food Chem.*, 60, 3700, 2012.)

8.8.6 Potential Applications, Impact

The plant by-products may be a significantly impressive source of compounds with high nutritional value whose recycling has technological, nutritional, and environmental interests. Applied for this purpose are the extractive techniques with solvents that allow obtaining some fractions with food-related potential interest (e.g., polyphenols), valorizing by-products. On the contrary, the alternative processes, represented by the use of extractive innovative techniques such as ultrasound- and microwave-assisted techniques or bioconversion processes by enzymes, need to be further developed and optimized. The possibility of converting them to "co-products" should be considered one of the most promising prospects for their recovery and for the rationalization of production processes, because this application completely valorizes the by-product so reducing the costs of production and/or disposal. The "biorefinery" approach will assure a sustainable method to this project, allowing also new green/white biotechnology strategies useful to obtain novel bio-based products. Because synthetic oligosaccharides are rare and expensive, there is an urgent need to find alternative sources from which to obtain sufficient amounts to perform clinical studies and examine their potential for use in human nutrition. Taking this into consideration, the significant number of by-products generated during vine cultivation and the wine making process appear to be a valuable low-cost source of these oligosaccharides, and purification may increase their availability in the food industry. Enzyme-assisted extraction, but also other different innovative techniques, such as ultrasound- and microwave-assisted extraction, represents a potential strategy to enhance extraction yield for the production of oligosaccharides with high stability for pharmaceutical and food applications. Developing new technological approaches as the immobilization of enzymes offers several improvements for enzyme applications because, if compared to the free enzymes, the reusability of immobilized enzymes represents a great advantage (also from the economic point of view). Particularly important it will be to assess the impact of the extracted oligosaccharides, with potential prebiotic activity, on the growth promotion of well-known probiotic bacteria, as well as improving their shelf life after lyophilization. Future legislation regarding industrial waste

FUTURE PERSPECTIVES   317

will become even more demanding, thus increasing the cost of its management. In that direction, the recovery of high value-added compounds from by-products is part of a new process of sustainable agriculture. Most of these biorefinery processes are being developed individually, but have the potential to be more efficient and economical when combined in multiprocess crossover regimens using waste/by-products from one process to produce high-value chemicals.

8.9 Spent Grape Pomace

8.9.1 Potential of By-Products

Conventionally, most crops have been focused on obtaining a single high-value product (e.g., wine, oil, etc.) discarding the remaining parts of the plant or fruit, or exploiting them to obtain low-value products (sometimes highly polluting and, therefore, undesirable products). A rational use of these residues would create a source of additional wealth, which in many cases could have a value even higher than that traditionally given, thus presenting interesting perspectives for the design of new industries. Residues from the plant contain compounds of interest to the pharmacological, nutraceutical, cosmetics, and food industries. Approximately 1.4–2.0 tons of shoots can be obtained per hectare of vine crop. Since the vineyard-cultivated areas in the world are around 8 million ha, an estimated 11.2–16 million tons of vine shoots are produced each year. On the other hand, the most abundant residue of the wine making process is the solid waste remaining after grape pressing: the grape pomace. Since 20%–25% of the weight of processed grapes remain as pomace and about 80% of the world grape output is used to make wine, around 10.5–13.1 million tons of grape pomace are produced in the world each year. Exploitation of residues could encompass three gradational states, namely: (1) Residues from which compounds of high added value have been recognized as such by the European Food Safety Authority (EFSA), as is the case with hydroxytyrosol, extracted from grape oil. This compound is claimed to reduce an emerging risk factor of atherosclerosis (oxidized LDL cholesterol) (EFSA 2011) and maintain normal blood HDL-cholesterol levels. (2) Products from the residues in (1), the beneficial effects of which are not enough

318 VALORIZATION OF WINE MAKING BY-PRODUCTS

supported at present for EFSA acceptance, but widely used in the food and cosmetics fields. (3) Residues, the potential of which has not been (or has been poorly) recognized and exploited. There is a growing demand for exploitation of agricultural residues. One of them is the present pressing need to obtain natural dyes able to substitute existing artificial dyes, particularly in the food industry. The need for more natural colorants began with a study, in 2009, which reported an experiment with six artificial colorants that caused hyperactivity and allergy on children subjected to the test. This situation has pushed the search for, and application of, natural colorants to replace synthetic colorants. The importance of residues from both the crops and products of one of the major industries in the Mediterranean basin (wine). The exploitation of these residues can provide benefits to the food industry thanks to the high content in antioxidants with proved healthy, cosmetics and nutritional properties.

8.9.2 Potential Exploitation of Residues from Vineyards and Wine Production

8.9.2.1 Vine Shoots Despite the fact that vine shoots constitute an abundant agricultural residue, their present economic value is very small as they are mostly used as fuel or fertilizer. Most of the scant research on vine shoots has been focused on the production of paper pulp and ethanol, the former requiring in-depth studies to improve production as vine shoots provide pulp of lower quality than other agricultural residues such as wheat straw. The composition of vine shoots is characterized by three main fractions: cellulose, hemicellulose, and lignin; the content of total cellulose being around 68% and that of lignin around 20% (dry weight). Lignin, a well-known component of secondary cell walls, is a high molecular mass cross-linked polymer. These monomeric units (namely, *trans-p*-coumaryl [4-hydroxycinnamyl], coniferyl [4-hydroxy-3-methoxycinnamyl, forming guayacyl units], and sinapyl [3,5-dimethoxy-4-hydroxycinnamyl, forming syringyl units]) are characterized by a phenolic structure. Lignins can be hydrolyzed to release aromatic phenolic compounds such as low molecular mass alcohols, aldehydes, ketones, or acids; therefore, vine shoots can be a suitable source of phenols (Bordiga et al. 2013b, 2014). The abundance and richness of vine shoots make their exploitation highly interesting in economic terms.

FUTURE PERSPECTIVES **319**

8.9.2.2 Vine Leaves This field is a small investigated area with respect to senescent leaves. In this state, vine leaves have a color ranging from yellow to red and brown, depending on the variety, which reveals the presence of flavonoids, especially flavonols, anthocyanins/anthocyanidins, and carotenoids with very different characteristics. Senescent vine leaves are, therefore, a potential raw material for the production of natural colorants that do not require an experimentation period prior to regulating their use.

8.9.2.3 Grape Skins Most of the research carried out thus far has been devoted to red grape skins directly from grapes, representing a significant cost of raw material. Several studies have demonstrated that the grape skin from the making of red wine has a number of and concentration of colorants that make this residue profitable as a raw material for colorant production. The need to remove the high moisture content of this raw material (an expensive step) prior to its use is something that should certainly be taken into account in the design of an industrial scaling.

8.9.2.4 Grape Seeds Grape seeds are the most studied and exploited residues from wine making due to the large variety of compounds in this raw material. Between 38% and 52% of a grape is made by seeds, which gives an overall idea of the amount of residues generated in each harvest. The high content in phenols, mainly flavonols and proanthocyanidins, endows them with a high antioxidant capacity. Proanthocyanidins, compounds responsible for astringency, are recognized by their beneficial health effects (anti-inflammatory, cardioprotective, radioprotective, antihyperglycemic, antitumor, and antigenotoxic). In addition to the components mentioned before, the lipid fraction of grape seeds has a high commercial value, which has been only partially exploited. Grape-seed oil is made up of 90% of poly- and mono-unsaturated fatty acids, particularly linoleic acid, followed by oleic acid, which are responsible for its nutritive value as edible oil. This oil only contains low amounts of saturated fatty acids (10%). The high content of unsaturated fatty acids makes grape-seed oil a high-quality nutritional oil, which exhibits properties for prevention of thrombosis, inhibition of cardiovascular diseases, reduction of cholesterol in serum, dilation of blood vessels, and regulation of autonomic nerves.

320 VALORIZATION OF WINE MAKING BY-PRODUCTS

8.9.2.5 Vinification Lees The residue formed at the bottom of wine containers after fermentation, during the storage or after authorized treatments, and the residue from filtering or centrifuging this product have a key role in wine making. For example, these are responsible for the removal of undesirable compounds, interaction with the volatile fraction, biogenic amines and phenolic compounds, and the release of mannoproteins (natural stabilizer of colloidal charge). Their present use is reduced to the recovery of tartaric acid, which is employed to correct the pH of must prior to fermentation. However, the two most promising uses of lees are as raw material for extraction of mannoproteins and colorants, both irreplaceable materials for wine clarification. The most valuable finding of a recent study was that the adsorptive capacity of lees makes the concentration of colorants in them up to ten times higher than in the skin of red grapes.

8.9.2.6 Residue from Residues An exhaustive exploitation of residue could lead to the use of "residues from residues." An example of this potential exploitation would be the residue remaining after ethanol distillation from the wine pomace. Grape skins and seeds from a distillery have been subjected to extraction with ethanol and water in equal proportions to obtain the largest possible number of compounds from these residues, which had so far been used only as a heat source. Despite the degradation caused by the drastic conditions of the distillation process, many interesting compounds exist in the extracts from the waste of this process, which make it a useful matter for more than as a heating source. The variety of identified compounds in the extracts makes them exploitable as additives in the food industry (either as colorants, as flavor modifiers or as antioxidants), and in the cosmetics and nutraceuticals industries.

8.10 Potential Case Study II

8.10.1 Spent Grape Pomace: Still a Potential By-Product

The grape (*Vitis vinifera* L.) is one of the world's largest fruit crops with an annual production of about 60 million metric tons (FAO 2014). Grape growing plays a major role in the worldwide fruit production, with an international acreage of approximately 8 million ha

FUTURE PERSPECTIVES   321

(OIV 2013). Wine industry waste accounts for almost 30% of the grapes used for wine production (Rondeau et al. 2013). Grape pomace, a remnant of the wine making process, is one of the most important residues of the wine industry. It consists of different amounts of grape, skin, pulp, seeds, and stems if not removed (Fontana et al. 2013; Yu and Ahmedna 2013). During the vinification process, phenolic compounds move from solid parts of the grape cluster into wine. The rate of transfer depends on various factors including phenolic concentration of grapes, level of pressing, maceration time, fermentation contact time, temperature, and alcohol levels. These waste materials contain biodegradable organic matter; however, their disposal generates huge amounts of industrial waste and creates serious environmental problems (Gonzalez-Paramas et al. 2004). The waste loads at the processing plants could be significantly reduced through by-product usage. Some industrial uses currently under investigation for wine pomace waste include use as animal feed, as possible nutritive ingredients for value-added products, in the production of citric acid, and the use of anthocyanins from grape skins as colorants (Basalan et al. 2011; Molina-Alcaide et al. 2008). According to conventional practice, grape pomace is also used in the production of distilled alcoholic beverages (Da Porto 1998). Spirits from fermented grape pomace are very popular in Mediterranean countries. Grappa is the typical Italian spirit of commercial, cultural, and historical importance, which is obtained by processing grape pomace or marc at the end of alcoholic fermentation. It is well known that these wine making by-products, mainly consisting of grape pomace, still contain a significant amount of phenolics with beneficial health-related effects, at different concentrations and chemical structures depending on the grape variety considered (Sagdic et al. 2011; Torres et al. 2002; Laufenberg et al. 2003). Available studies regarding these phenolic compositions are focused on pomace deriving both from red and white grape varieties (Deng et al. 2011; Sanchez et al. 2009; Rockenbach et al. 2011). The aim of this study was to demonstrate that not only the waste from the wine making industry can be exploited to obtain valuable compounds, but that these compounds—and even others resulting from the drastic conditions to which grape pomace is subjected for ethanol distillation—can be obtained from the spent waste of this industry. This research consisted of the determination of several phenolic compounds and fatty acid content grape seeds and skins obtained from PRE and POST distilled pomace. Considering the greater time of high temperature exposure, the discontinuous methods have been utilized in this study in order to characterize the pomace degradation caused by the drastic conditions of the distillation process. The identification and quantification of flavan-3-ol composition by HPLC-DAD, the determination of the mean degree of polymerization (mDP) of the condensed tannins, and the estimation of their antiradical activity by DPPH procedures have been also realized.

8.10.2 Distillation Process

One kg of red winegrape pomace (cv. Nebbiolo) matrix was steam-distilled by adding 250 mL of distilled water in a traditional copper 5 L alembic, with natural gas as heating source. Before heating, the alembic was hermetically closed in order to prevent any steam leakage. When the temperature reached 80°C–90°C, the liquid spirit started to run from the funnel, and the distillation products were collected (total overall volume V = 180 mL). Two rates of pomace (100 g of PRE and POST distillation process), randomly selected from the collected samples of each step, were used for the study (Figures 8.6 and 8.7). Grape seeds and skins were manually separated from the pomace.

Figure 8.6 PRE distillation process pomace.

Figure 8.7 POST distillation process pomace.

8.10.3 Extraction of Fatty Acids and Phenolic Compounds

Grape seeds and skins were ground to a powder using liquid N_2. Four grams of the seeds powder was extracted with Soxhlet apparatus (Büchi Universal Extraction System B-811, Germany) using 150 mL hexane for 9 h to obtain total fat fraction. The solvent was then evaporated to dryness (vacuum, 40°C) and the dry extract was stored at –20°C until use. Subsequently, ground de-fatted seeds (1.0 g) or skins (1.0 g) were extracted using 50 mL of acetone/water (70:30, v/v) under constant magnetic stirring (2 h), in the dark at room temperature (22°C) to obtain phenolic compounds fraction. The process was repeated three times for an exhaustive extraction and performed under a nitrogen atmosphere. Extracts were finally combined and evaporated to dryness (vacuum, 40°C) and the dry extract was stored at –20°C until use.

8.10.4 Phenolic Compound Analysis

The phenolic compound analysis was performed according to the OIV method (OIV-MA-AS315-11) reported by resolution Oeno 12/2007, after minor modifications (OIV 2007). A Shimadzu LC-20A Prominence chromatographic system equipped with a diode array detector (DAD detector SPD-M20A) was used. Separation

324 VALORIZATION OF WINE MAKING BY-PRODUCTS

was performed on a reversed-phase Synergi™ 4 μm Max-RP 80 Å, LC Column (250 × 4.6 mm, Phenomenex, Torrance, CA) protected by a guard column containing the same material, at 30°C. Eluent A was water/acetonitrile/formic acid 87:3:10 v/v and Eluent B water/acetonitrile/formic acid 40:50:10 v/v. The flow rate was kept constant at 0.5 mL/min. The elution program used was as follows: 6%–20% B linear from 0 to 20 min, 20%–40% B linear from 20 to 35 min, 40%–60% B linear from 35 to 40 min, 60%–90% B linear from 40 to 45 min, 90% B isocratic from 45 to 50 min, 90%–6% B linear from 50 to 51 min, and equilibration of the column from 51 to 63 min under initial gradient conditions. DAD detection was performed at 280, 330, and 520 nm. The injection volume was 5 μL. The anthocyanins were identified by comparison to their retention times with those of the authentic standards: 3-monoglucosides of delphidine, cyaniding, petunidin, peonidin, and malvidin.

8.10.5 Proanthocyanidin Analysis

A Shimadzu LC-20A Prominence chromatographic system equipped with a diode array detector (DAD detector SPD-M20A) was used. Separation was performed on a reversed-phase Supelcosil TM LC-318 (250 × 4.6 mm, 5 μm, Supelco, Bellefonte, PA) column, protected by a guard column containing the same material, at 35°C. Eluent A was water/formic acid 0.1% v/v and Eluent B acetonitrile/formic acid 0.1% v/v. The flow rate was kept constant at 1 mL/min. The elution program used was as follows: 3% B isocratic from 0 to 5 min, 3%–8% B linear from 5 to 9 min, 8% B isocratic from 9 to 15 min, 8%–9% B linear from 15 to 16 min, 9% B isocratic from 16 to 22 min, 9%–14% B linear from 22 to 25 min, 14%–20% B linear from 25 to 35 min, 20%–40% B linear from 35 to 46 min, 40%–97% B linear from 46 to 47 min, 97% B isocratic from 47 to 50 min, 97%–3% B linear from 50 to 51 min, and equilibration of the column from 51 to 56 min under initial gradient conditions. DAD detection was performed at 280 nm. The injection volume was 5 μL. PAs composition was determined by reversed-phase HPLC after acid-catalyzed degradation in the presence of phloroglucinol (Figure 8.8). A solution of 0.1 N HCl in methanol, containing 50 g/L phloroglucinol and 10 g/L ascorbic acid, was prepared. Five milligrams of polymeric proanthocyanidin extract were dissolved in 1 mL of the reagent solution

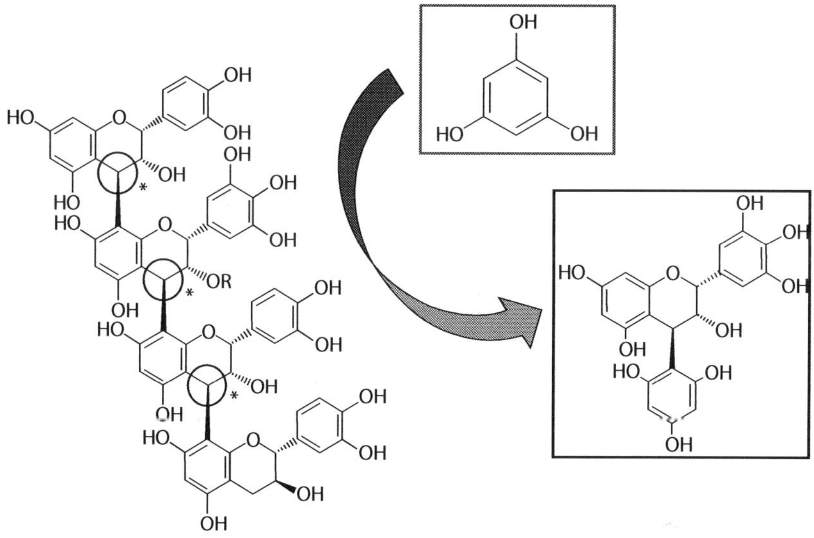

FUTURE PERSPECTIVES

Figure 8.8 Acid-catalyzed degradation of proanthocyanidins in the presence of phloroglucinol. (From Bordiga, M. et al., *Food Chem.*, 127, 180, 2011.)

and the reaction was performed at 50°C for 20 min. PAs cleavage products were estimated using their response factors relative to (+)-catechin (quantitative standard). The results of phloroglucinolysis provided information on PAs subunit composition and mean degree of polymerization (mDP) (Bordiga et al. 2013a). The mDP was determined by summing the terminal and extension subunit amounts (in moles) and dividing by the terminal subunit amount (in moles).

8.10.6 Fatty Acid Analysis

Fatty acid methyl esters (FAMEs) were obtained by trans-esterification of triglycerides (extracted as total fat using hexane in a Soxhlet apparatus for 9 h) with a 1.5% sodium methylate in methanol solution (Bonetta et al. 2008). The reaction was carried out into a closed vial at 80°C with periodic stirring. The content of each vial was then washed with water and diethyl ether until the organic phase was clear and transparent. After water elimination with Na_2SO_4, the solvent was removed and 1 mL of CH_2Cl_2 was added. FAMEs were analyzed on a Shimadzu gas chromatograph 17-A, equipped with a DB23 column (J&W Scientific) (30 m × 0.25 mm i.d. × 0.25 μm film thickness). Hydrogen was used as carrier and a flame ionization

326 VALORIZATION OF WINE MAKING BY-PRODUCTS

detector (FID) as detector. The temperature program was as follows: Separation was performed starting at 90°C, held for 5 min, temperature increased at a rate of 10°C/min to 220°C and finally held for 15 min, with constant flow of 1.6 mL/min.

8.10.7 Antiradical Activity

The DPPH radical scavenging assay was performed to evaluate the antiradical activity. Seven hundred microliters of samples diluted in MeOH (final concentration: 25 µg extract/mL) or MeOH (control) were added to the same volume of methanolic solution of a 100 µM DPPH (Locatelli et al. 2009). Mixtures were shaken vigorously and left to stand in the dark at room temperature for 20 min and then absorbance was read at 515 nm, using a Kontron UVIKON 930 Spectrophotometer (Kontron Instruments, Milan, Italy). Antiradical activity was expressed as inhibition percentage (I%) and calculated using the following equation:

$$\text{Inhibition percentage (I\%)} = (\text{Abs}_{control} - \text{Abs}_{sample})/\text{Abs}_{control} \times 100$$

All tests were performed in triplicate.

8.10.8 Statistical Analysis

Results were expressed as mean ± standard deviation (SD) of at least three independent experiments. Differences among samples were estimated by analysis of variance (ANOVA) followed by Tukey's "Honest Significant Difference" test. The statistical significance level was set to 0.05. All statistical analyses were performed using the free statistical software R 2.10.0 version (R Development Core Team 2008).

8.10.9 Results

8.10.9.1 Phenolic Compounds The contents of phenolic compounds detected in grape seeds and skins, previously and subsequently the distillation process, are reported in Table 8.1. Twenty-six phenolic compounds were identified and quantified in samples, including hydroxybenzoic acids (5), hydroxycinnamic acids (5), flavan-3-ols (5), flavanols (6), and anthocyanins (5) compounds. The skins' total

FUTURE PERSPECTIVES 327

Table 8.1 Phenolic Compounds Content (Expressed on a Dry Matter, DM), Detected in Grape Seeds and Skins, Previously and Subsequently to the Distillation Process

mg/kg DRY MATTER	SKINS PRE	SKINS POST	SEEDS PRE	SEEDS POST
Hydroxybenzoic acids				
Gallic acid	nd	19.3 ± 0.14	21.9 ± 0.50	18.3 ± 0.20
Protocatechuic acid	nd	5.24 ± 0.41	11.4 ± 0.23	2.19 ± 0.08
p-OH-benzoic acid	nd	1.68 ± 0.08	nd	10.1 ± 0.09
Syringic acid	nd	3.73 ± 0.10	6.16 ± 0.28	6.27 ± 0.21
Ellagic acid	1.19 ± 0.11	2.22 ± 0.16	1.16 ± 0.03	1.66 ± 0.09
Hydroxycinnamic acids				
p-Coumaric acid	nd	0.77 ± 0.01	0.84 ± 0.03	0.73 ± 0.01
Ferulic acid	0.44 ± 0.02	0.50 ± 0.01	0.37 ± 0.01	0.31 ± 0.01
Sinapic acid	1.45 ± 0.04	0.78 ± 0.03	0.42 ± 0.02	1.85 ± 0.03
Flavan-3-ols				
Procyanidin B1	nd	nd	8.60 ± 0.11	11.8 ± 0.17
Catechin	9.40 ± 0.15	146 ± 1.39	438 ± 0.31	392 ± 1.63
Procyanidin B2	39.6 ± 0.64	15.0 ± 0.27	53.6 ± 0.87	59.2 ± 1.39
Epicatechin	20.4 ± 0.71	67.4 ± 1.12	213 ± 1.30	214 ± 0.67
Epicatechin gallate	0.44 ± 0.01	10.5 ± 0.19	0.75 ± 0.04	15.3 ± 0.26
Flavonols				
Quercetin-3-*O*-galactoside	25.9 ± 0.33	23.4 ± 0.39	1.80 ± 0.10	6.50 ± 0.08
Quercetin-3-*O*-glucoside	2.29 ± 0.04	8.93 ± 0.16	nd	2.77 ± 0.12
Kaempferol-3-*O*-glucoside	1.11 ± 0.06	2.15 ± 0.07	nd	nd
Myricetin	2.86 ± 0.19	1.07 ± 0.04	0.73 ± 0.08	1.15 ± 0.07
Quercetin	5.71 ± 0.12	30.4 ± 0.55	1.32 ± 0.07	14.4 ± 0.33
Kaempferol	2.29 ± 0.09	2.33 ± 0.20	2.67 ± 0.19	2.58 ± 0.26
Tartaric esters of hydroxycinnamic acids				
Caftaric acid	0.95 ± 0.01	0.40 ± 0.03	7.61 ± 0.10	2.33 ± 0.09
Cutaric acid	0.60 ± 0.01	4.72 ± 0.04	2.46 ± 0.03	2.50 ± 0.04
Anthocyanins				
Delphinidin-3-*O*-glucoside	18.6 ± 0.23	1.45 ± 0.04	—	—
Cyanidin-3-*O*-glucoside	7.91 ± 0.18	1.33 ± 0.02	—	—
Petunidin-3-*O*-glucoside	29.6 ± 0.31	2.06 ± 0.03	—	—
Peonidin-3-*O*-glucoside	35.8 ± 0.38	4.53 ± 0.04	—	—
Malvidin-3-*O*-glucoside	89.4 ± 0.44	9.91 ± 0.06	—	—
Total	296 ± 4.07	366 ± 5.59	774 ± 4.30	766 ± 5.83

Source: From Bordiga, M. et al., *J. Agric. Food Chem.*, 60, 3700, 2012.

328 VALORIZATION OF WINE MAKING BY-PRODUCTS

amount of phenolic compounds was 296 ± 4.07 mg/kg dry matter in PRE sample and 366 ± 5.59 mg/kg dry matter in POST sample. Considering seeds, the total amount was 774 ± 4.30 mg/kg dry matter in PRE sample and 766 ± 5.83 mg/kg dry matter in POST sample.

8.10.9.2 Hydroxybenzoic Acids The total amount of hydroxybenzoic acids ranged from 1.19 ± 0.11 (PRE) up to 32.21 ± 0.90 (POST) mg/kg dry matter in skins and from 38.56 ± 0.68 (POST) up to 40.65 ± 1.05 (PRE) mg/kg dry matter in seeds. Considering skins, it must be pointed that, between the five compounds, ellagic acid was the unique molecule detected in PRE samples, thereby proving that distillation process was able to release these compounds from the matrix. On the contrary, in seeds, p-OH-benzoic acid was the only compound not present before distillation, resulting in confirming how skins are more effectively depleted of their bioactive components during wine making process. Moreover, protocatechuic acid showed a negative trend after distillation (about fivefold less), contrary to other molecules that generally presented similar values.

8.10.9.3 Hydroxycinnamic Acids and Tartaric Esters The total amount of hydroxycinnamic acids ranged from 1.89 ± 0.06 (PRE) up to 2.04 ± 0.05 (POST) mg/kg dry matter in skins and from 1.63 ± 0.06 (PRE) up to 2.89 ± 0.04 (POST) mg/kg dry matter in seeds. While tartaric esters of hydroxycinnamic acids (caftaric and cutaric acids) ranged from 1.55 ± 0.02 (PRE) up to 5.12 ± 0.07 (POST) mg/kg dry matter and from 4.83 ± 0.13 (POST) up to 10.07 ± 0.13 (PRE) mg/kg dry matter in skins and seeds, respectively. p-Coumaric acid was the only compound not detected in skins before distillation but generally the total values appeared similar. In seeds, sinapic acid increased about fourfold its content after process causing a slightly positive trend. Cutaric acid significantly increased in POST skins sample but it remained substantially the same in the PRE and POST seed samples. Both in skins and in seeds caftaric acid showed a negative trend after distillation, but this decrement was more pronounced in the latter.

8.10.9.4 Flavan-3-Ols Seeds are the main source of this monomers and dimers (procyanidin B1 and B2) in grape, their presence in skins usually being quite lower. As expected, the content of flavan-3-ols in skins was low, ranging from 69.97 ± 1.51 (PRE) up to 239.90 ± 2.98 (POST) mg/kg dry matter. With respect to content in seeds, these compounds ranged from 692.72 ± 4.12 (POST) up to 715.26 ± 2.63 (PRE) mg/kg dry matter and the singular molecule values substantially appeared similar before and after distillation. Conversely in skins the distillation process strongly affected the release of these substances, proving in an about 3.5-fold higher total value, particularly due to the contribution of catechin which increases its value from about 9 to 146 mg/kg dry matter. Catechin was more abundant than epicatechin in both seed (PRE and POST) and skin (POST) samples analyzed, generally about twofold higher. This monomer showed in seeds particularly great values 392.22 ± 1.63 (POST) and 438.65 ± 0.31 (PRE) mg/kg dry matter while epicatechin 213.59 ± 1.30 (PRE) and 214.08 ± 0.67 (POST) mg/kg dry matter, respectively. Epicatechin gallate content showed to be apparently affected by distillation process and this positive trend was detected in both matrices with an average increment of about 23-fold for skins and 20-fold for seeds.

8.10.9.5 Flavonols Six main flavonols were detected in skin and seed samples, which are as follows: quercetin-3-*O*-galactoside, quercetin-3-*O*-glucoside, kaempferol-3-*O*-glucoside (detected only in skins), myricetin, quercetin, and kaempferol. Total flavonols content ranged from 40.19 ± 0.83 (PRE) up to 68.35 ± 1.41 (POST) mg/kg dry matter in skins and from 6.51 ± 0.44 (PRE) up to 27.44 ± 0.86 (POST) mg/kg dry matter in seeds. The main flavonols found were quercetin-3-*O*-galactoside and quercetin in all samples analyzed. Moreover, the latter content seemed to be greatly affected by distillation process, in fact it was highlighted an increment of 10-fold and 6-fold for seeds and skins, respectively. Same positive trend was reported for quercetin-3-*O*-galactoside but in this case only in the seeds while skins showed a slightly negative progress after distillation. Even quercetin-3-*O*-glucoside, but less pronounced, showed a positive trend in PRE and POST samples specially in seeds considering that this molecule was not detected before distillation.

8.10.9.6 Anthocyanins The identification of the five different compounds was achieved by comparison of both retention time and

330 VALORIZATION OF WINE MAKING BY-PRODUCTS

the absorption spectra obtained for each eluted peak with those obtained for the standards (delphinidin 3-O-glucoside, cyanidin 3-O-glucoside, petunidin 3-O-glucoside, peonidin 3-O-glucoside, and malvidin 3-O-glucoside). The use of eluents A and B at pH 1.5 is necessary for efficient separation of anthocyanins as well as maximum sensitivity for detection at 520 nm (flavylium form). The major anthocyanin in grape skins was malvidin 3-O-glucoside (about 51%). The remaining four anthocyanins contents of the grape skins showed the following values: about 4% for cyanidin, 10% for delphinidin, 16% for petunidin, and 19% for peonidin 3-O-glucoside derivatives. However, conversely to other phenolic compounds, anthocyanins content was strongly affected by distillation (generally 10-fold lower in POST skins), proving that the high temperature applied on the matrix caused both a degradation and a less extractability of these molecules in grape skin.

8.10.9.7 RP-HPLC-Phloroglucinolysis Analytical methods based on acid-catalyzed cleavage of the polymer combined with HPLC separation enable quantification of individual polymer subunits and calculation of mean degree of polymerization, as well as determination of total flavan-3-ols. Table 8.2 shows the evolution of the extractable fraction isolated by phloroglucinolysis from seeds and skins before and after the distillation process. The results, obtained after acid-catalyzed degradation in the presence of phloroglucinol, pointed out that the sample characterized by the lowest PAs content was skin PRE with a catechin equivalent value of 1.28 g/kg dry matter whereas skin POST showed a value of 1.69 ± 0.06 g/kg dry matter. On the contrary, seed POST showed the lower value, with 7.70 g/kg dry matter and seed PRE the higher one, with 14.34 g/kg dry matter, respectively. Epigallocatechin–phloroglucinol (EGC-P) (detected only in the skins), (–)-epicatechin–phloroglucinol (EC-P), (+)-catechin–phloroglucinol (C-P), and epicatechin-3-O-gallate–phloroglucinol (ECG-P) were identified as extension units, whereas (+)-catechin (C), (–)-epicatechin (EC) and (–)-epicatechin-3-O-gallate (ECG) as a terminal unit in the samples (Bordiga et al. 2011; Pastor del Rio and Kennedy 2006). By quantifying these molecules, we estimated both mean degree of polymerization (mDP), percentage of galloylation (%G), and percentage of prodelphinidin units (%P) (solely for skins). In the seeds (PRE), the EC-P

Table 8.2   Fatty Acid Composition of the Grape-Seed Oil Samples

	FATTY ACID COMPOSITION (%)								
SAMPLE	OIL CONTENT (%)	MYRISTIC (C14:0)	PALMITIC (C16:0)	PALMITOLEIC (C16:1)	STEARIC (C18:0)	OLEIC (C18:1)	LINOLEIC (C18:2)	LINOLENIC (C18:3)	ARACHIDIC (C20:0)
PRE	9.72 ± 0.41	0.08 ± 0.01	7.35 ± 0.25	0.12 ± 0.01	4.56 ± 0.24	16.1 ± 0.96	71.3 ± 4.15	0.39 ± 0.02	0.17 ± 0.03
POST	10.56 ± 0.37	0.10 ± 0.01	7.40 ± 0.31	0.14 ± 0.02	4.35 ± 0.28	16.2 ± 1.14	71.2 ± 3.98	0.42 ± 0.03	0.19 ± 0.01

Source: Created using information from Bordiga, M. et al., *J. Agric. Food Chem.*, 60, 3700, 2012.

extension unit was the most abundant (about 71%) followed by the ECG-P extension unit (percentage values was about 26%) (Moreno et al. 2008). Catechin–phloroglucinol (C-P) extension units showed the lower percentage value (about 2%). Concerning the values of terminal units, epicatechin showed a percentage slightly greater (about 39%) than the catechin monomer that was within 36%. After the distillation process, concerning the extension units, the percentages resulted substantially similar even if EC-P value was slightly higher (about 73%) whereas ECG-P value was slightly lower (about 25%). On the contrary, concerning terminal units, the distillation caused a decrease in C value (about 20%) and consequently an increase in ECG value (about 47%). In the skins (PRE), the EC-P extension unit was the most abundant (about 75%) followed by the EGC-P extension unit (Moreno et al. 2008; Pastor del Rio and Kennedy 2006). Regarding epigallocatechin–phloroglucinol (EGC-P) and epicatechin-3-O-gallate-phloroglucinol (ECG-P) extension units, the percentage values were about 11% and 10%, respectively. Concerning the values of terminal units, catechin showed a greater percentage (about 53%) than the other two, especially of EC that was within 19%. After the distillation process, concerning the extension units, the percentages resulted substantially similar even if EC-P value was slightly lower (about 72%), whereas ECG-P and EGC-P values were slightly higher (about 12% and 13%, respectively). On the contrary, concerning terminal units, the distillation caused a decrease in C value (about 40%) and consequently an increase in ECG value (about 41%). As shown in Table 8.2, the mDP of seed proanthocyanidins increased after the distillation process reaching an average value of about 20 in POST sample starting from a detected value of about 9 in PRE sample. Similarly, the mDP of skin proanthocyanidins reached after the distillation a value of about 27 in POST sample starting from a value of about 19 in PRE sample.

8.10.9.8 Fatty Acid Content The oil contents and the fatty acid composition of grape seeds are given in Table 8.3. The oil contents ranged from 9.72% (PRE) to 10.56% (POST) (Gokturk Baydar and Akkurt 2001; Beveridge et al. 2005). The major fatty acid in grape-seed oil was linoleic acid (about 71%). The fatty acid contents of the grape-seed oils showed the following values: about 0.10% for myristic, 7.40% for palmitic, 0.12% for palmitoleic, 4.40% for stearic, 16.10% for oleic,

Table 8.3 Oligomeric and Polymeric Proanthocyanidins (PAs) Content (Expressed on a Dry Matter, DM) of Seed and Skin Samples

SAMPLE	PAS CONCENTRATION g/kg DRY MATTER	TERMINAL UNITS (%)			EXTENSION UNITS (%)						
		C	EC	ECG	EC	C	ECG	EGC	mDP	%G	%P
Seed PRE	14.34	35.8	39.5	24.7	71.5	2.3	26.2	—	8.67	26.3	—
Seed POST	7.70	20.6	31.8	47.6	73.0	2.0	25.0	—	19.60	14.1	—
Skin PRE	1.28	53.3	18.6	28.1	75.3	3.2	11.5	10.0	19.04	12.1	9.4
Skin POST	1.69	40.1	18.4	41.5	71.9	2.9	12.0	13.2	27.35	13.3	12.7

Source: From Bordiga, M. et al., *J. Agric. Food Chem.*, 60, 3700, 2012.
Structural characteristic and composition of oligomeric and polymeric proanthocyanidins (mDP, mean degree of polymerization; C, catechin; EC, epicatechin; ECG, epicatechin gallate; EGC, epigallocatechin; %G, percentage of galloylation; %P, percentage of prodelphinidin units).

334 VALORIZATION OF WINE MAKING BY-PRODUCTS

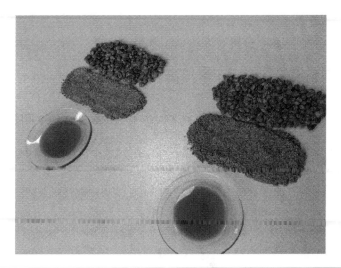

Figure 8.9 Seed oil extracted from PRE (left) and POST (right) samples.

71.30% for linoleic, 0.40% for linolenic, and 0.18% for arachidic acid. These percentage values were reported in both oils (from PRE and POST seeds) showing no significant differences. In this case, considering data reported, the distillation process did not seem to significantly affect the fatty acid composition of extracted oils (Figure 8.9).

8.10.9.9 Antiradical Activity Antiradical activity of grape seeds and skins extracts was evaluated as inhibition percentage of DPPH radical. The lower inhibition percentage highlighted in skin samples was 21.86 ± 1.97, reported for the PRE; the highest one was 33.96 ± 1.42 for the POST. Among seeds, the lowest percentage of inhibition was 66.85 ± 3.71 for the POST sample and the highest was 87.95 ± 4.25 for the PRE sample.

8.11 Discussion

The high concentrations of bioactive compounds and the variation of their concentrations among different parts of grape pomace show the importance of analyzing in depth this wine making by-product as a good and economically viable source of natural molecules. The recovery of different bioactive compounds from grape pomace is an important challenge for the field-related scientists and the industry, resulting a complex approach that depends on several parameters

FUTURE PERSPECTIVES 335

(e.g., compounds variety and the chemical complexity of the matrices). In this perspective, this study was structured and thus evaluating the potential degradation grade of bioactive compounds caused by the drastic conditions of the distillation process toward grape pomace. There are two typical distillation methods: the first, continuous, finds its application in the industrial production of grappa and the second, discontinuous, used by small distillers. In the continuous distillation method, the pomace is constantly introduced in a big column passed through, from the bottom to the top, by a flow of hot steam at a constant temperature. The steam rising to the top becomes rich in alcohol and, when condensing, generates an alcoholic mixture. The distillation process results substantially realized under the same conditions, so achieved a standard product with good chemical characteristics. On the contrary, the discontinuous method implies an intermittent alimentation of the distiller with a pre-established quantity of pomace and, consequently, resulting in an intermittent separation and extraction of the distillate. The pomace is warmed up with increasing temperatures from the beginning to the end of operations. The management of the distillation is completely different in the two systems earlier described, especially considering the time during which the pomace remains exposed to high temperatures. The continuous method generally takes 10–15 min at 90°C–95°C, while the discontinuous one takes 30–40 min reaching the same temperatures for the distillation of each batch loaded into the boiler. Considering this, we decided to utilize the discontinuous methods in this study as the greater time of high temperature exposure in order to characterize the pomace degradation caused by the drastic conditions of the distillation process. Overall, looking at the values detected and reported in Table 8.1, distilling at 95°C for 40 min resulted in about 1% reduction for seeds and about 23% increase for skins in extractable phenolic compounds, respectively. The matrix of grape by-products (seeds or skins) might also influence the effect of heat treatment on grape bioactive compounds. About seeds, protocatechuic acid, catechin, and caftaric acid showed the greatest reduction. The decrease for different compounds obtained could be justified since these phenolics are present in a free form and, therefore, more susceptible to the distillation process (Davidov-Pardo et al. 2011; Travaglia et al. 2011). Moreover, considering oligomeric and polymeric seeds flavan-3-ols, the drastic

336 VALORIZATION OF WINE MAKING BY-PRODUCTS

conditions of the distillation process have reduced by approximately 50% the extractable content of this group of compounds, thus reducing also the antioxidant activity of grape seeds POST (Table 8.2) (Kim et al. 2006). In the case of skins, heating process increased catechin (>1500%), epicatechin (>300%), and epicatechin gallate (>2300%) concentrations (Table 8.1). Moreover, gallic acid was detected after the distillation (also other hydroxybenzoic acids), thereby proving that its release may come from the excision of the gallate group attached to the C-ring of the flavonoids or from the hydrolysis of gallotannins present in grape pomace. These modifications indicate that distillation process seems to change from more highly polymerized molecules to relatively less polymerized molecules because of the breakdown of polymerized flavan-3-ols during the treatment. This effect could be due to the ability of heating process in releasing these molecules generally bound to various cell wall components such as arabinoxylans and proteins (Pinelo et al. 2006). On the other side, proving the complexity of these matrices, previously studies have also demonstrated that polymeric flavan-3-ols may bind irreversibly to cell wall polysaccharides through hydrogen bonding and/or hydrophobic interactions, and therefore, not all compounds may get converted into monomers and lower-level oligomers (Renard et al. 2001; Nunan et al. 1998). Between seeds and skins, a similar behavior emerged about epicatechin gallate and quercetin content: in both cases, heating process significantly increased the content of these two compounds. Anthocyanins chemical stability is the core theme of many recent researches due to their several potential applications, beneficial effects, and their use as alternative to artificial colorants in foods. Heating process strongly decreased anthocyanins concentrations (e.g., malvidin-3-O-glucoside reduced its concentration of about 90%). Current knowledge indicates that in general high temperature treatments can affect the levels of anthocyanins in fruit and vegetable food products and both magnitude and duration of heating has a strong influence on stability (Sadilova et al. 2006; Patras et al. 2009). However, conversely to other phenolic compounds, anthocyanins content decreased after distillation (generally 10-fold lower in POST skins), proving that the high temperature applied on the matrix caused both a degradation and a less extractability of these molecules in grape skins. Eight fatty acids were detected in the grape seed samples (Table 8.3). The percentage values detected in both oils

FUTURE PERSPECTIVES **337**

(from PRE and POST seeds) did not show significant differences. In this case, considering data reported, the distillation process applied did not seem to significantly affect the fatty acid distribution of extracted oils. However, to better evaluate the thermal impact toward the oil, the oxidation index (OI) was determined (Gordon and Roedig-Penman 1998). In addition, in this case, PRE and POST samples did not show significant differences with a value of 0.087 ± 0.002 and 0.090 ± 0.003, respectively. Distillation process changed the antiradical activity of both seeds and skins. Heating significantly decreased the seeds activity (even remaining higher than that of the skins) by resulting in about 25% reduction. Conversely, skins activity increased to around 50% compared to PRE sample. Looking at related previous studies, there are debatable results in this aspect. A reduction in the antioxidant activity has been reported drying at temperature higher than 100°C and after sterilization of seed extracts (Kim et al. 2006). Conversely, an increase on the radical scavenging activity has been reported in roasted seeds (Kim et al. 2010). Due to different interactions, polyphenols may behave differently in mixture than when they occur individually. Defining the apparent thermal stability results is not easy, because there are many compounds involved in and some of them may be increased, while the others decreased. However, this behavior could partially be related to the flavan-3-ols content detected in the samples. In seeds, the process has reduced by approximately 50% (about 7 g/kg DM) the extractable content of this group of compounds, thus proving a lower antiradical activity of POST sample. In skins, gallic acid, catechin, epicatechin, and epicatechin gallate strongly increased their concentrations, so proving a higher activity.

8.12 Conclusions

The conclusion from this study is that despite the degradation caused by the drastic conditions of the distillation process, many interesting, valuable compounds are still present in the extract from the waste of this process, which make it a useful matter for a better exploitation than as fertilizer or for producing renewable energy. The variety of tentatively identified compounds in the extracts makes them exploitable as additives in the food industry (either as colorants, as flavor modifiers or as antioxidants), and in the cosmetics and nutraceuticals industries.

References

Avila-Fernandez, A., Galicia-Lagunas, N., Rodrıguez-Alegrıa, M.E., Olvera, C., Lopez-Munguıa, A. (2011). Production of functional oligosaccharides through limited acid hydrolysis of agave fructans. *Food Chemistry*, 129, 380–386.

Ayestaran, B., Guadalupe, Z., León, D. (2004). Quantification of major grape polysaccharides (*Tempranillo* v.) released by maceration enzymes during the fermentation process. *Analytica Chimica Acta*, 513(1), 29–39.

Barile, D., Tao, N., Lebrilla, C.B., Coisson, J.D., Arlorio, M., German, J.B. (2009). Permeate from cheese whey ultrafiltration is a source of milk oligosaccharides. *International Dairy Journal*, 19(9), 524–530.

Barreteau, H., Delattre, C., Michaud, P. (2006). Production of oligosaccharides as promising new food additive generation. *Food Technology and Biotechnology*, 44, 323–333.

Basalan, M., Gungor, T., Owens, F.N., Yalcinkaya, I. (2011). Nutrient content and in vitro digestibility of Turkish grape pomaces. *Animal Feed Science and Technology*, 169, 194–198.

Belleville, M.P., Williams, P., Brillouet, J.M. (1993). A linear Arabinan from a red wine. *Phytochemistry*, 33(1), 227–229.

Beveridge, T.H.J., Girard, B., Kopp, T., Drover, J.C.G. (2005). Yield and composition of grape seed oils extracted by supercritical carbon dioxide and petroleum ether: Varietal effects. *Journal of Agricultural Food and Chemistry*, 53, 1799–1804.

Boehm, G., Stahl, B. (2003). Oligosaccharides. In: Mattila, T., Saarela, M. (Eds.), *Functional Dairy Products*. Cambridge, England: Woodhead Pub. Ltd., pp. 203–243.

Bonetta, S., Coïsson, J.D., Barile, D., Bonetta, S., Travaglia, F., Piana, G., Carraro, E., Arlorio, M. (2008). Microbiological and chemical characterization of a typical Italian cheese: Robiola di Roccaverano. *Journal of Agricultural Food and Chemistry*, 56, 7223–7230.

Bordiga, M., Coïsson, J.D., Locatelli, M., Arlorio, M., Travaglia, F. (2013a). Pyrogallol: An alternative trapping agent in proanthocyanidins analysis. *Food Analytical Methods*, 6, 148–156.

Bordiga, M., Piana, G., Coïsson, J.D., Travaglia, F., Arlorio, M. (2014). Headspace solid-phase micro extraction coupled to comprehensive two-dimensional with time-of-flight mass spectrometry applied to the evaluation of Nebbiolo-based wine volatile aroma during ageing. *International Journal of Food Science & Technology*, 49, 787–796.

Bordiga, M., Rinaldi, M., Locatelli, M., Piana, G., Travaglia, F., Coïsson, J.D., Arlorio, M. (2013b). Characterization of Muscat wines aroma evolution using comprehensive gas chromatography followed by a post-analytic approach to 2D contour plots comparison. *Food Chemistry*, 140, 57–67.

Bordiga, M., Travaglia, F., Locatelli, M., Coïsson, J.D., Arlorio, M. (2011). Characterisation of polymeric skin and seed proanthocyanidins during ripening in six *Vitis vinifera* L. cv. *Food Chemistry*, 127, 180–187.

FUTURE PERSPECTIVES 339

Bordiga, M., Travaglia, F., Meyrand, M., German, J.B., Lebrilla, C.B., Coïsson, J.D., Arlorio, M., Barile, D. (2012). Identification and characterization of complex bioactive oligosaccharides in white and red wine by a combination of mass spectrometry and gas chromatography. *Journal of Agricultural and Food Chemistry*, 60, 3700–3707.

Bradburn, D.M., Mathers, J.C., Gunn, A., Burn, J., Chapman, P.D., Johnston, I.D. (1993). Colonic fermentation of complex carbohydrates in patients with familial adenomatous polyposis. *Gut*, 34(5), 630–636.

British Nutrition Foundation. (1990). Complex carbohydrates in foods: Report of the British Nutrition's Task Force. London, UK: Chapman & Hall.

Chalier, P., Angot, B., Delteil, D., Doco, T., Gunata, Z. (2007). Interactions between aroma compounds and whole mannoprotein isolated from *Saccharomyces cerevisiae* strains. *Food Chemistry*, 100(1), 22–30.

Chen, J., Liang, R., Liu, W. et al. (2013). Pectic-oligosaccharides prepared by dynamic high-pressure microfluidization and their in vitro fermentation properties. *Carbohydrate Polymers*, 91, 175–182.

Coelho, E., Rocha, M.A.M., Saraiva, J.A., Coimbra, M.A. (2014). Microwave superheated water and dilute alkali extraction of brewers' spent grain arabinoxylans and arabinoxylo-oligosaccharides. *Carbohydrate Polymers*, 99, 415–422.

Da Porto, C. (1998). Grappa and grape-spirit production. *Critical Reviews in Biotechnology*, 18, 13–24.

Davidov-Pardo, G., Arozarena, I., Marin-Arroyo, M.R. (2011). Stability of polyphenolic extracts from grape seeds after thermal treatments. *European Food Research and Technology*, 232, 211–220.

Deng, Q., Penner, M.H., Zhao, Y. (2011). Chemical composition of dietary fiber and polyphenols of five different varieties of wine grape pomace skins. *Food Research International*, 44, 2712–2720.

Doco, T., Brillouet, J.M. (1993). Isolation and characterization of a rhamnogalacturonan-II from red wine. *Carbohydrate Research*, 243(2), 333–343.

Doco, T., Williams, P., Cheynier, V. (2007). Effect of flash release and pectinolytic enzyme treatments on wine polysaccharide composition. *Journal of Agricultural and Food Chemistry*, 55(16), 6643–6649.

Doco, T., Williams, P., Pauly, M., O'Neill, M.A., Pellerin, P. (2003). Polysaccharides from grape berry cell walls. Part II. Structural characterization of the xyloglucan polysaccharides. *Carbohydrate Polymers*, 53, 253–261.

Du, B., Song, Y., Hu, X., Liao, X., Ni, Y., Li, Q. (2011). Oligosaccharides prepared by acid hydrolysis of polysaccharides from pumpkin (*Cucurbita moschata*) pulp and their prebiotic activities. *International Journal of Food Science and Technology*, 46, 982–987.

Ducasse, M.A., Williams, P., Meudec, E., Cheynier, V., Doco, T. (2010). Isolation of Carignan and Merlot red wine oligosaccharides and their characterization by ESI-MS. *Carbohydrate Polymers*, 79(3), 747–754.

340 VALORIZATION OF WINE MAKING BY-PRODUCTS

EFSA. (2011). Scientific Opinion on the substantiation of health claims related to polyphenols in olive and protection of LDL particles from oxidative damage (ID 1333, 1638, 1639, 1696, 2865), maintenance of normal blood HDL cholesterol concentrations (ID 1639), maintenance of normal blood pressure (ID 3781), "anti-inflammatory properties" (ID 1882), "contributes to the upper respiratory tract health" (ID 3468), "can help to maintain a normal function of gastrointestinal tract" (3779), and "contributes to body defences against external agents" (ID 3467) pursuant to Article 13(1) of Regulation (EC) No 1924/2006. *EFSA Journal* 9(4), 2033.

Egert, D., Beuscher, N. (1992). Studies on antigen-specificity of immuno-reactive arabinogalactan proteins extracted from *Baptisia-Tinctoria* and *Echinacea-Purpurea. Planta Medica*, 58(2), 163–165.

Englyst, H., Hay, S., Macfarlane, G. (1987). Polysaccharide breakdown by mixed populations of human faecal bacteria. *FEMS Microbiology Letters*, 45(3), 163–171.

FAO. (2014). FAOSTAT. http://faostat.fao.org/. Accessed on November 8, 2014.

Fontana, A.R., Antonilli, A., Bottini, R. (2013). Grape pomace as a sustainable source of bioactive compounds: Extraction, characterization, and biotechnological applications of phenolics. *Journal of Agricultural and Food Chemistry*, 61, 8989–9003.

Gerbaud, V., Gabas, N., Laguerie, C., Blouin, J., Vidal, S., Moutounet, M., Pellerin, P. (1996). Effect of wine polysaccharides on the nucleation of potassium hydrogen tartrate in model solutions. *Chemical Engineering Research and Design*, 74(A7), 782–790.

Gibson, G.R., Roberfroid, M.B. (1995). Dietary modulation of the human colonic microbiota—Introducing the concept of prebiotics. *Journal of Nutrition*, 125(6), 1401–1412.

Gobinath, D., Madhu, A.N., Prashant, G., Srinivasan, K., Prapulla, S.G. (2010). Beneficial effect of xylo-oligosaccharides and fructo-oligosaccharides in streptozotocin-induced diabetic rats. *British Journal of Nutrition*, 104, 40–47.

Göktürk Baydar, N., Akkurt, M. (2001). Oil content and oil quality properties of some grape seeds. *Turkish Journal of Agriculture and Forestry*, 25, 163–168.

Gonzalez-Paramas, A., Esteban-Ruano, S., Santos-Buelga, C., Pascual-Teresa, S., Rivas-Gonzalo, J. (2004). Flavanol content and antioxidant activity in winery byproducts. *Journal of Agricultural Food and Chemistry*, 52, 234–238.

Gordon, M.H., Roedig-Penman, A. (1998). Antioxidant activity of quercetin and myricetin in liposomes. *Chemistry and Physics of Lipids*, 97, 79–85.

Guggenbichler, J.P., De Bettignies-Dutz, A., Meissner, P., Schellmoser, S., Jurenitsch, J. (1997). Acidic oligosaccharides from natural sources block adherence of *Escherichia coli* on uroepithelial cells. *Pharmaceutical and Pharmacological Letters*, 7(1), 35–38.

FUTURE PERSPECTIVES 341

Gullon, B., Gullon, P., Sanz, Y., Alonso, J.L., Parajo, J.C. (2011). Prebiotic potential of a refined product containing pectic oligosaccharides. *LWT— Food Science and Technology*, 44, 1687–1696.

Harmsen, H.J.M., Wildeboer-Veloo, A., Raangs, G.C., Wagendorp, A.A., Klijn, N., Bindels, J.G., Welling, G.W. (2000). Analysis of intestinal flora development in breast-fed and formula-fed infants by using molecular identification and detection methods. *Journal of Pediatric Gastroenterology and Nutrition*, 30(1), 61–67.

Hopkins, M.J., Macfarlane, G.T. (2003). Nondigestible oligosaccharides enhance bacterial colonization resistance against *Clostridium difficile* in vitro. *Applied and Environmental Microbiology*, 69(4), 1920–1927.

Hu, K., Liu, Q., Wang, S., Ding, K. (2009). New oligosaccharides prepared by acid hydrolysis of the polysaccharides from *Nerium indicum* Mill. and their anti-angiogenesis activities. *Carbohydrate Research*, 344, 198–203.

Kang, O.L., Ghani, M., Hassan, O., Rahmati, S., Ramli, N. (2014). Novel agaro-oligosaccharide production through enzymatic hydrolysis: physicochemical properties and antioxidant activities. *Food Hydrocolloids*, 42, 304–308.

Kim, S.Y., Jeong, S.M., Park, W.P., Nam, K.C., Ahn, D.U., Lee, S.C. (2006). Effect of heating conditions of grape seeds on the antioxidant activity of grape seed extracts. *Food Chemistry*, 97, 472–479.

Kim, T.J., Silva, J.L., Kim, M.K., Jung, Y.S. (2010). Enhanced antioxidant capacity and antimicrobial activity of tannic acid by thermal processing. *Food Chemistry*, 118, 740–746.

Kunz, C., Rudloff, S., Baier, W., Klein, N., Strobel, S. (2000). Oligosaccharides in human milk: Structural, functional, and metabolic aspects. *Annual Review Nutrition*, 20, 699–722.

Laufenberg, G., Kunz, B., Nystroem, M. (2003). Transformation of vegetable waste into value added products: (A) the upgrading concept, (B) practical implementations. *Bioresource Technology*, 87, 167–198.

Locatelli, M., Gindro, R., Travaglia, F., Coïsson, J.D., Rinaldi, M., Arlorio, M. (2009). Study of the DPPH-scavenging activity: Development of a free software for the correct interpretation of data. *Food Chemistry*, 114, 889–897.

Macfarlane, G.T., Steed, H., Macfarlane, S. (2008). Bacterial metabolism and health-related effects of galacto-oligosaccharides and other prebiotics. *Journal of Applied Microbiology*, 104(2), 305–344.

Molina-Alcaide, E., Moumen, A., Martín-García, A.I. (2008). By-products from viticulture and the wine industry: Potential as sources of nutrients for ruminants. *Journal of the Science of Food and Agriculture*, 4, 597–604.

Moreno, J., Cerpa-Calderon, F., Cohen, S., Fang, Y., Qian, M., Kennedy, J. (2008). Effect of postharvest dehydration on the composition of Pinot Noir grapes (*Vitis vinifera* L.) and wine. *Food Chemistry*, 109, 755–762.

Moure, A., Gullon, P., Dominguez, H., Parajo, J.C. (2006). Advances in the manufacture, purification and applications of xylo-oligosaccharides as food additives and nutraceuticals. *Process Biochemistry*, 41, 1913–1923.

342 VALORIZATION OF WINE MAKING BY-PRODUCTS

Mussatto, S.I., Mancilha, I.M. (2007). Non-digestible oligosaccharides: A review. *Carbohydrate Polymers*, 68, 587–597.

Nabarlatz, D., Ebringerova, A., Montane, D. (2007). Autohydrolysis of agricultural by-products for the production of xylo-oligosaccharides. *Carbohydrate Polymers*, 69, 20–28.

Newburg, D.S. (2005). Innate immunity and human milk. *Journal of Nutrition*, 135(5), 1308–1312.

Newburg, D.S., Ruiz-Palacios, G.M., Altaye, M., Chaturvedi, P., Guerrero, M.L., Meinzen-Derr, J.K., Morrow, A.L. (2004). Human milk alpha 1,2-linked fucosylated oligosaccharides decrease risk of diarrhea due to stable toxin of *E. coli* in breastfed infants. *Advances in Experimental Medicine and Biology*, 554, 457–461.

Ninonuevo, M.R., Park, Y., Yin, H., Zhang, J., Ward, R.E., Clowers, B.H., German, J.B., Freeman, S.L., Killeen, K., Grimm, R. (2006). A strategy for annotating the human milk glycome. *Journal of Agricultural and Food Chemistry*, 54(20), 7471–7480.

Nunan, K.J., Sims, I.M., Bacic, A., Robinson, S.P., Fincher, G.B. (1998). Changes in cell wall composition during ripening of grape berries. *Plant Physiology*, 118(3), 783–792.

OIV. (2007). Resolution Oeno 12/2007. Amendment to resolution Oeno 22–2003—Limit of detection and limit of quantification. Assemblee generale, Paris, France, June 17, 2007.

OIV. (2013). State of the Vitiviniculture World Market, statistics. http://www.oiv.int/oiv/info/enconjoncture. Accessed on October 5, 2014.

O'Neill, M.A., Warrenfeltz, D., Kates, K., Pellerin, P., Doco, T., Darvill, A.G., Albersheim, P. (1996). Rhamnogalacturonan-II, a pectic polysaccharide in the walls of growing plant cell, forms a dimer that is covalently cross-linked by a borate ester. *Journal of Biological Chemistry*, 271(37), 22923–22930.

Pastor del Rio, J., Kennedy, J.A. (2006). Development of proanthocyanidin in *Vitis vinifera* L. cv. Pinot Noir grapes and extraction into wine. *American Journal of Enology and Viticulture*, 57, 125–132.

Patras, A., Brunton, N.P., Gormely, T.R., Butler, F. (2009). Impact of high pressure processing on antioxidant activity, ascorbic acid, anthocyanins and instrumental colour of blackberry and strawberry puree. *Innovative Food Science and Emerging Technologies*, 10(3), 308–313.

Pellerin, P., O'Neill, M., Pierre, C., Cabanis, M., Darvill, A., Albersheim, P., Moutounet, M. (1997). Lead complexation in wines with the dimers of the grape pectic polysaccharide rhamnogalacturonan 2. *Journal International des Sciences de la Vigne et du Vin* (France), 31, 33–41.

Penn, S.G., Cancilla, M.T., Green, M.K., Lebrilla, C.B. (1997). Direct comparison of matrix-assisted laser desorption/ionisation and electrospray ionisation in the analysis of gangliosides by Fourier transform mass spectrometry. *European of Mass Spectrometry*, 3(1), 67–79.

Pinelo, M., Arnous, A., Meyer, A.S. (2006). Upgrading of grape skins: significance of plant cell-wall structural components and extraction techniques for phenol release. *Trends in Food Science & Technology*, 17, 579–590.

FUTURE PERSPECTIVES 343

Qiang, X., Yonglie, C., Qianbing, W. (2009). Health benefit application of functional oligosaccharides. *Carbohydrate Polymers*, 77, 435–441.

R Development Core Team. (2008). *R: A Language and Environment for Statistical Computing.* Vienna, Austria: R Foundation for Statistical Computing.

Renard, C.M., Baron, A., Guyot, S., Drilleau, J.F. (2001). Interactions between apple cell walls and native apple polyphenols: Quantification and some consequences. *International Journal of Biological Macromolecules*, 29, 115–125.

Riou, V., Vernhet, A., Doco, T., Moutounet, M. (2002). Aggregation of grape seed tannins in model wine—Effect of wine polysaccharides. *Food Hydrocolloid*, 16(1), 17–23.

Rockenbach, I.I., Rodrigues, E., Gonzaga, L.V., Caliari, V., Genovese, M.I., Gonçalves, A.E.S.S., Fett, R. (2011). Phenolic compounds content and antioxidant activity in pomace from selected red grapes (*Vitis vinifera* L. and *Vitis labrusca* L.) widely produced in Brazil. *Food Chemistry*, 127, 174–179.

Rondeau, P., Gambier, F., Jolibert, F., Brosse, N. (2013). Compositions and chemical variability of grape pomaces from French vineyard. *Industrial Crops and Products*, 43, 251–254.

Ruiz-Palacios, G.M., Cervantes, L.E., Ramos, P., Chavez-Munguia, B., Newburg, D.S. (2003). *Campylobacter jejuni* binds intestinal H(O) antigen (Fuc alpha 1, 2Gal beta 1, 4GlcNAc), and fucosyloligosaccharides of human milk inhibit its binding and infection. *Journal of Biological Chemistry*, 278(16), 14112–14120.

Sadilova, E., Stintzing, F.C., Carle, R. (2006). Thermal degradation of acylated and nonacylated anthocyanins. *Journal of Food Science*, 71, C504–C512.

Sagdic, O., Ozturk, I., Ozkan, G., Yatim, H., Ekici, L., Yilmaz, M.T. (2011). RP-HPLC-DAD analysis of phenolic compounds in pomace extracts from five grape cultivars: Evaluation of their antioxidant, antiradical and antifugal activities in orange and apple juices. *Food Chemistry*, 126, 1749–1758.

Samanta, A.K., Jayapal, N., Kolte, A.P., Senani, S., Sridhar, M., Dhali, A., Suresh, K.P., Jayaram, C., Prasad, C.S. (2014). Process for enzymatic production of xylooligosaccharides from the xylan of corn cobs. *Journal of Food Processing and Preservation*. doi:10.1111/jfpp.12282.

Sánchez, M., Franco, D., Sineiro, J., Magariños, B., Nuñez, M.J. (2009). Antioxidant power, bacteriostatic activity, and characterization of white grape pomace extracts by HPLC–ESI–MS. *European Food Research and Technology*, 230, 291–301.

Shoaf, K., Mulvey, G.L., Armstrong, G.D., Hutkins, R.W. (2006). Prebiotic galactooligosaccharides reduce adherence of enteropathogenic *Escherichia coli* to tissue culture cells. *Infection and Immunity*, 74(12), 6920–6928.

Torres, J.L., Varela, B., García, M.T., Carilla, J., Matito, C., Centelles, J.J., Cascante, M., Sort, X., Bobet, R. (2002). Valorization of grape (*Vitis vinifera*) by-products. Antioxidant and biological properties of polyphenolic fractions differing in procyanidin composition and flavonol content. *Journal of Agricultural and Food Chemistry*, 50, 7548–7555.

344 VALORIZATION OF WINE MAKING BY-PRODUCTS

Travaglia, F., Bordiga, M., Locatelli, M., Coisson, J.D., Arlorio, M. (2011). Polymeric proanthocyanidins in skins and seeds of 37 *Vitis vinifera* L. cultivars: A methodological comparative study. *Journal of Food Science*, 76, C742–C749.

Urashima, T., Saito, T., Nakamura, T., Messer, M. (2001). Oligosaccharides of milk and colostrum in non-human mammals. *Glycoconjugate Journal*, 18(5), 357–371.

Vicens, A., Fournand, D., Williams, P., Sidhoum, L., Moutounet, M., Doco, T. (2009). Changes in polysaccharide and protein composition of cell walls in grape berry skin (Cv. Shiraz) during ripening and over-ripening. *Journal of Agricultural and Food Chemistry*, 57(7), 2955–2960.

Vidal, S., Courcoux, P., Francis, L., Kwiatkowski, M., Gawel, R., Williams, P., Waters, E., Cheynier, V. (2004). Use of an experimental design approach for evaluation of key wine components on mouth feel perception. *Food Quality & Preferences*, 15(3), 209–217.

Warrand, J., Janssen, H.-G. (2007). Controlled production of oligosaccharides from amylose by acid-hydrolysis under microwave treatment: Comparison with conventional heating. *Carbohydrate Polymers*, 69, 353–362.

Yu, J., Ahmedna, M. (2013). Functional components of grape pomace: Their composition, biological properties and potential applications. *International Journal of Food Science and Technology*, 48, 221–237.

Index

A

Acetone, butanol, and ethanol mixture (ABE), 133
Acetone, for phenolic extraction, 148
Acid hydrolysis, oligosaccharide production by, 301–302
Active dry yeasts (ADY), 51
Advanced oxidation processes (AOP), 63, 105
Aftertaste of wines, 53
Agarooligosaccharides, 298
Aging, wine, 56
Agriculture
  sustainability, 232
  use of grape waste in, 129–130
  use of wine waste in, 118
  waste, 3
Agriculture raisonnée, 248–249
Albumin-based fining, 60
Alcoholic beverages, distilled, 321
Alcoholic fermentation
  by-products, 54–55
  control, 48–49
  management, 49–51
  yeast, 47–48
Alcoholic solvents, for phenolic extraction, 148
Algal bloom, in water, 219
Amine fluoride (Fluorinol®), 162
Amine solutions, absorption of CO_2 with, 101–102
Ammonium salts, 50
Ammonium sulfate, 50
Anaerobic digestion, 133, 177
Animal feed, 124–129
Animal health, 128–129
Anthelmintic drugs, 129
Anthocyanins, 46, 51, 54, 58, 77, 122, 136, 156, 329–330
  and lees, 94
  structure of, 159
Antiadenoviral activity, of wine residues, 164
Antibacterial activity, of phenolics, 162–164
Antifungal activity, of phenolics, 162–164

345

346 INDEX

Anti-influenza virus activity, of
phenolics, 165
Antinutrient compounds in wine
by-products, 126
Antioxidant(s), 122
activity of grape leaves, 176
activity of phenolics, 160, 162
dietary fibers, 166
grape-seed extracts as, 156
grape stem extracts as, 175
Antiviral activity, of phenolics,
164–166
AOP, see Advanced oxidation
processes
Appellations d'Origine Contrôlées
(AOC) system, 235
Aquatic organisms, effect of waste
on, 219
Aqueous waste, 200
Arabic gum, 61
Arabinogalactans, 294–295
Aroma of wines
effect of malolactic fermentation
on, 53
effect of temperature on, 49–50
Aspergillus, 4
A. niger, 174–175
A. oryzae, 175
Australia
sustainable winegrowing in,
233–234
Sustaining Success
strategy, 233
waste management in, 219
environmental monitoring,
218, 220–221
Environment Protection
Authority, 218,
220–221, 233–234
reporting of data, 221
Australian/New Zealand
Standards (AS/NZS
5667 1998), 220

Autohydrolysis, 301
Autoimmune diseases, grape-seed
extracts for, 167
Autolysis, yeast, 54, 56, 91–92, 94

B

Barrel, wine aging in, 56–57, 245
Beef cattle finishing, grape pomace
in, 127
Bentonite fining, 59
Berries, grape, 35–37, 86, 89
β-cyclodextrin, 154
Bifidobacterium, 296
Bioactives
in grape canes, 81–82
in grape leaves, 176
novel conventional
technologies for
extraction of, 151–154
site-based changes in, 177
Bioactivity of phenolics, 160
Biochemical oxygen demand
(BOD), 104, 202, 219
Biodiesel, 131–132
Biodiversity, vineyard, 230, 234,
261, 263–264
Biodiversity and Wine Initiative
(BWI), 255
Biodynamic winegrowing, 230
Bioenergy, 250
Biofuel
biomass for, 131–133
conversion of grape oil to, 132
pellets, 131
Biogas, 133
Biological treatment of winery
wastewater, 63, 104
Biomass
for biofuel, 131–133
degradation of, 174–175
refining, 43
Biomphalaria alexandrina, 173

INDEX

Biorefinery
cell wall polysaccharides, 304
chemical characterization,
310–313
cryoprotectant agents, 311
environmental and
biotechnological impact, 305
enzymatic-driven digestions,
306, 314
enzyme treatments, 310
fine chemical characterization,
314–315
food-related potential
applications, 316
freeze-drying, 311
gas chromatography
technique, 308
green technology extraction, 307
insoluble dietary fiber, 304
probiotic bacteria activity, 306
recovery yield, 310
residual polyphenols, 308
seeds and skins, 309
solid-phase extraction, 314
soluble dietary fiber, 305
stabilization processes, 308
ultrasounds and microwaves
assisted, 306
vanillin biotechnological
production, 309
vegetable matrix
by-products, 309
Biosurfactants, 176
Bio-waste, treatment of, 216
Bipolar membrane
electrodialysis, 100
Black table grapes, antiviral activity
of, 164
Bloating, 128
Boards, use of grape skins in
making, 174
BOD, *see* Biochemical oxygen
demand

Botrytis cinerea, 38, 81, 164
Bottling, wine, 56
Brettanomyces, 57
Brick production, grape pomace
in, 174
Butanol production, grape pomace
in, 133
Butyrivibrio
B. fibrisolvens, 128
B. proteoclasticus, 128
BWI, *see* Biodiversity and Wine
Initiative

C

Cabernet Sauvignon, 75,
127, 164
Calcium tartrate, 60, 99, 166
California
sustainable winegrowing in, 233
waste management in, 203–206
solid waste, 205–206
WRAP, 204
California Association of Winegrape
Growers (CAWG), 233,
256, 258–259
California Code of Sustainable
Winegrowing Practices
workbook, 205, 233,
258–259
California Sustainable
Winegrowing Alliance
(CSWA), 258
California Sustainable Winegrowing
Program (CSWP),
256–260
Canada, wine market, 20
Candida albicans, 162
Cane pruning, 80–81
Canes, grape, 81–82
Canopy management, 37
CAP, *see* Common Agricultural
Policy

348   INDEX

Carbon dioxide (CO_2), 101–103
  in alcoholic fermentation, 49
  capture/separation technologies, 102
  emissions
    of food waste, 200
    of wineries, 256
  supercritical fluid extraction, 151,
    171–172
Carbonic maceration, 42
Carboxymethylcellulose, 61
Carmènere, 164
Castilla y León (Spain),
    environmental
    sustainability of wine sector
    in, 249–250
Catechins, 36, 46, 81, 329
CAWG, *see* California
    Association of Winegrape
    Growers
CCSW, *see* Certified California
    Sustainable Winegrowing
    program
CCVT, *see* Central Coast Vineyard
    Team
Cellulose, 84
Cell wall polysaccharides, 304
Central Coast Vineyard Team
    (CCVT), 254
Certification
  and communication, 234
  ISO 14001, 234, 259, 264
  organic, 129
  programs, 251, 254, 258
  protocols, 250
Certified California Sustainable
    Winegrowing program
    (CCSW), 258
Charcoal briquettes, grape, 130
Chardonnay, 41, 59, 75, 133
Chemical composition of wine
    making by-products, 77–79
Chemical oxygen demand (COD),
    104–105, 202, 219

Chemical stabilization, 58
China
  wine market, 20
  wine production and
    consumption in, 20–21
Chromatographic separation,
    oligosaccharides, 304
Cisplatin A-induced nephropathy,
    proanthocyanidin for, 168
Climate
  change
    greenhouse gas
      emissions, 256
    impact on vineyards, 250
  for growing grapes, 33
Clostridium saccharobutylicum, 133
CMO, *see* Common
    Market Organization
    for Wine
CO_2, *see* Carbon dioxide
COD, *see* Chemical oxygen
    demand
Cold soaking, 51
Cold stabilization, 60
Colloidal instability, of wines, 58
Columella, Lucius, 5
Combustion, of grape waste,
    130–131
Commercial recycling, 204
Commission Directive 98/15/
    EC, 211
Commission Directive 2007/68/EC,
    213–214
Commission Regulation (EC)
    415/2009, 213
Commission Regulation (EC)
    555/2008, 216
Commission Regulation (EC)
    607/2009, 212
Commission Regulation (EC)
    753/2002, 213
Commission Regulation (EC)
    884/2001, 213

INDEX

Commission Regulation (EC)
981/2008, 213
Commission Regulation (EC)
1507/2006, 213
Commission Regulation (EC)
2868/95, 213
Common Agricultural Policy
(CAP), 212
Common Market Organization for
Wine (CMO), 212
Compost
food waste, 201
marc/sludge, 207
pomace, 204
wine waste, 118, 129–130
Concord seed oil, 91
Consumers Union, 251
Continuous belt presses, 43
Continuous extraction systems, 147
Cooperative Extension grape
programs, 260
Cosmetics, 122
Council Directive 75/106/EEC, 214
Council Directive 76/211/EEC, 214
Council Directive 76/464/EEC, 210
Council Directive 80/232/EEC, 214
Council Directive 91/271/EEC, 211
Council Directive 91/676/EEC, 211
Council Directive 1999/31/EC,
211, 214
Council Directive 2005/25/EC, 214
Council Directive 2006/12/EC,
211, 215
Council Regulation (EC)
3/2008, 213
Council Regulation (EC) 40/94,
213–214
Council Regulation (EC) 479/2008,
76, 212–213, 235
Council Regulation (EC) 491/2009,
212, 235
Council Regulation (EC)
606/2009, 213

Council Regulation (EC)
607/2009, 235
Council Regulation (EC)
1234/2007, 213
Council Regulation (EC)
1493/1999, 213, 216
Council Regulation (EC)
1782/2003, 213
Council Regulation (EEC) 822/87,
212, 235
Council Regulation (EEC) 823/87,
212, 235
Council Regulation (EEC)
1567/89, 135
Cross-flow microfiltration, 62
Crude oils, conversion into
biofuels, 132
Crude wine, 61
Crushing, 41, 83, 147
Cryogenic distillation for CO_2
recovery, 102–103
CSWA, *see* California
Sustainable Winegrowing
Alliance
CSWP, *see* California
Sustainable Winegrowing
Program
Culex pipiens, 173
Culinary ingredients, vine leaves
as, 176
Cultivation of vines, 34–35
Cyclodextrins, 154

D

DAD, *see* Diode array detector
Dairy cattle feed, grape pomace
in, 128
Deacidification, *see* Malolactic
fermentation
Dekkera, 57
De-oiling/de-fatting of grape seeds,
143, 147

350 INDEX

Depolymerization, polysaccharides, 297
De-stemming, 83
Diammonium phosphate, 50
Diatomaceous earth filtration, 62
Dietary fibers, 46, 166
Dietary supplements, 75, 81, 122
Diethyl ether, for grape-seed oil extraction, 171
Diode array detector (DAD), 323–324
Directive 2008/98/EC, 305
Directive 2000/13/EC, 213–214
Directive 2000/60/EC, 211
Directive 2006/118/EC, 211
Directive 2007/45/EC, 214
Directive 2008/1/EC, 211–212
Directive 2008/98/EC, 215
Disease forecasting models, 243
Disposal fees, for wine making by-products, 76
Distillation of wine by-products, 132–135
  fractions, 134
  grape pomace
    methods, 325–336
    process, 322–323
Docosahexaenoic acid, 167
Drainage capacity, of vineyard soil, 32
Dregs, *see* Lees
Drenching, cattle, 129
D-(-) tartaric acid, 99

E

EC, *see* Electrical conductivity of water
Eco-Management and Audit Scheme (EMAS II), 234
Effluents, treatment of, 241–242
EFSA, *see* European Food Safety Authority

EGC-P, *see* Epigallocatechin–phloroglucinol
Eisenia andrei, 130
EIU, *see* Environmental Impact Unit
Electrical conductivity (EC) of water, 220
Electrodialysis, 60, 100
EMAS II, *see* Eco-Management and Audit Scheme
Emergency procedures flipchart, 208
Energy management, 201, 250, 253
Enocyanin, 122, 160
Ensiling, 127
Environment
  biodiversity conservation, 263–264
  impacts
    of wine production, 232–234
    of winery waste constituents, 219
  sustainability assessment of winegrowing activity in France, 248–249
  and waste management challenges for, 3–5
  pollution, 118
Environmental Conservation Act (South Africa), 217
Environmental Impact Unit (EIU), 251
Environmental monitoring program, 218, 220
Environmental risk assessment, 240
Environment Protection Act (Australia), 218
Environment Protection Authority (EPA), Australia, 218, 220–221, 233–234
Environment protection policies (EPPs), 218

INDEX

351

Enzymes, 4
  for grape-seed oil extraction, 173
  in wine making, 42
EPA, *see* Environment
    Protection Authority,
    Australia
Epicatechin, 46, 81
Epigallocatechin–phloroglucinol
    (EGC-P), 330
EPPs, *see* Environment protection
    policies
Escherichia coli, 162
ESP, *see* Exchangeable sodium
    percentage
Ethanol
  for grape-seed oil extraction,
    171, 173
  level in supercritical fluid
    extraction, 151
  oxidative damage of, 168
  for phenolic extraction, 148–149
  production, 132–134
Ethyl acetate, for phenolic
    extraction, 149
Ethyl esters, 132
EU, *see* European Union
Europe
  wine policy, 2008 reform, 240
  wine regulations, 234–236
European Commission
  legislation related to wastewater
    management, 210–214
  purpose of, 211
European Community
    regulations, 135
European Food Safety Authority
    (EFSA), 317–318
European Union (EU)
  environmental regulations, 200
  vineyards in, 9
  waste management
    bio-waste, 216
    community legislation, 214–217

    Council Resolution of 24
      February 1997, 214
    turning waste into resource,
      209–210
    waste hierarchy, 215
    withdrawal of by-products,
      216–217
European Waste Catalogue
    and Hazardous Waste
    List, 216
Eutrophication, 219
Exchangeable sodium percentage
    (ESP), 220
Exporters, wine, 16–17

F

Fatty acid methyl esters (FAMEs),
    325–326
Fatty acids, 90–91, 93
  contents, 332–335
  extraction, 323
  in grape seeds, 131–132
  in sherry wine lees, 93
Fermentation, 177; *see also* Alcoholic
    fermentation
  by-products, 174–175, 284–286
  lees after, 257
  malolactic, 52–54
  solid-state, 132, 174–175
Fertilizers, 129–130, 243, 256
  vineyard, 34
  wine solid residues as, 118
Filtration, 61–62
Finger Lakes Grape
    Program, 260
Fining, 59–60
Flavan-3-ols, 46, 77, 156, 328–329
  polymers, and anthocyanins, 94
  structures of constitutive units
    of, 159
Flavonols, 77, 136, 156, 329
Flavor of wines, 53

352   INDEX

FNTV, *see* National Federation TerraVitis, France
Foliage management, vines, 34
Food
  applications of wine making by-products
    grape-seed oil, 171–173
    health applications, 167–169
    ingredients, 166
    meat products, 169–171
    phenolics, 155–166
  colorants, 160
  processing
    by-products, 1–2
    in Europe, 238
  waste, 4–5, 199–202
FOS, *see* Fructooligosaccharides
Fourier transform ion cyclotron resonance (FTICR) method, 297
France
  environmental sustainability assessment of winegrowing activity in, 248–249
  wine regulations in, 235
Frankfurters, use of grape-seed oil in, 170
Freeze-drying, 311
Fructooligosaccharides (FOS), 298, 301
Fructose, 37
Fruit and vegetable waste (FVW), 202
Fruit processing, 4
FTICR, *see* Fourier transform ion cyclotron resonance method
Fungi, filamentous, 4
Fusarium, 4
FVW, *see* Fruit and vegetable waste

G

Galactooligosaccharides (GOS), 294–295, 298
Gas separation membranes, for CO_2 recovery, 102
Gastrointestinal parasitism, 129
Gelatin-based fining, 60
Germany
  sustainability schemes in, 234
  wine market, 19–20
Glass recycling/reuse, 207
Glucose, 37, 176
GOS, *see* Galactooligosaccharides
GP, *see* Grape pomace (GP)
Grape and Wine Research and Development Corporation (GWRDC), 264–265
Grape juice, antiviral activity of, 164
Grape pomace (GP), 38–39, 45–46, 75, 84–86, 155
  adhesive applications of, 174
  as animal feed, 124
  anthocyanins, 329–330
  antimicrobial activity of, 162
  antiradical activity, 326, 334
  biosorbent activity of, 173–174
  composting, 130, 204
  conventional solvent extraction methods for extraction of polyphenols from, 140–143
  dietary fibers in, 166
  digestibility of, 124, 127
  distillation
    methods, 325–336
    process, 322–323
  distilled alcoholic beverages, 321
  ensiling of, 127
  ethanol production from, 132–134
  exploitation, 317–318, 320

INDEX   **353**

extracts, phenolic profile of, 87–88

FAMEs, 325–326

fatty acids
  contents, 332–335
  extraction, 323

flavan-3-ols, 328–329

flavonols, 329

grape skins, 319

HPLC phloroglucinolysis, 330–332

hydroxybenzoic acid, 328

hydroxycinnamic acids, 328

nutritional value of, 124, 127

phenolic compounds
  contents, 326–328
  extraction, 323–324

proanthocyanidin analysis, 324–325

production, 320

red grapes, 124, 150–151

role in butanol production, 133

seeds, 319

spirits, 321

statistical analysis, 326

tartaric esters, 328

use in brick production, 174

use in meat products, 170

utilization of, 119–121

vine leaves, 319

vine shoots, 318

vinification lees, 320

white grapes, 124, 135

wine industry waste, 321

Grapes
  average yield, 272
  by-products
    biological importance, 277
    fermentation, 284–286
    pomace, 84–86, 277–278
    seeds (*see* Grape-seed extracts Grape-seed oil; Seeds, grapes)
    skins (*see* Skins, grape)
    stems, 83–84, 284
    tartaric acid, 284
  composition, 273–274
  consumption, 271–272
  natural antioxidant, 272
  phenolics in different fraction of, 157–158
  prices, 276
  uses, 272
  Vitis vinifera
  for wine making
    berries, 35–37
    by-products, 38–39
    climate, 33
    good practices, 33–35
    growing, 30–31
    pruning waste, 37–38
    pulp structure and components, 44
    slope, 33
    soil, 32
    terroir, 31
  world production, 271, 274–276

Grape-seed extracts (GSEs), 156
  antimicrobial activity of, 162, 164
  as food ingredients, 166
  health applications of, 167–169
  molluscidal/insecticidal activities of, 173
  use in meat products, 170–171

Grape-seed oil, 46, 86–91, 158, 171–173
  cosmetic applications of, 174
  extraction yield of, 131
  health applications of, 168
  oil content of seeds, 90
  use in meat products, 169–170

Grappa, 134–135, 155

Green campaigns, 230

Greenhouse gas emissions, of wineries, 256

Green production, 4, 119, 201

354 INDEX

GSEs, *see* Grape-seed extracts
GWRDC, *see* Grape and
    Wine Research
    and Development
    Corporation

H

Haemonchus contortus, 128
Harvest, grape, 40–41, 244
Hawke's Bay Winegrowers
    Association, 252
Health applications, of by-products,
    167–169
Health supplements and extracts,
    156–160
Heavy lees, 92
Hemicellulosic sugars, 38
Heterocyclic aromatic amines, in
    beef patties, 170–171
Hexane, for grape-seed oil
    extraction, 171–172
High voltage electric discharge
    (HVED), 152–153
Hills, vineyards in, 33
H_2O_2, 105
Honest Significant Difference
    test, 326
Horizontal membrane presses, 43
Horizontal plate presses, 43–44
Horseradish peroxidase (HRP),
    105–106
Hot water extraction, 148
Household waste, reuse and
    recycling of, 216
HPLC phloroglucinolysis,
    330–332
HRP, *see* Horseradish peroxidase
HVED, *see* High voltage electric
    discharge
Hydro-alcoholic solvents, for
    phenolic extraction, 148
Hydrogen peroxide, 105–106

Hydroxybenzoic acid, 328
Hydroxycinnamic acids, 35–36, 328

I

Importers, wine, 17–19
Inactivated dry yeasts (IDY), 50–51
INDIGO® method, 249
Industrial biotechnology, 2
Industrial economy, and wine
    making, 75–76
Inoculation, 50
Insoluble dietary fiber, 304
Integrated farming, 248–249
Integrated pest management (IPM),
    251, 255
Integrated Production of Wine
    (IPW) scheme, South
    Africa, 217–218, 234,
    255–256
International Organization for
    Biological Control (IOBC),
    238, 254
International Organization of Vine
    and Wine (OIV), 6, 9,
    235–237
    mission, 237–238
    resolution CST 1/2004,
        238–239, 246
    resolution CST 1/2008, 238–245
    Scientific and Technical
        Committee, 238, 246
    Strategic Plan (2012–2014), 245
International Standard Organization
    (ISO) certification, 264
Inulin, 300
IOBC, *see* International
    Organization for Biological
    Control
Ion exchange, 100
Ion exchange resins, 60
IPM, *see* Integrated pest
    management

INDEX

IPW, *see* Integrated Production of Wine scheme, South Africa
Irrigation techniques, 243, 256
ISO, *see* International Standard Organization (ISO) certification
ISO 14001, 234, 259, 264
Italian spirit production, 134
Italian wine industry, 246–248
    Tergeo program, 247
    V.I.V.A. Sustainable Wine project, 247

J

Joint stakeholder platforms, 3

K

Kinetics, fermentation, 49–51

L

LAB, *see* Lactic acid bacteria
Laccase, 156, 158, 175
Lactic acid, 175–176
Lactic acid bacteria (LAB), 52–53, 133
Lactobacilli, 54
Lactobacillus spp., 52
    L. brevis, 134
    L. pentosus, 175–176, 284
    L. plantarum, 134
    L. rhamnosus, 55
Lactococcus lactis, 175
Lactosucrose, 299
Lactulose, 299
Lake Erie Regional Grape Program, 260
Landfills, 214–215
Landscape, vineyard, 232, 241, 250

LCA, *see* Life cycle assessment
Leaves
Leaves, vine, 176, 319
Lees, 54–56, 75–77, 121
    after fermentation, 257
    anthelmintic activity of, 128–129
    elemental composition and physicochemical characteristics, 92
    ethanol recovery from, 133
    grappa production from, 134
    yeast
        protein compounds and phenolic compounds, 94
        soluble components, 92–93
Life cycle assessment (LCA), 201
LIFE HAproWINE project, 249–250
Light lees, 92
Lignins, 43, 46, 83–84, 318
Lignocellulose biorefinery, 305
Linoleic acid, 91
Lipids
    addition during fermentation, 50
    in lees, 93
    oxidation, 156, 168, 170–171, 278
Liposomal encapsulation, of grape-seed extract, 170
Liquid–liquid extraction, 164
Liquid-to-solid (L/S) ratio, in phenolic extraction, 147
Liquor Products 1989 (South Africa), 255
LISW, *see* Long Island Sustainable Winegrowing
Lithium compounds, adsorption of CO_2 with, 102
LIVE, *see* Low Input Viticulture and Enology program
Livestock production, waste, 3
Lodi Rules program, 251–252

356 INDEX

Lodi Winegrape Commission (LWC), 251–252
Long Island Sustainable Practices Workbook, 260
Long Island Sustainable Winegrowing (LISW), 260–261
Lori-Bakhtiari lamb feed, grape pomace in, 127
Low Input Viticulture and Enology (LIVE) program, 254
L/S, *see* Liquid-to-solid ratio, in phenolic extraction
L-(+) tartaric acid, 99
LWC, *see* Lodi Winegrape Commission
Lymnea cailliaudi, 173

M

Maceration, 51–52, 86
MAE, *see* Microwave-assisted extraction
Magnesium, role in fermentation, 51
MALDI, *see* Matrix-assisted laser desorption ionization
Malic acid, 35, 37, 44
Malolactic fermentation (MLF), 52–54
Mannoproteins, 48, 56, 60–61, 94
Manto Negro grape by-products, 84, 166
Manure, 3
Marc, grape, 45–46, 75–77
  cell wall, 45
  composts, 207
  conventional solvent extraction methods for polyphenols extraction, 140–143
  grappa production from, 134
  protein content of, 128, 175
Marcus Cato, 5

Mash cooler, 42
Mass spectrometry, 296
Matrix-assisted laser desorption ionization (MALDI), 297
Maturation, wine, 56
McLaren Vale Grape Wine and Tourism Association (MVGWTA), 262
  Pest and Disease Code of Conduct, 262
  Soil Management, Water Management, and Preservation of Biodiversity Codes, 262
McLaren Vale Sustainable Winegrowing Australia (MVSWGA) program, 262–264
Meat products, 169–171
Mechanical pressing, of grape seeds, 171
Mediterranean basin (wine), 318
Merlot, 75, 136
Metal ions, role in fermentation, 51
Metatartaric acid, 61
Methane
  emissions, 128, 200
  production, 132–133
Methanol
  content in spirits, 135
  for grape-seed oil extraction, 171, 173
  for phenolic extraction, 148–149
Methoxypyrazine compounds, 37
Methyl esters, 132
Microbiological stabilization, 58
Microfiltration, 58
Microorganisms, cultivation from food waste, 4
Micro-oxygenation, 56, 58

INDEX

Microwave-assisted extraction (MAE), 153–154
Microwave radiation, 302
MLF, *see* Malolactic fermentation
Moisture content of lees, 92–93
Monascus purpureus, 175
Monensin, 128
Monounsaturated fatty acids, 90
Mulching, 38
Musts, 41–42, 44, 49
MVGWTA, *see* McLaren Vale Grape Wine and Tourism Association
MVSWGA, *see* McLaren Vale Sustainable Winegrowing Australia program

N

National Association of Testing Authorities (NATA), 220
National Environmental Management Act (South Africa), 217
National Environment Protection (Assessment of Site Contamination) Measure 1999, 220
National Federation TerraVitis (FNTV), France, 248
National Water Act (South Africa), 217
Natural preservatives, phenolics as, 164
Natural resource management, 240–241
Nebbiolo, 257
Nerium indicum, 301
Neurospora, 4
New York State's sustainable viticulture program, 260
New Zealand

SWNZ, 206–209, 233, 252–254
waste management in, 206
emergency procedures flipchart, 208
glass recycling, 207
wastewater management, 208–209
New Zealand Government Waste Minimization Act, 206
New Zealand Winegrowers Code of Practice for Management of Winery Waste, 206
Nitrogen, in lees, 92
NMR, *see* Nuclear magnetic resonance technique
Nonanthocyanin phenolics, 94
Nondigestible disaccharides, 297
Non-flavonoids, 77, 80, 154
Non-food applications, of by-products, 173–174
Nuclear magnetic resonance (NMR) technique, 296
Nutrients
addition during fermentation, 50
composition, in wine making by-products, 77–80
content of wine waste used for animal feed, 125
NYS Agricultural Environmental Management (AEM) worksheets, 260

O

Oak barrel aging technology, 56–57
Oak chips/shavings, 51
Obesity, grape-seed extracts for, 167
Oenococcus oeni, 52
Oil, *see* Grape-seed oil
OIV, *see* International Organization of Vine and Wine

358   INDEX

Oleuopein, 176
Oligosaccharides; *see also*
     Polysaccharides
   animal milks, 296
   definition, 296
   food industry, 297–298
   FOS, 298–300
   glycosyl and linkage analyses, 296
   GOS, 298–300
   human milks, 296
   lactosucrose, 299
   lactulose, 299
   MALDI, 297
   nondigestible carbohydrates,
     298–299
   physiological effects, 30
   prebiotic activity, 296 (*see also*
     Prebiotic oligosaccharides)
   production
     acid hydrolysis, 301–302
     enzymatic hydrolysis, 302–303
     physical hydrolysis, 303
   properties, 298
   purification, 303–304
   saccharide compounds, 297
   structure–function
     properties, 296
   wine, 294–295
   XOS, 300
Organic acids, 166
Organic compounds in wastewater,
   103–104
Organic farming
   definition, 228
   parasitism in, 129
Organic solvents, for phenolic
   extraction, 148
Organic winegrowing, 228
Organization Internationale De
     La Vigne Et Du Vin, *see*
     International Organization
     of Vine and Wine (OIV)
Oxygen, for fermentation, 50

P

Packaging materials, recycling
   of, 245
PACs, *see* Proanthocyanidins
Parasitic gastroenteritis, 129
PEAS, *see* Pesticide Environmental
     Assessment System
Pecticoligosaccharides (POS),
     298, 301
Pectin methylesterase, and methanol
     concentration, 135
Pectin polysaccharides, 84
Pectins, 166
Pediococci, 54
Pediococcus spp., 52
PEF, *see* Pulsed electric field
Penicillium, 4
   P. chrysogenum, 174–175
   P. citrinum, 174–175
Pesticide Environmental Assessment
     System (PEAS), 251
Pest management, vines, 35
Pharmaceuticals, 122
Phenolics, 77, 84
   antibacterial/antifungal activities
     of, 162–164
   anti-influenza virus
     activity, 165
   antioxidant activity of, 122,
     160, 162
   antiviral activity of, 164–166
   bioactivity of, 160
   biological properties, 161
   contents, 326–328
   in different fraction of grapes,
     157–158
   extraction of, 51–52, 136,
     323–324
     novel conventional
      technologies for bioactives
      extraction, 151–154
   sample pretreatment, 143, 147

INDEX

size reduction, 147
solvent-to-sample ratio, 147
time and temperature,
149–151
types of solvents, 148–149
health supplements and extracts,
156–160
inhibition effects against
microorganisms, 163
in lees, 94
profile of Pinot Noir lees extracts,
93, 95–96
profile of Riesling lees extracts,
93, 97–98
in wastewater, 103–104
and yeast lees, 94
Phomopsis viticola, 38
Physical stabilization, 58
Physicochemical treatment of
winery wastewater, 104
Phytoalexins, 81
Phytochemicals, 5, 81, 155, 158
Phytosanitary protection,
243–244
Pinot Gris, 75
Pinot Meunier, 85
Pinot Noir, 81, 85–86, 89
grape pomace, 124
lees, 93, 95–96, 128–129
Pint Noir, 75
Pkatsiteli grape-seed oil, 91
Plant-based waste, 2–3
Plug seedling production, use of
grape waste in, 130
Plunging, in red wine
making, 51–52
Pollution
fertilizer-induced, 34
food waste-oriented, 4
water, 210
Polyethylene glycol, 127
Polyphenols, 46, 55, 85, 89, 94, 124,
294; *see also* Phenolics

antiviral activity of, 165
extraction
emerging technologies,
145–146
from grape pomace/marc,
140–143
from grape seeds, 137–139
from grape skins/stalks, 144
high voltage electric discharge,
152–153
microwave-assisted extraction,
153–154
pulsed electric field, 152–153
supercritical fluid extraction,
151–152
ultrasound-assisted
extraction, 152
health applications
of, 167–168
hyperglycemic activity of, 169
stability of, 151–152
Polysaccharides, 56, 77, 92,
154, 294
enzymatic depolymerization, 302
extraction, natural sources,
300–301
nondigestible
carbohydrates, 298
wines, 294
Polyunsaturated fatty acids, 90
Pomace, *see* Grape pomace
Pork sausage, use of grape-seed oil
in, 169–170
POS, *see* Pecticoligosaccharides
Positive Points System (PPS), 254
Potassium bitartrate, 60, 99
Potassium hydrogen tartrate, 60
Potassium metabisulfite, 135
PPS, *see* Positive Points System
Prebiotics, definition, 296
Pre-fermentation, 41–43
by-products, 42–43
maceration, 51

360   **INDEX**

Pressing
  by-products, 45–46
  methods, 43–45
Proanthocyanidins (PACs), 46, 77
  analysis, 324–325
  antiviral activity of, 165
  extraction of, 149, 151
  health applications of, 167–169
Procyanidins, 46, 160, 175, 278
  extraction of, 148
  health applications of, 167–168
Prodelphinidins, 175
Protease A, 91–92
Protected Harvest, 251
Protein-based fining, 59–60
Protein compounds, in lees, 94
Proteins, 294
Pruning, vine, 34, 80–82, 129, 207
  waste, 37–38
  winter, 80, 243
Pullulan, 122
Pulsed electric field (PEF),
  152–153
Pump overs, in red wine making,
  51–52
Pyrazines, 37

R

Racking, 92–93
Recycled water, definition, 209
Red grape pomace, 124, 150–151
Reduce, reuse, and recycle,
  203, 207
Red wine
  antiviral activity of, 164
  making
    anthocyanins, 94
    flowchart, 40
    maceration, 51
    malolactic fermentation, 52–54
    pressing, 43
    reactions, 57

  stabilization, 59
  temperature conditions, 50
Refining, crude oil, 134
Regulation (EC) 1924/2006, 213
Residual polyphenols, 308
Resources
  natural resource management,
    240–241
  turning waste into, 209–210
Response surface
  methodology, 152
Resveratrol, 80–81
  antiviral activity of, 165–166
  health applications of, 167
Reused water, definition,
  208–209
Riesling, 75, 93, 97–98
Ripeness, of grapes, 40–41
Ripening of grape berries,
  36–37
Romanian Spotted cow feed, grape
  pomace in, 128
Rootstocks, vine, 34
Ruby Red, 91
Rumen acidosis, 128

S

Saccharomyces
  S. bayanus, 132
  S. cerevisiae, 47, 91, 132,
    134, 303
Salinity of drinking water, 220
Sand bioreactors, 63
SAR, *see* Sodium adsorption ratio
Saturated fatty acids, 90
Sauvignon Blanc, 75
Sauvignon musts, 50
SCFE, *see* Supercritical fluid
  extraction
Secoiridoids, 176
Secondary fermentation, *see*
  Malolactic fermentation

INDEX

Seeds, grape, 46, 75, 85–91, 319
 antimicrobial activity, 278
 conventional solvent extraction
  methods for polyphenols
  extraction, 137–139
 de-oiling of, 143, 147
 dietary intake, 278
 nutraceutical market, 281
 nutritional value of, 124
 oil, 278–280
 phytochemicals, 282
 potential income, 280
 procyanidins, 278
 separated from pomace, 131
 structures of constitutive units of
  flavan-3-ols in, 159
 tannins in, 136
 utilization of, 121
 world production, 281
Sherry wine lees, fatty acid
  composition of, 93
Shiraz, 75, 128
Single-cell protein, 128
SIP, *see* Sustainability in Practice
  program
Skin contact, in wine making, 42, 51
Skins, grape, 46, 75, 85–86, 120,
  122, 282–284, 319
 conventional solvent extraction
  methods for extraction of
  polyphenols from, 144
 extracts, health applications,
  168–169
 for making boards, 174
 nutritional value of, 124
 phenolics in, 160
 phytochemicals in, 158
 structures of constitutive units of
  flavan-3-ols in, 159
 tannins in, 136
Slope, for growing grapes, 33
Sludge, wine, 129, 207
Sodicity, 220

Sodium adsorption ratio
  (SAR), 220
Soil, vineyard, 32, 34, 243
Soil erosion, prevention of,
  255–256
Solid–liquid extraction, 152
Solid-state fermentation, 132,
  174–175
Solid-state fungal fermentation, 46
Solid waste
 management, in California,
  205–206
 treatment of, 241–242, 256
Solid wine by-products, 121
Solvents
 extraction, 100
 for phenolic extraction, 148–149
Solvent-to-sample ratio, for phenolic
  extraction, 147
South Africa
 Integrated Production of Wine
  program, 234
 sustainable winegrowing in, 234
 waste management in, 217–218
Soxhlet extraction, 164, 173
Spanish Law 10/1998, 76
Sparkling wine, 9, 16
Spirits, 135, 321
Spontaneous fermentation, 47
Spur pruning, 80–81
Squalene, 173
SSs, *see* Suspended solids
Stalks, 43, 75, 121, 128, 174–176
 conventional solvent extraction
  methods for polyphenols
  extraction, 144
 nutritional value of, 124
 tannins in, 136
 use in composts, 129–130
Staphylococcus aureus, 162
Stems, grape, 39, 83–84, 119
Sterols, 51
Stilbenoid compounds, 166

362   **INDEX**

Straw, 3
Streptococcus mutans, 162
Stuffed grape leaves (dish), 178
Summer pruning, 80
Supercritical fluid extraction
    (SCFE), 151–152, 171
Suspended solids (SSs), 200, 202
Sustainability
    CSWP, 256–260
    definition, 228
    environmental impact of wine
      production, 232–234
    European wine regulations,
      234–235
    France, environmental
      sustainability assessment in,
      248–249
    IPW scheme, 255–256
    Italian wine industry, 246–248
    LIFE HAproWINE project,
      249–250
    LISW, 260–261
    LWC and Lodi Rules program,
      251–252
    MVSWGA program, 262–264
    New York State's sustainable
      viticulture program, 260
    OIV, 235–238
    principles, 231, 258
    SWNZ, 252–254
    UKVA, 264–265
    VineBalance, 260
    WOC Sustainability Program,
      261–262
Sustainability in Practice (SIP)
    program, 254
Sustainable vitiviniculture
    definition, 238
    development of, 238–239
    guidelines for, 239–245
      environmental risk
        assessment, 240
      harvesting, 244

    natural resource management,
      240–241
    phytosanitary protection,
      243–244
    recycling of packaging
      materials, 245
    vineyards, 242–243
    waste management, 241–242
Sustainable waste management, 2
Sustainable water
    management, 253
Sustainable winegrowing, 228–232;
    see also Sustainability
    best practices, 230
    definition, 231
    programs, 233
Sustainable Winegrowing
    New Zealand (SWNZ),
    206–209, 233, 252–254
Syrah, 164

T

TA, *see* Tartaric acid
Tannase, 127
Tannins, 36–37, 77, 86, 118, 124,
    127, 158, 175, 294
    oxidation of, 105–106
    role in animal health, 128
Tartaric acid (TA), 35, 42, 55,
    99–101
    in lees, 93
    recovery, 100
Tartaric esters, 328
Tartaric stabilization, 60–61
Tartrates, 166
TDS, *see* Total dissolved salts,
    in drinking water
Temperature
    fermentation, 49–50
    for growing vines, 33
    of phenolic extraction, 149–151
    pre-fermentation, 42

INDEX 363

Tergeo program, 247
Tergeo Scientific Committee, 247
TerraVitis®, 248
Terroir, 31
Thermal insulation boards, 174
Thermophilic anaerobic digestion, 133
Time, of phenolic extraction, 149–151
TOC, *see* Total organic carbon
Tocols, in grape-seed oil, 172–173
Tocopherols, 91
Total dissolved salts (TDS), in drinking water, 220
Total organic carbon (TOC), 104, 219
Total suspended solids (TSS), 220
Tourist-oriented approach, in marketing, 230
Trametes
    T. pubescens, 175
    T. versicolor, 43
Transesterification, biofuel, 132
Trellising system, of vines, 34
Triacetin, 129
Trichoderma
    T. harzianum, 174–175
    T. reesei, 175
Trilinolein, 91
Triple-bottom line approach, 233, 263
Tsigai lamb feed, grape pomace in, 127–128
TSS, *see* Total suspended solids
Turbidity of wine, 50, 61
Tyrosol, 80

U

Ultrasound-assisted extraction (UAE), 152, 173
Unione Italiana Vini (UIV), 247

United Kingdom
    UKVA, 264–265
    wine market, 19
United Kingdom Vineyard Association (UKVA), 264–265
United Nations Commission on Environment and Development, "Our Common Future" report, 232
United States, wine market, 19
University of Montpellier (France), Integrated Wine Production study, 234

V

Vaccinium vitis-idaea, 165
Vanillin biotechnological production, 309
Vegetable processing, 4
Vermicomposting, 130
Vertical presses, 43
VineBalance's New York Guide to Sustainable Viticulture Practices Grower Self-Assessment Workbook, 260
Vines, 5
    cultivation of, 34–35
    leaves, 176, 319
    pest management, 35
    pruning, 34, 37–38, 80–82, 129, 207, 243
    rootstocks, 34
    shoots, 318
    temperature for growing, 33
    trellising system, 34
Vine training system, 243
Vineyards
    establishment of, 242–243
    waste management, 207

364 INDEX

Viniferin, 81
Viticulture, 5
Vitis
  V. berlandieri, 29
  V. labrusca, 89, 168
  V. riparia, 29
  V. rupestris, 29
  V. vinifera, 29, 35, 51, 75, 81,
    89, 271
Vitiviniculture
  leading producers, 7
  production potential
    European Union, 9–10
    outside European
      Union, 9–11
  sustainable (*see* Sustainable
    vitiviniculture)
  world market, 6–9
V.I.V.A. Sustainable Wine
  project, 247
Volatile solids (VSs), 202

W

WAO, *see* Wet air oxidation
Waste hierarchy, 215
Waste management, 249
  environmental/economic
    challenges, 3–5
  food, 201–202
  sustainable, 2
Waste Reduction Award's Program
  (WRAP), 204
Wastewater, 103–106
  chemical parameters and
    composition, 103
  management, 250,
    253, 256
    European Commission
      legislation related to,
      210–214
    in New Zealand, 208–209
  winery, 62–63

Water
  algal bloom in, 219
  consumption in wine
    making, 229
  electrical conductivity of, 220
  input for vineyards, 243
  management, 250, 253
  permits, 217
  pollution, 210
  quality, 3
  recycled water, 209
  retention, of vineyard soil, 32
  reused water, 208–209
  salinity of drinking
    water, 220
  sustainable water
    management, 253
Welch, 260
Wet air oxidation (WAO), 105
White grape pomace, 124, 135
White-rot fungi, 43
White wine
  aging, 56
  making
    fermentation, 49–50
    fining, 59
    flowchart, 40
    malolactic fermentation, 54
    pressing, 43–45
    stabilization, 59
    temperature conditions,
      49–50
Whole cluster fermentation, 42
WIETA, *see* Wine and Agricultural
    Industry Ethical Trade
    Association
Wine
  aging, 56
  consumption, 8–9, 14–15,
    20–21
  exporters, 16–17
  importers, 17–19
  largest markets, 19–20

INDEX

production, 5, 75–76
  in China, 20–21
  European Union, 11–12
  outside European
    Union, 12–13
  trade, global trends in, 15–16
Wine and Agricultural Industry
    Ethical Trade Association
    (WIETA), 255
Wine Institute, California, 233, 256,
  258–259
Wine making process
  alcoholic fermentation (*see*
    Alcoholic fermentation)
  filtration, 61–62
  fining, 59–60
  flowchart, 40
  grape harvest and
    documentation, 40–41
  grapes for (*see* Grapes, for wine
    making)
  maceration, 51–52
  malolactic fermentation, 52–54
  post-fermentation
    operations, 55
  prebiotic oligosaccharides (*see*
    Biorefinery)
  pre-fermentation, 41–43
  pressing
    by-products, 45–46
    methods, 43–45
  process, by-products, and
    potential utilization, 74
  reactions, 57–58
  stabilization, 58
    red wine, 59
    tartaric, 60–61
    white wine, 59
  vines, 29
  winery wastewater, 62–63
  wood, 56–57
Wine of Origin (WO) scheme, 255

Wines of Chile (WOC)
    Sustainability Program,
    261–262
Wine stability, definition, 58
Winter pruning, 80, 243
WO, *see* Wine of Origin
    (WO) scheme
WOC, *see* Wines of Chile (WOC)
    Sustainability Program
Wood, for wine making, 56–57
  barrels, 56–57, 245
  toasting intensities, 57
World Bank, Biodiversity & Wine
    Initiative, 234
WRAP, *see* Waste Reduction
    Award's Program

X

Xyloglucans, 294–295
Xylooligosaccharides (XOS),
  298, 300
Xylose, 176

Y

Yeast, 47–48
  active dry yeasts, 51
  autolysis, 54, 56, 91–92, 94
  inactivated dry yeasts, 50–51
  lees
    protein compounds and
      phenolics, 94
    soluble components, 92–93
  role in ethanol production, 132

Z

Zygosaccahromyces
  Z. bailii, 164
  Z. rouxii, 164